Risk Methodologies for Technological Legacies

NATO Science Series

A Series presenting the results of scientific meetings supported under the NATO Science Programme.

The Series is published by IOS Press, Amsterdam, and Kluwer Academic Publishers in conjunction with the NATO Scientific Affairs Division

Sub-Series

I. **Life and Behavioural Sciences**	IOS Press
II. **Mathematics, Physics and Chemistry**	Kluwer Academic Publishers
III. **Computer and Systems Science**	IOS Press
IV. **Earth and Environmental Sciences**	Kluwer Academic Publishers
V. **Science and Technology Policy**	IOS Press

The NATO Science Series continues the series of books published formerly as the NATO ASI Series.

The NATO Science Programme offers support for collaboration in civil science between scientists of countries of the Euro-Atlantic Partnership Council. The types of scientific meeting generally supported are "Advanced Study Institutes" and "Advanced Research Workshops", although other types of meeting are supported from time to time. The NATO Science Series collects together the results of these meetings. The meetings are co-organized bij scientists from NATO countries and scientists from NATO's Partner countries – countries of the CIS and Central and Eastern Europe.

Advanced Study Institutes are high-level tutorial courses offering in-depth study of latest advances in a field.
Advanced Research Workshops are expert meetings aimed at critical assessment of a field, and identification of directions for future action.

As a consequence of the restructuring of the NATO Science Programme in 1999, the NATO Science Series has been re-organised and there are currently five sub-series as noted above. Please consult the following web sites for information on previous volumes published in the Series, as well as details of earlier sub-series.

http://www.nato.int/science
http://www.wkap.nl
http://www.iospress.nl
http://www.wtv-books.de/nato-pco.htm

Risk Methodologies for Technological Legacies

edited by

Dennis C. Bley

Buttonwood Consulting, Inc.,
Oakton, VA, U.S.A.

James G. Droppo

Battelle,
Pacific Northwest National Laboratory,
Richland, WA, U.S.A.

and

Vitaly A. Eremenko

ICES, Department of Risk Analysis & Management,
Moscow, Russia

Technical editor:

Regina Lundgren

Independent Technical Consultant,
Kennewick, WA, U.S.A.

Kluwer Academic Publishers

Dordrecht / Boston / London

Published in cooperation with NATO Scientific Affairs Division

Proceedings of the NATO Advanced Study Institute on
Risk Assessment Activities for the Cold War Facilities and Environmental Legacies
Bourgas, Bulgaria
2–11 May 2000

A C.I.P. Catalogue record for this book is available from the Library of Congress.

ISBN 1-4020-1257-8 (HB)
ISBN 1-4020-1258-6 (PB)

Published by Kluwer Academic Publishers,
P.O. Box 17, 3300 AA Dordrecht, The Netherlands.

Sold and distributed in North, Central and South America
by Kluwer Academic Publishers,
101 Philip Drive, Norwell, MA 02061, U.S.A.

In all other countries, sold and distributed
by Kluwer Academic Publishers,
P.O. Box 322, 3300 AH Dordrecht, The Netherlands.

Printed on acid-free paper

Dedication

This book is dedicated to the families of those living with the legacies of the Cold War. May our work contribute to lasting solutions and a better life for us all.

Table of Contents

List of Figures

List of Tables

List of Contributors

Editors
Dr. Dennis C. Bley, President, Buttonwood Consulting, USA
Dr. James G. Droppo, Senior Scientist, Pacific Northwest National Laboratory, USA
Dr. Vitaly A. Eremenko, General Director, Educational Center "TRAOMD," Russia

Technical Editor
Ms. Regina Lundgren, Independent Technical Consultant, USA

Contributors
Ms. Rosanne L. Aaberg, Senior Scientist, Pacific Northwest National Laboratory, USA (Chapter 4)
Dr. Vladilena N. Abramova, Head of Psychology, Obninsk Institute of Nuclear Power Engineering, Russia (Chapter 13)
Mr. Svetoslav Andonov, Deputy Director, Civil Protection Service, Bulgaria (Chapter 11)
Mr. William Andrews, Senior Program Manager, Pacific Northwest National Laboratory, USA (Chapter 4)
Dr. Olga Aneziris, Researcher, National Centre for Scientific Research "Demokritos," Greece (Chapter 15)
Dr. Lynn R. Ansbaugh, Professor, University of Utah, USA (Chapter 9)
Ms. Susan Bayley, Senior Engineer, Science Applications International Corporation, USA (Chapter 8)
Dr. Allan S. Benjamin, Senior Risk Consultant, ARES Corporation, USA (Chapter 10)
Dr. Dennis C. Bley, President, Buttonwood Consulting, USA (Chapters 1, 2 and 18)
Dr. Esko Blokker, Head of International Department, DCMR Environmental Agency, the Netherlands (Chapters 1 and 2)
Mr. Mark E. Bollinger, Program Manager, U.S. Department of Energy Center for Risk Excellence, USA (Chapter 4)
Dr. Ilko Bonev, Director, Bourgas Copper Mines, Bulgaria (Chapter 7)
Mr. Gary Boyd, Senior Safety Engineer, Science Applications International Corporation, USA (Chapter 8)
Mr. John W. Buck, Senior Scientist, Pacific Northwest National Laboratory, USA (Chapter 14)
Mr. Karl J. Castleton, Scientist, Pacific Northwest National Laboratory, USA (Chapter 14)

Dr. M. O. Degteva, Scientist, Urals Center for Radiation Medicine, Russia (Chapter 9)

Dr. James G. Droppo, Senior Scientist, Pacific Northwest National Laboratory, (Chapters 1, 2, 6 and 18)

Dr. Vitaly A. Eremenko, General Director, Educational Center "TRAOMD," Russia (Chapters 1, 2, 17, 18 and Appendix C)

Mr. Steven Fogarty, Senior Risk Consultant, ARES Corporation, USA (Chapter 10)

Ms. Gariann M. Gelston, Scientist/Engineer, Pacific Northwest National Laboratory, USA (Chapter 14)

Mr. Yuri Gorlinsky, Director RTC Systems Analysis, Russian Research Centre, Kurchatov Institute, Russia (Chapter 3)

Ms. Bonnie L. Hoopes, Senior Technical Specialist, Pacific Northwest National Laboratory, USA (Chapter 14)

Dr. Hristo Hristov, Senior Regulatory Toxicologist, Ontario Ministry of the Environment, Canada (Chapter 16)

Dr. Jordan Jordanov, Engineer, Bourgas Cooper Mines, Bulgaria (Chapter 7)

Ms. Olga A. Juharyan, Country Coordinator, International Risk Assessment Network, Armenia (Chapter 11)

Dr. Daniela Kolarova, Professor, Sofia University, Bulgaria (Chapter 5)

Dr. V. P. Kozheurov, Scientist, Urals Research Center for Radiation Medicine, Russia (Chapter 9)

Mr. Gerald F. Laniak, Scientist, Environmental Protection Agency, USA (Chapter 14)

Ms. Judith Lowe, Policy Advisor, Parkman Environment, Great Britain (Chapter 2)

Ms. Regina Lundgren, Independent Technical Consultant, USA (Chapter 2)

Dr. Lyubcho Lyubchev, Associate Professor, Department of Safety in Industry, University, Bulgaria (Chapter 1)

Dr. Georgiy Lysychenko, Senior Scientist, State Scientific Centre of Environmental Radiogeochemistry, Ukraine (Chapter 12)

Mr. Wilson McGinn, Scientist, Argonne National Laboratory, USA (Chapter 4)

Mr. Bruce Napier, Senior Scientist, Pacific Northwest National Laboratory, USA (Chapter 9)

Ms. Kathryn M. Naasan, Senior Risk Consultant, ARES Corporation, USA (Chapter 10)

Dr. I. A. Papazoglou, Researcher, National Centre for Scientific Research "Demokritos," Greece (Chapter 15)

Mr. Mitch Pelton, Senior Technical Specialist, Pacific Northwest National Laboratory, USA (Chapter 14)

Mr. Kurt Picel, Scientist, Argonne National Laboratory, USA (Chapter 4)

Mr. Aleksey Ryabushkin, Deputy Director, International Radioecology Laboratory, Ukraine (Chapters 11 and 17)

Dr. Simeon Simeonov, Director, Regional Environmental Inspectorate-Bourgas, Bulgaria (Chapter 1)

Dr. Robert D. Stenner, Senior Scientist, Pacific Northwest National Laboratory, USA (Chapter 4)

Mr. Greg St. Pierre, Engineer, U.S. Army Program Manager for Chemical Demilitarization, USA (Chapter 8)

Dr. E. I. Tolstykh, Scientist, Urals Research Centre for Radiation Medicine, Russia (Chapter 9)

Dr. M. I. Vorobiova, Scientist, Urals Research Centre for Radiation Medicine, Russia (Chapter 9)

Dr. Gene Whelan, Chief Engineer, Pacific Northwest National Laboratory, USA (Chapter 14)

Mr. Michael K. White, Senior Scientist, Pacific Northwest National Laboratory, USA (Chapter 4)

Dr. Alvin Young, Director, Department of Energy Center for Risk Excellence, USA (Chapters 4 and 17)

Preface

The Cold War Era left the major participants, the United States and the former Soviet Union (FSU), with large legacies in terms of both contamination and potential accidents. Facility contamination and environmental degradation, as well as the accident-vulnerable facilities and equipment, are a result of weapons development, testing, and production. Although the countries face similar issues from similar activities, important differences in waste management practices make the potential environmental and health risks of more immediate concern in the FSU and Eastern Europe. In the West, most nuclear and chemical waste is stored in known contained locations, while in the East, much of the equivalent material is unconfined, contaminating the environment.

In the past decade, the U.S. started to address and remediate these Cold War legacies. Costs have been very high, and the projected cost estimates for total cleanup are still increasing. Currently in Russia, the resources for starting such major activities continue to be unavailable.

It is now clear that even the large budgets provided to the U.S. Department of Energy and Department of Defense (DOE and DOD, respectively) cannot cover the cleanup activities. The high cost projections in the U.S. have resulted in a movement toward risk-based decision making for setting priorities among these activities. The knowledge and experiences of the U.S. in these initial cleanup efforts are seen as important information in many North Atlantic Treaty Organization (NATO) Partner countries, where the environmental problems are more severe and the cleanup budgets more limited.

This situation created the need for an Advanced Study Institute (ASI) on "Risk Assessment Activities for the Cold War Facilities and Environmental Legacies." This high-level course was held in Bourgas, Bulgaria, May 2-11, 2000. The objective of the ASI was to provide information to facilitate and enable decision-making activities affecting the environment and human populations in the NATO and Partner countries. Specifically, the ASI provided a forum to communicate the current status of risk analysis and management methodologies and their appropriate application. It addressed scientific approaches and application experiences from the initial U.S. risk assessment activities. In addition, integrated approaches that have only recently been developed were documented and made available.

This book is the direct product of the ASI. The power of the text lies in linking information on legacies with an integrated view of controlling the risk of those legacies. Chapters 1 and 2 expand these ideas to explain how all the topics in the book are related.

Risk can only be effectively controlled by proper balance of three central concepts: risk analysis, risk perception, and risk management. The editors were first drawn together by the joint recognition that risk analysis methods had matured over the past 30 years in several fields, relatively independent of each other. It was time to integrate all these forms of risk analysis under one framework, identifying the reasons for the seemingly disparate approaches and the gains to be reaped by bringing them together. The second key issue in the control of risk is the recognition that risk perception is the product of many factors in our lives and that cultural differences between the East and West can have significant impact on how we view risk and measures to control it.

This aspect of effective risk control leads to the third concept: risk management. What factors must the decision maker consider in selecting among alternative options? How do cultural factors influence these decisions? How can better information be provided to these decision makers in the East and West to help them make the best decisions for their people? Two previously alternative approaches receive focus–facility risk management (i.e., use of risk analysis to control the risk to facilities) and human-centred risk management (i.e., use of risk analysis to control the risk to people in the surrounding areas and within facilities). Part I of this book gives detailed information on the three concepts and gives further definition to facility-centred and human-centred approaches to risk analysis and risk management.

The striking extent of Cold War legacy problems needs to be understood if decision makers in the East and West are to be able to relate to each other's problems and assist each other. The information presented at the ASI surprised many participants. Participants from the West learned the extent of contamination that exists in the East and the resulting current health problems. Participants from Europe learned of the massive amounts of hazardous materials currently stored in the U.S. Analysts who have studied contamination problems learned of the likelihood of possible accidents and those focused on accident analysis learned of the extent of contamination. Part II of this book gives extensive information on the legacies, our perception of the risk associated with them, and, in some cases, tools for analysing that risk.

Part III of the book relies heavily on applications as a means of presenting detailed information on risk assessment programs and methodologies. Applications were selected that illustrate the strengths and limitations of different methodologies for assessments of military and Cold War legacy facilities in NATO and Partner countries. The concept is to communicate how specific needs have been met by the various methodologies and stress the need for an integrated view of risk assessment.

Finally, Part IV provides details on future activities that were spawned at the ASI.

Part I carries the central message of the ASI and the book. The rest of the book gives examples and extensions for some of the ideas developed in Part I. Because these examples were developed before the unifying ideas of Part I were completely developed and published, they cannot hope to fully convey the integrated and human-centred message proposed there.

Dennis C. Bley, Buttonwood Consulting, Inc.
U.S.A., bley@ieee.org
James G. Droppo, Pacific Northwest National Laboratory
U.S.A., james.droppo@pnl.gov
Vitaly A. Eremenko, Education Center on "Technologies for Risk Analysis for Optimisation of Management Decisions,"
Russian Federation, vitaly@vitaly.msk.ru

Acknowledgments

This book would not be possible without the contributions of many organizations and individuals. The Security-Related Civil Science Committee of the North Atlantic Treaty Organization (NATO) Science Programme served as primary sponsor of this NATO Advanced Study Institute and this book. Buttonwood Consulting, the Pacific Northwest National Laboratory, and the U.S. Department of Energy provided active and important co-sponsorship of these activities.

However, it was truly the participants who made this institute a success. The lecturers provided pertinent and quality content. Participants showed a high level of interest and enthusiasm. The nearly full attendance at all sessions was impressive.

The untiring staff of the Greener Bourgas Foundation and their Risk Training Centre worked days, evenings, weekends, holidays, and more to support this institute. We asked for a lot, and they always delivered. Special thanks go to Esko Blokker, who provided invaluable help in organizing the meeting. The editors also thank Leslie Bowen, Buttonwood Consulting, for her important organizational efforts.

The local organizers and supporters, particularly the Bourgas Mayor and Vice-Mayors, made the Institute run smoothly. The facilities, service, and cultural events that they arranged were excellent and appreciated by all. We believe we speak for all participants when we say we really enjoyed having the institute in Bourgas.

Organizations that graciously supported additional aspects of this activity include the U.S. Army, Program Manager for Chemical Demilitarization; U.S. Department of Energy, Center for Risk Excellence; U.S. Department of Energy under Contract DE-AC06-76RLO 1830; Commission of the European Union DGXII through contract ENV4-CT96-0243 (DG12-DTEE); U.S. Department of Energy, Office of International Health Studies; Federal Department of the Ministry of Health of the Russian Federation; U.S. Environmental Protection Agency; U.S. National Aeronautics and Space Administration (NASA); and Coffey International.

The editors of this book also thank the lecturers for their prompt preparation of their draft materials and their willingness to allow the editors to reorganize those materials into chapters for this textbook. And finally, our thanks go to Regina Lundgren, who was invaluable as the technical editor for this book and to Wil Bruins, whose advice, encouragement and assistance guided us through the final preparation of the book for publication.

1.   Introduction

This book is primarily aimed at decision makers in the East and West, who must manage technological risks. It will also be useful to the technical experts who hope to advise decision makers. The book defines what is meant by risk, how it can be analysed, and how that analysis can be used within decision-making processes in government agencies, public organizations, and private companies operating under different value structures. The objective is to provide information to facilitate and enable decision-making activities affecting the environment and human populations in the NATO Member and Partner countries.

The three pillars of risk science are *risk management, risk analysis*, and *risk perception*. All the chapters of this book deal with one or more of these aspects.

Risk management involves identifying options and balancing risks against resources and preferences. To manage risks, we must first understand them. That is the purpose of risk analysis. In addition, the risk manager must select alternatives in light of the risks they pose, the resources they demand, and the values controlling organization and culture.

Risk analysis has become a well-defined process for analysing the likelihood and consequences of operations, accidents, or the spread of pollutants. Unfortunately, practitioners in different fields have defined their processes quite differently. The time has come to provide a framework that integrates all existing approaches, showing how they are related and the purposes they serve. Experts in many fields should understand what others are doing and why. In this way they can learn from each other and provide more useful advice in their own areas.

Risk perception describes how various individuals and societies view risk in light of their culture, values, and understanding of the risk. The editors hold the view that all risk is perceived risk. We never know exactly what the future will bring; we just have differing levels of sophistication in our understanding of the range and likelihood of possible events and their consequences.

Part I lays the foundations for an integrated approach for balancing risks and preferences. It focuses on two ideas:

Risk management. Risk managers must consider varied risks from both existing contamination and potential accidents and releases–economic risks, potential damage to valuable facilities, acute and chronic human health effects, and risk to the environment. They must balance these risks against available resources in light of existing preference structures–cultural mores in their society, value systems of those affected by their decisions, regulatory and legal requirements, and the judgments of their own organizations.

Unification of risk methods. All the varied approaches toward risk analysis can be organized under structures that describe:

- The risk itself (i.e., the parameters to be calculated and the process that is involved)
- The aspects of the risk that are to be managed.

The risk management approach can involve compliance assurance or a full characterization of the risk. Each offers advantages and disadvantages in terms of cost, understanding, technical and public communication, and flexibility to deal effectively with unanticipated problems, within the context of the cultural milieu under which the decision maker operates.

This textbook flows out of a multiyear involvement of the editors. After meeting through several international efforts to deal with specific hazards that flowed out of the Cold War legacies, the ideas introduced above began to jell. The NATO Science Programme supported an Expert Visit to bring the editors together to develop the ideas into a coherent framework. That work led to the NATO Advanced Study Institute, "Risk Assessment Activities for the Cold War Facilities and Environmental Legacies," held in Bourgas, Bulgaria, May 2-11, 2000. The institute brought an internationally prominent group of lecturers together to work with students from more than 20 countries. Finally, NATO sponsored the publication of this textbook to describe the high-level course offered at the institute. The U.S. Department of Energy, a co-sponsor of the institute, also co-sponsored the publication of the textbook and is using it in courses and conferences on related topics.

The book is intended as an aid to decision makers who must make key risk management decisions in complicated situations, with sometimes conflicting analyses and claims and value structures. The book should also serve risk scientists, providing an integrated description of many analysis techniques, advising them of alternative approaches, and alerting them to the issues of risk perception and preferences that must be part of the decision maker's basis. It urges all concerned to attempt to separate the issues of science and value structures as much as possible.

While the importance of separating issues of science and values has been well recognized, for example in the guidance of the U.S. National Research Council, it is especially important in international activities. For example, in countries of the former Soviet Union (FSU), local authorities and populations had no say in the siting of government owned hazardous facilities. However in new facilities (for example, private facilities built to treat or move toxic materials), those local populations are asking questions. Why here? Why that way? Decision makers need to develop rational tools to support and defend their own decision-making processes.

The power of the book lies in its three parts:

- Part I provides a unified view of risk analysis and risk management for decision makers; it shows how to balance risks and preferences and unifies the many risk analysis methodologies
- Part II provides information on legacies in the both the East and the West; in some cases it includes analysis of the associated risks
- Part III provides examples of risk analysis and risk management programs that illustrate aspects of the approach outlined in Part 1.

Part IV outlines future activities that arose as a result of the ASI.

PART I: UNIFYING RISK MANAGEMENT AND ANALYSIS FOR DECISION MAKERS

Never in my life did I think I would have to set about writing memoirs when I had only just passed my 50th birthday. But such events happened [in my oversight role at Chernobyl], on such a scale and involving people with such contradictory interests, with so many different interpretations of how it happened, that it is surely my duty, to some extent, to write about what I know, how I saw the events that occurred...I must say that at the time it did not enter my head that we were moving toward an event on a planetary scale, an event which would apparently go down forever in mankind's history, like the eruption of famous volcanoes, the destruction of Pompeii.
Soviet Prof. Valery A. Legasov, First Deputy Director of the Kurchatov Institute, "Memoirs," *Pravda*, November 1988, shortly before he took his own life

No decision maker should be placed in the position in which Prof. Legasov found himself, facing catastrophic consequences and great uncertainty. Decision makers must have a rational approach to balance the many types of risk associated with sites and facilities under their control against the available resources, in light of the preferences dictated by their cultures and their organizations. They need to understand what their risk analysts can do for them so that they can fully understand the risks they face, before disaster strikes, and develop plans to eliminate or mitigate risks.

2. Complementary Risk Management: A Unified View for Decision Makers

50% of the problems in the world result from using the same word with different meanings. The other 50% comes from people using different words with the same meaning.
Stan Kaplan
PSAM II Short Course
"Risk Assessment/Risk Management Fundamentals"
San Diego 1994

The goal of this book and the NATO Advanced Study Institute (ASI) that spawned it is to provide a unified view of risk methodologies for decision makers and their experts. Often some combination of approaches is necessary to meet the needs of decision makers. We call the effective and appropriate use of these methodologies the Complementary Risk Management approach. The Complementary Risk Management approach balances flexibility within specific applications.

2.1. A Short History of Risk Assessment Traditions

The revolutionary idea that defines the boundary between modern times and the past is the mastery of risk: the notion that the future is more than a whim of the gods and that men and women are not passive before nature.

Peter L. Bernstein
Against the Gods: The Remarkable Story of Risk
Wiley, New York, 1996

The modern history of risk assessment begins just following the industrial revolution. Failures of new machines led to the use of redundancy to improve reliability. For example, ocean liners were equipped with redundant power plants and rudder gear, and later came the introduction of multiengine aircraft. During World War II, German V-1 missiles could not accommodate redundant parts and were built following the weak link theory (i.e., a chain is no stronger than its weakest link). However, improving the weakest part did not help; the rockets were still completely unreliable. The solution comes from understanding that the problem is very different from the simple chain analogy, where, if all links of a chain see the same stress, the weakest link will fail. The V-1 was a complex system that could fail if any of its components should fail from random causes. Robert Lusser, a German engineer, showed that the overall reliability is the product of the reliabilities of the individual series components:

$$R_S = \prod_{i=1}^{N} R_i$$

This is "Lusser's Law" and shows that, for a series system with many components, the reliability of every component must be very high. Reliability theory continued to develop throughout the remainder of the 20th century. Calculations can now be made for systems of high complexity, with many series and parallel components.

In 1957, WASH-740 was published in the U.S. It provided judgmental estimates of the frequency of stylised accidents in nuclear power plants. The accidents involved very conservative ("incredible" combinations of events) estimates of the consequences. At this time, it did not seem possible to analyse such complex systems adequately.

During the 1950s, 1960s, and 1970s, the aerospace, defence, chemical, and nuclear industries in the West promoted development and application of quantitative reliability and availability algorithms. For example, in the 1960s, the C5-A airplane program developed the automatic reliability mathematical model (ARAM), the first computerized reliability model for large complex systems. The U.S. Atomic Energy Commission [later to become the U.S. Nuclear Regulatory Commission (NRC)] supported research to develop new quantitative safety analysis methods and applied them to selected issues. Intercontinental ballistic missiles and the man-rated rocket programs (Mercury and Gemini) instituted reliability requirements. New systems analysis techniques were developed to support these requirements including reliability block diagrams and fault tree analysis.

In the early 1970s, the NRC's Reactor Safety Study[1] used logic models (event trees and fault trees) to estimate the risks to the public from potential accidents in large nuclear power plants. This seminal work, the first probabilistic risk assessment (PRA, also known as probabilistic safety analysis, PSA), broke new ground in many areas. Saul Levine and Norman Rasmussen, the study's directors, developed a new modelling tool, the event tree on critical safety functions, that permitted the organisation of the massive logic model in a way that was tractable and could be reviewed and quantified. The study attempted to quantify the uncertainty in its results. It moved beyond the assumptions of independence, which had led to absurd results in many previous reliability calculations. It addressed accidents that were well beyond the design basis of the plants. It modelled the phenomenology of severe accidents more thoroughly than past studies for a wide variety of existing conditions. It pioneered the development of methods for the analysis of the probability of human error. It treated atmospheric dispersion probabilistically. It developed new computer codes to systematize the complex analyses.

Breaking so much new ground, the Reactor Safety Study greatly extended our ability to analyse complex systems, but it also identified many problems for future studies to examine further. The 1980s and 1990s saw the application and maturing of PSA. There came an understanding of the importance of modelling uncertainty more formally, which led to the refinement of Bayesian methods. A great deal of work on human cognitive processing and human reliability analysis was performed and continues today. Work on dependent failures was greatly expanded including examination of "common-cause initiating events" such as earthquakes, winds, and fire, as well as parametric modelling of common-cause failures of components. The interactions between "level 1" PSA (the calculation of accident frequencies), "level 2" PSA [the characterization of radioactive releases (source terms)], and "level 3" PSA (the consequence analysis that tracks the release through the environment to ultimate receptors and the impacts on those receptors) were studied and methods for treating them were refined.

During this same time period, the methods for tracking releases through the environment and their impacts ("fate and transport" models) began to be applied to in situ pollution and routine releases from incinerators and other chemical processes. Accident PSA methods were adapted to defence, aerospace, chemical processes, and other industries.

Finally the 1990s saw the development of risk management techniques and risk-informed regulation in the U.S., Europe, and Asia.

2.2.   Defining Risk Analysis

The evolution of risk analysis in various industries and applications occurred relatively independent of each other. What has ensued is bit like the Tower of Babel: analysts working in different fields have not stayed abreast of developments in other fields. Their language is often quite different, and their methods often appear quite different. They use different words to mean the same thing and the same words to mean different things. And so they can appear inept to each other. The Society for Risk Analysis was formed in 1981 to bring these diverse groups together, but they tend to remain isolated from each other, even at international conferences, where practitioners often interact mostly with their colleagues. The time has come to learn from each other, to agree on a common language, and to integrate the available approaches.

The NATO ASI brought analysts from diverse fields together, and we found much common ground. Let us begin here with a general framework for risk analysis as shown in Figure 2.1.

Conceptually risk analysis identifies a simple triplet[2]:

S_i – the scenario (i.e., what can go wrong)

ℓ_i – the likelihood of the scenarios occurring

X_i – the consequences of the complete scenario

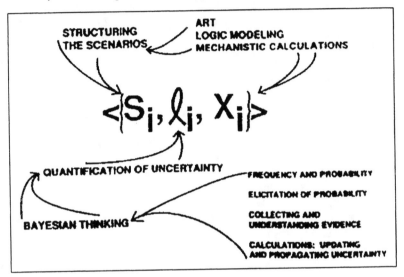

Figure 2.1 General framework for risk analysis

Then the risk analysis is the assembly of all possible such triplets

$$\{\!\}S_i, \ell_i, X_i\{\!\}_C$$

The art of risk analysis comes in structuring the search for scenarios, S_i, and in organizing the structure of the scenarios in a way that facilitates analysis. This can mean effectiveness of search, ease of calculation, clarity of presentation, etc. The science comes in the detailed analysis of the identified scenarios and their consequences. And tying it all together is the structure for identifying, quantifying, and explaining the uncertainty in the elements of the analysis.

Structuring the scenarios is both an engineering art requiring experience and a nice sense of analysis, and a process that draws on the techniques of logic modelling and traditional engineering and scientific mechanistic calculations. Next, no matter how finely we partition the space of scenarios, it is important to recognize that each scenario really represents a group of similar subscenarios. All members of each group must lead to the same consequence. If they do not, the group should be broken into smaller subgroups until that is the case. The calculation of the frequency of each scenario must be based on considering all possible members of the group (i.e., all possible conditions that might exist under each scenario). The calculation of the consequences, the X_i, relies on traditional, mechanistic calculations from the engineering disciplines but is distinguished in that consequences from many more cases are calculated as compared to other approaches. The mechanistic calculations include thermal-hydraulic calculations, electric circuit analysis, neutronic calculations, chemical process analysis, atmospheric dispersion analysis, and so on. The logic modelling required to structure the scenarios traditionally draws on fault trees and event trees for accident PSA, but other approaches, including digraphs and Markov models, are often used. In some cases, other tools that bridge the gap between logic and mechanistic calculations, such as simulation models, are especially appropriate.

Under the formulation already described, we incorporate the ideas of uncertainty into our calculation of the frequency for each individual scenario group. In addressing the uncertainty of frequency, it is important to adopt a coherent and consistent approach. The Bayesian model provides just such an approach, and under its umbrella, we address the issues of frequency and probability, elicitation of probability, collection and understanding of evidence, and calculations.

Clarity of thought regarding the difference between what we call frequency and probability provides a philosophical framework for understanding a consistent treatment of uncertainty. The two concepts are often confused in the literature of probability, both being called probability. Let us say here that frequency is simply the result of an experiment, be it a real experiment or a *gedanken* experiment in which we simply count the number of times the event in question occurs out of the total number of possible trials or expired time. Probability, then, represents our state of knowledge about the real world frequency. In the literature, what we call probability has gone under various names, including subjective probability, state of knowledge probability, and prevision[3,4]. Probability, as a measure of what is in our heads rather than a property of the physical world, is a measure of what we know and what we do not know—our complete state of knowledge.

If probability is a personal state of knowledge, how then do we determine probabilities to use in risk assessments? Let us consider two cases. In the first case, our state of knowledge comes directly from information that has been collected for other applications; for example, we have collected a wide range of equipment failure data from a variety of power plants around the world. From these collected data, we have existing curves showing the plant-to-plant variability of, say, the failure rate from motor-operated valves. This plant-to-plant variability curve shows the variation in frequency of failure as we move from plant to plant in a large population. When we now consider the probability of failure of motor-operated valves at a new plant, our probability distribution for the failure rate is numerically identical to the plant-to-plant variability curve or the frequency variability curve.

In other cases, no such plant-to-plant variability curve is available. Therefore, we must elicit the probability from the best experts available to our work. How one obtains the information from experts and builds a probability distribution is the subject of a large body of literature. Probability is often not elicited in risk assessments or is not elicited well. The reasons it is not elicited well have been documented by Hogarth[5] and others, and include biases built into the human thinking process such as anchoring, overconfidence, and selective interpretation of new data. Careful techniques must be used to avoid these problems[6].

The last two elements in determining the probability of frequency of each scenario (collecting and understanding the evidence, and running calculations using Bayes' theorem for updating probability distributions and propagating uncertainty) are now fairly well established and have been covered in other papers and reports (for example, Pickard, Lowe and Garrick, Inc.[7]). The structured language of PSA provides a powerful model for addressing safety and uncertainty involved in engineered facilities and in situ pollution sites. It provides a framework for organizing a wide variety of standard mathematical and engineering models to address safety issues directly. Note that PSA is more than a set of tools for analysing large systems and calculating a risk parameter. It is a process for understanding the safety status of a facility, identifying contributions of people and specific equipment to safety problems, and evaluating potential improvements. At a deeper level, PSA is really a language for addressing uncertainty in all engineering applications.

Decision makers are sometimes confused by the wide range of analysis methods and endpoints that can be the product of a risk analysis or assessment. The needs of specific applications largely define what end points are most appropriate. Even analysts from different disciplines have become confused and believed their counterparts from other industries were guilty of poor practice. A recent National Research Council report[8] from the U.S. addresses this issue:

[A risk analysis] based on a conservative analysis acceptable for regulatory decision making, such as whether to grant a permit, lacks many essential details. If efforts to control risk are based on [such a risk analysis], they could mistakenly be focused on areas that have been artificially inflated in the conservative analysis. Problems that could arise from using an [analysis] performed for regulatory compliance in communicating with other interested parties are listed below:

- *The [risk analysis] may be assumed to describe actual releases rather than upper-bound results. Thus, the [facility operator] could be accused of releasing more [hazardous materials] than are actually being released.*
- *Attempts to correct "conservative" assumptions could be interpreted as a cover-up.*
- *Risk management is likely to be focused on aspects of the [risk analysis] with the most pessimistic assumptions, rather than those with the most impact.*
- *The scenarios required for the [analysis] may not reflect the most serious facility risks.*

Problems could also arise from using an [analysis] intended to be a risk management tool in communicating with other interested parties for the following reasons:

- *It contains complex results that acknowledge uncertainties.*
- *It does not include simple worst-case scenarios based on point-estimate analyses, and results may be more difficult to interpret and explain.*
- *Because it is site-specific, it does not necessarily follow established generic screening guidance for compliance-oriented* [risk analyses], *which may compromise the credibility of the results.*

To better understand the range of endpoints possible in risk analyses, Figure 2.2 illustrates the range of such endpoints and how they are related. Each of the listed "products" may use any, or all, of the indicated sequential steps. Starting at the top of the table product #1 is a description of the properties of the material; inventory, mass, radioactivity of the constituents are typical properties that are used to define the hazard. This site or source characterization can also be the first step for evaluating the other endpoints, which include sequentially the occurrence of some event, release to the environment, environmental concentrations, receptor exposure, uptake by receptors, and impact to receptors. The "examples" column provides some of the typical endpoints for each of "products."

Intermediate endpoints, usually interpreted by standards, norms, etc., often have their origin in protecting human health and the environment. In other cases, the implied impacts of the intermediate endpoint may be so severe that an explicit analysis is considered unnecessary. Thus, even for immediate risk analysis endpoints, the "health endpoint" often will be indirectly considered as part of the analysis.

Different endpoints for different media/hazards are often used in regulations, norms, and standards. This situation makes it very difficult, if not impossible, to compare the impacts using the risk analysis products tied directly to regulations and standards. Several papers in this book refer to risk analysis studies that go consider the gamut, from event occurrence to environmental or health impacts, with outputs of health risks for all contaminants, all media, and all exposure pathways. In that way, a common basis is used to compare the relative impacts.

The figure shows that the range of different risk analysis endpoints represents a continuum of ways of characterizing potential hazards. The appropriate endpoint for a risk analysis will depend on the needs of an application. When many hazards are to be compared, an endpoint is needed that will allow that comparison. When a single hazard is being evaluated, it may well be sufficient to evaluate what the form of the material is and will be in the future. If the risk analysis study is focused on meeting standards, norms, or regulations, then these define the logical endpoints for the study.

One of the first questions that arise when decision makers consider conducting an risk analysis concerns the type and amount of data required. Figure 2.2 shows why it is difficult to define the data until the scope of the risk analysis has been clearly defined. Moving left to right in the figure, each column feeds data to the next column. That is, the outputs of one column are input data required by the next column. The situation is further complicated by the fact that the risk analysis may start in any column. For example, if monitored environmental concentrations are available, the analysis may start

14

Risk Analysis Endpoints		Sequential Steps in Evaluating Various Risk Analysis Endpoints						
Product	Examples	Hazard Description	Source Designation	Release of Hazard	Transport Processes	Exposure	Uptake	Health Effects
1. Material description	Radioactive, carcinogenic, explosive, etc.	Based on properties of material						
2. Event occurrence	Drop during handling; Systems failure	→	Occurrence Frequency					
3. Release to environment	Accident; Vent or stack	→	→	Rate (amount per time)				
4. Environmental concentrations	Standards for monitored concentrations	→	→	→	Concentration in air, soil, water, meat, fruit, vegetables			
5. Dose to receptors	Used in radiation protection	→	→	→	→	Dose (time integrated exposure)		
6. Receptor uptake	Organ concentrations	→	→	→	→	→	Concentration	
7. Health or environmental receptor impacts	Cancer Risk, other diseases; Ecological Impacts	→	→	→	→	→	→	Impact, Risk (probability)

Figure 2.2 Alternative risk assessment endpoints

with the input of these concentrations—and the risk analysis will not explicitly need the input data for that column or the other columns to the left.

Quite different input data are needed in each column. Material description, for example, requires data such as the identity of constituents; their chemical, physical, and radioactive properties; their volume; their location; and their inventory. Examples of the types of data typically required in other steps include:

1. Event Occurrence—potential events of concern, failure rates, event frequencies
2. Release—potential modes of release (accidents including fires, spills and explosions; routine releases such as stack/vent releases, pipe discharges, suspension/volatilisation of materials from ponds, contaminated soils), containment properties, release barriers, release mechanisms, properties of released materials
3. Environmental concentrations—environmental properties of the media in which the released materials are transported, including dispersion, degradation, deposition, and transport rates, or parameters used to define these rates
4. Dose to receptors—definition of receptors as well as timing and duration of exposures, including demographic information, agricultural activities and production rates, timing and duration of recreational activities, and local dietary habits
5. Uptake, concentration in receptors—uptake and retention rates in people and food (crops, farm animals, and wild game)
6. Health or environmental receptor impacts—toxicology data to define the potential impacts as well as information on the makeup of the population, including potentially more sensitive population segments.

In setting up a risk analysis that will start with the source material and go to some direct or indirect health or environmental impact, it is important to develop a conceptual model that includes all analysis steps. Although these steps are conducted in sequence, there are important data dependencies in both directions across the table. For example, the data on potential environmental receptors can define the event and media to be considered.

Data in any of these steps may have great uncertainty that may translate to increased uncertainty in the risk results. Sensitivity and uncertainty studies are very useful in identifying the most critical parameters and understanding the uncertainty of the risk estimates.

2.3. Risk Management

We Athenians, in our own persons, take our decisions on policy and submit them to proper discussion. The worst thing is to rush into action before the consequences have been properly debated. And this is another point where we differ from other people. We are capable at the same time of taking risks and of estimating them beforehand. Others are brave out of ignorance; and when they stop to think, they begin to fear. But the man who can most truly be

accounted brave is he who best knows the meaning of what is sweet in life and what is terrible, and he then goes out undeterred to meet what is to come.
Pericles' Funeral Oration in Thucydides'
History of the Peloponnesian War

Risk analysis was originally done to understand what the risk was. Once we reached that goal, the next step was risk management: creating things with less inherent risk and controlling the risk.

Risk analysis can help managers approach legacy issues and manage them more effectively. The purpose of this book is to present information to facilitate and enable decision-making activities affecting the environment and human populations in the NATO member and partner countries, regardless of their cultural underpinnings. The book reorganizes information studied during the ASI from a unified point of view that was only possible after the broad ranging material had been assimilated.

The book is unique in that it recognizes that risk perception is the product of many factors in our lives and that cultural differences between the East and West can have significant impact on how we view risk and measures to control it. Taken as a whole, it provides answers to a number of key questions about risk management:

- What factors must the decision maker in each of these cultures consider in selecting among alternative options?
- How do cultural factors influence these decisions?
- How can better information be provided to decision makers in the East and in the west to help them make the best decisions for their people?

Two previously alternative approaches receive focus–facility-centred risk management (i.e., use of risk analysis to control the risk to facilities) and human-centred risk management (i.e., use of risk analysis to control the risk to people in the surrounding areas and within facilities). The resulting viewpoint, Complementary Risk Management, integrates the previous approaches, seeking a balance that best serves each community.

To better appreciate the two previous approaches, consider Figure 2.3. If we consider the kinds of situations we might want to analyse, they can be grouped into four main categories:

- Accidents in operating and storage facilities
- Releases during routine operations and mild upset
- Releases from in situ contamination
- Project cost and schedule.

In each case, direct local management (facility management) can be applied. Management activities are shown in **bold** type. Many different levels of management are possible to support a wide variety of goals. For example, the decision maker may want to reduce the frequency of accidents; reduce the risk of equipment damage, worker injury, or offsite release (i.e., either frequency or consequences); optimise costs and schedules; or minimize negative criticism (from bosses, regulator, or public).

Depending on the source of risk (the four categories above), various consequences and

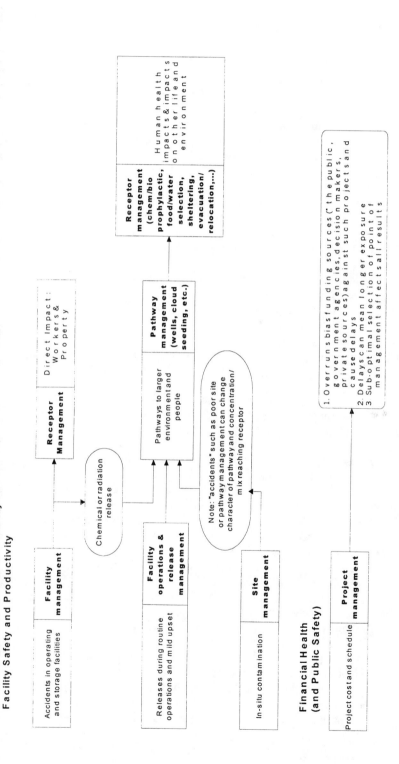

Facility Safety and Productivity

Financial Health
(and Public Safety)

Figure 2.3 Facility- and human-centred risk management

opportunities for further management structures are possible. Possible endpoint consequences are shown in **distinctive** type.

In the case of accident risk, if an accident occurs, there can be direct damage (i.e., the accident can directly damage equipment and injure workers at the scene). Local risk management ("receptor" management) can limit the damage. In addition, contaminants could be released to the environment. Once there is a release, the pathways to the environment and the public will be similar to those from other types of releases, although the energy and composition of the release may be different.

In the case of releases during routine operations and from in situ contamination, the release does not require special failures or accidents, but the risk analysis must characterize the release and again track its pathways to receptors. Here, the source is generally modelled through analysis of data and sampling techniques, compared to the systems analysis tools required for low-frequency accidents. In all release cases, analysis of the pathways and impact on receptors is required. There are opportunities for both management of the pathways and for protection of the receptors. One early distinction was that, when the risk comes from single facilities, it is often most effective to emphasize facility management; when receptors receive insult from many sources, management at the receptor location may be the only cost-effective approach. This is one difference between facility-centred risk management and human-centred risk management.

A societal/values distinction has also been used. Facility-centred risk management may focus on protecting the facility (the investment), while human-centred risk management aims at protecting the public. Often, in the West, the two are heavily intermingled, with a focus on the facility (say the core damage frequency for a nuclear reactor) being used as a surrogate for the off-site risk to humans. If the core does not melt, no significant release can occur. As risk analysis and management methods are adopted in the former Soviet Union (FSU) and Eastern Europe, where different societal value structures exist and a greater variety of hazards are already affecting public health, there is concern that focus on facility-centred risk management could work to the detriment of nearby populations.

The level and purpose of the various risk management activities outlined in Figure 2.3 are closely linked with the goals of the decision maker, be it a regulator, owner, or politician. For example, the U.S. Environmental Protection Agency has supported a compliance-oriented, conservative point estimate risk analysis approach focused on routine and mild upset conditions, while the NRC has worked toward a risk management approach focused on accident risk. Recently, efforts in the Army's chemical demilitarisation program have sought a combined approach that considers worker risk as well as public risk from accidents and routine releases, as described in Chapter 8. Finally we mention the case of project cost and schedule risk. On the surface, this seems a very different type of risk. It may be surprising to find that many of the analytical tools used in accident risk analysis and the general risk framework of Figure 2.1 apply and can be used effectively. In addition, there can be close coupling between cost and schedule and human health risk. Especially in the case of Cold War legacy sites, delay in cleanup and demilitarisation activities can expose workers and the public to substantially more risk than activities associated with processing the waste. It lengthens the time of exposure to routine releases and the exposure time to accidents.

The NATO ASI and this book dedicated significant focus to relevant societal, technical, and management problems in the East. Rather than just a compendium of methods developed elsewhere, the participants and the material considered carefully how the existing methods could be adapted and how they should be used to support unique problems. We even found that such a simple thing as language needed a surprising amount of thought and discussion. Hence we have included the Russian language "glossary" in Appendix C. A simple example will suffice. Regulation in the West speaks of a facility "site" and associated site boundaries. It is not just a matter of picking the right word to translate "site." The very notion of a site as separate from the surrounding community requires a fairly lengthy explanation to make clear the meaning and its application to risk issues.

Specific situations in the East must be considered when laying out a scheme for risk management. To date, the demand for basing decisions on risk analysis has been meagre and applications few. Several reasons are apparent. First the risk methodology was developed in the West, based on a western outlook. In the East, other scientific approaches were applied to resolve known technical issues. Until recently these techniques were not known and their value was not clear to FSU and Eastern European managers. One of the goals of this textbook is to bring these methods to the fore.

Second, other severe problems forced attention on economic problems that threaten survival of institutions and people. Attention to safety of the people and the state of the environment has been necessarily postponed as other crises were addressed. Therefore, one of the purposes of this book is to demonstrate that addressing these public health and environmental problems is necessary, for survival in the future, in spite of the absence of economic resources. Tomorrow it will cost even more. The book makes it clear that methods are available now that can help and can be applied in a step-wise approach of increasing sophistication, making real improvements beginning with a modest commitment of resources.

Finally the lack of use of risk methods can also be attributed to the fact that the democratic procedures for deciding important social issues are not yet well-formed in Eastern Europe. Participants noted that, where we observe a lowered degree of risk management activities, democratic relations between authorities and the public are less well developed. Some observed that the degree of consideration of public opinion is high only during election campaigns. Institutional systems that give a strong voice to the people are developing gradually.

So how is the decision maker to proceed? The approach must balance among the competing needs of the society. An approach that goes beyond self-interest or narrow technical criteria is needed. The Complementary Risk Management approach is illustrated in Figure 2.4.

The challenge is for risk analysts from the FSU and East European countries to carry out analyses relative to risks to workers and the population. These applications should satisfy any established national norms and standards of an acceptable risk. Successful applications will result in optimal risk-based decisions, taking into account available domestic resources and social factors.

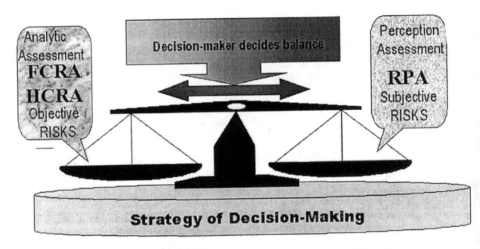

Figure 2.4 Complementary Risk Management: balancing the needs of decision makers

Risk analyses, recommended for decision makers of Eastern Europe, cannot be limited to a single risk analysis methodology. The ASI identified a general set of methodologies that must be considered as part of a Complementary Risk Management effort. These methodologies share many of the same factors as Western types of analysis but differ in the purpose for the effort:

- Facility-centred risk analysis is used mainly to define or demonstrate acceptable risk-based operating parameters for facilities.
- Human-centred risk analysis is used mainly to study and understand human exposures and risk for environmental contamination. These

studies normally are based on fixed operating parameters for hazardous facilities.

- Risk perception assessment analyses the perception of risks by the involved parties (decision makers and local populations).

Misunderstandings about the roles of these analyses can lead to apparent inappropriate competitive views. In fact, all three are needed as part of a Complementary Risk Management approach.

Case studies in the following chapters show the flexibility of using risk analysis methodologies to address different situations in quite different, but appropriate ways. For such a complementary approach to be effective, decision makers must clearly define in advance exactly what issues are being addressed. Experience has shown that clear definitions of the products and their application are essential before starting an applied risk analyses, if results are to be meaningful in the context of the decisions to be made.

The proactive consideration of the many aspects of risk is a relatively new development, even in the West. The trend for the future is clearly away from using single measures of risk and simple upper bounds as input to decision makers. As much as a single number is an appealingly simple approach, decision makers must consider many aspects of risk—and make decisions as a balance of the different types of risk. Furthermore, a single number can, at best, offer a vague comfort, if the number is low. It provides no understanding of the causes of risk, the uncertainty in the results, or what can be done to control the risk.

After years of effort, characterization of accident risks from the legacy sites are only recently being completed in the U.S. Risk analyses at contaminated sites in the FSU are just beginning and can benefit from the risk methodologies developed previously. However, the application of accident risk modelling techniques to weapons handling is relatively new, even in the West. Note too that developing a "risk-informed" basis for regulations is just beginning to be applied in the West. Only a few regulators have promulgated risk-informed regulations. Many more are attempting to follow suit.

In the West, the cost of remediation and long-term management of legacy wastes has proved to be very high. Countries of the FSU cannot afford the magnitudes of costs being experienced in the West and thus must carefully invest what resources they can in keeping risks to a minimum. The Western approach using a balance of risk management, risk analysis, and risk perception is seen as a means of effectively directing priorities for management and cleanup efforts based on maximizing potential population safety.

Risk analysis results have been proposed to provide a basis for defining protective safety, remedial, or alternative actions. One of the most important proposals is to estimate incremental health treatment costs for populations as well as the size of appropriate insurance guarantees for those living in these zones. Such a use of risk analysis would be a departure from the Western view that the only acceptable risks are those with trivial risk levels. Another important proposal that does have an analogue in U.S. air emissions management is to use risk results to define optimal measures to protect the population—even if that new protection is not directly connected with the proposed new activity or facility. The idea is to reward the region that agreed to accept the new hazardous activity by reducing large current risks produced by other sources.

2.4.   Risk Perception

In both the East and West, many organizations have tried to manage risks using only science and policy as the underpinnings. However, both scientists and policy makers are beginning to realize that there is a third leg to the stool supporting risk management decisions[9]. This third leg involves the perception of the risk in question.

In general, risk perceptions vary among the key groups involved in analysing a risk—the scientific or expert component, the policy maker or manager component, and the stakeholder or public component. Expert perceptions of risk are generally grounded in scientific understanding of the phenomena and an appreciation for uncertainty factors. However, personal experiences can still colour expert views of science[10]. Management perceptions, on the other hand, can be informed by scientific understanding, if the manager is an expert or the experts effectively communicated results of their deliberations. However, management perceptions are also informed by the perceptions of their constituencies and other political motivators. Stakeholders—those who perceive themselves as being affected by the risk and its management decision—rely on a variety of sources to gather their information and form their perceptions. This information does not necessarily include information from experts, even when those experts communicate effectively[2].

Factoring these three perspectives into a risk assessment and risk management process is important for several reasons. First, particularly when the goal of risk management is to minimize harm, perception is key because it influences the body's response to environmental contamination. Researchers have found that people threatened by nuclear contamination in particular will experience anticipatory stress. That is, even if they are not exposed, their perception of radiation will result in psychological and neuroendocrine reactions. For example, the perception of risk by residents near the U.S. Three Mile Island nuclear facility influenced their performance, the number of psychosomatic symptoms they reported, and their pyschoneuroendocrine indices 3.5 years after the nuclear accident that occurred there, even when these residents were not actually exposed to any radiation. Similar responses were found associated with radioactive contamination in the FSU and

Brazil[12]. These perception-driven responses must be taken into account in risk management decisions.

Second, risk perceptions must be considered in risk analysis activities because only when perceptions are understood can behaviour be influenced. For example, particularly in the East, people must be encouraged to leave contaminated areas and remain outside them until the area is remediated. A failure to understand the perception of those evacuees can lead to people behaving in ways that endanger their lives and hinder effective remediation.

Second, perception influences communication, and communication is key not only to share expert risk information with stakeholders but to share it with decision makers. The inability to communicate risk information can result in faulty policy, poor remediation choices, and ultimately, lost lives and resources.

Finally, failure to factor perception into the scientific assessment of risk can make risk management more difficult. Policy makers cannot accept risk assessment activities that fail to consider the lifestyles and needs of their constituencies. Stakeholders cannot understand the results of risk assessments unless they can clearly see how their needs have been considered. Scientists studying risk from contaminated soil sites for the European Union found that some of the key questions that required additional research for effective risk assessment did not include algorithms or modelling per se but instead included a number of issues related to perception. These issues included such questions as:

- Is the risk in context?
- What is "acceptable"?
- Who decides?
- What information do they need?[13]

2.4.1. Dimensions of Perception

As mentioned, perception can vary depending on the role one plays in the risk analysis paradigm—expert, manager, or stakeholder. Extensive literature has pointed to the development of mental models—a detailed intellectual picture of what a risk entails, including source of risk, exposure pathways, transport mechanisms, pathogenic effects, and likelihood of exposure and resulting harm. People process new information within the context of their existing model[14]. Perception plays a strong role in developing and an even stronger role in revising these models. Indeed, evidence suggests that once perceptions have been embedded in the model, they are extremely resistant to change[13]. In general, the closer the models of all participants in the risk analysis process match, the easier it will be to assess, manage, and communicate the risks.

What frequently happens in the risk assessment process, however, is that the models developed by scientists and the models developed by stakeholders vary widely. A number of reasons have been posited as to why this discrepancy exists. Prevalent among these reasons is the general lack of understanding among the public of scientific principles. This reason does not explain, however, why scientists with similar educational credentials sometimes disagree on various aspects of risk, such as exposure pathway, harm engendered, or probability of occurrence[10]. While lack of scientific understanding can play a part, other reasons may have greater influence.

Primary Reasons for Differences Between Expert and Stakeholder Perceptions of Risk

The Concerted Action on Risk Assessment for Contaminated Sites in the European Union (CARACAS) study of 1996[13] sought to tackle the problem of contaminated land by bringing together academics, government representatives, and other experts from all European Union member states plus Norway and Switzerland. These experts found that the major difference between the perception of risk by experts and stakeholders was that the stakeholders tended to view the risk more intuitively. Stakeholders often viewed risks that are conspicuous, known from experience, recent, and occurring nearby as more likely to occur and more harmful than did their scientific counterparts. Differences of perception were also seen in the neglect of initial probability of certain phenomena and the willingness to extrapolate estimates of probability from analogies or other chance phenomena[13].

Another reason expert and stakeholder models often do not match is the existence of outrage factors. Sandman postulated that risk is actually made up of the hazard, which can be calculated through typical risk assessment methodologies, plus the outrage, or perception of the risk, which is more difficult to quantify[15]. Building on research from Slovic and others[16], researchers have identified over 40 outrage factors that can result in public over- or underestimating risk. These factors can include the potential for catastrophic consequences, the aerial size of the impact, the level of personal control, the level of personal experience, the equity of distribution of risk (those with benefits also are those at risk), voluntariness of exposure, level of associated dread, and visibility. These factors can explain why legacy contamination can elicit a high level of emotional response from stakeholders potentially exposed to it. Nuclear waste in particular scores highly on these outrage factors.

Beyond the outrage factors, the question of identity can also play a part in the disagreement of expert and stakeholder perceptions of risk. Changes in technologies or lifestyles as a result of risk can be viewed as a challenge to one's sociocultural identity[11]. For example, workers asked to be retrained when their industry has closed because of contamination may react with hostility because of a threat to their identity.

Secondary Reasons for Differences Between Expert and Stakeholder Perceptions of Risk

The lack of dialogue among experts, stakeholders, and managers has also been cited as a contributor to the disparate perceptions of risk. Certainly the ability and willingness to communicate, or lack thereof, can strongly influence how risk information is perceived[17]. However, unless existing perceptions and cultural differences are considered, dialogue can be ineffective and potentially disastrous[11].

The news media can also play a role in the development of perceptions. In general, public trust in institution and government has been decreasing in the U.S.[18], making it more likely that at least some stakeholders will rely on the news media as a primary source for informing their mental models of certain risks. However, it remains a question as to whether the media creates public furore or merely enlarges upon it. A certain amount of public interest must first be evident if the news media is to find a story worthy of coverage.

Key Differences in Risk Perceptions Between East and West

A number of factors affect how stakeholders in the East and West view risk issues in general, and those of legacy waste in particular. One such factor is culture. Researchers have identified an attitude prevalent in some countries of the FSU that appears to be tied to cultural views of the role in society of particular groups. Such groups are more likely to perceive themselves as victims and develop surprisingly effective coping strategies associated with living near hazardous materials[19]. In other cases, however, such an attitude erupts into social protest. The different reactions seemed to be tied to whether the risk is undertaken voluntarily (as for cleanup workers) or involuntarily (as in the population living in the contaminated area)[12,20]. Contrast this to the U.S. where involuntary exposure almost always results in increased outrage and vocal and physical action[15].

Another factor is the different role of the news media and politicians. Both are often viewed as less-than-trustworthy sources. Scientists, on the other hand, are more likely to be viewed as trustworthy. Contrast this to the U.S., where the scientists are often portrayed as villains and misguided fools in popular culture and discredited by the news media.

Another factor is the identification of sustainable development as a political force. The "Green" Party in many countries runs on a platform of environmental issues, whereas such issues are only tangentially represented in American politics. This enhanced visibility of environmental risk issues in Europe increases the likelihood that risk perceptions will be formed in an atmosphere of intensity.

2.4.2. Including Perceptions in Risk Assessment and Risk Management

For risk assessment and risk management to be effective, then, risk perceptions must be factored into the process. In past decades, risk perceptions often met the risk assessment process only at the end, where a head-on collision was often the result. In the last 10 years, however, models of risk analysis developed in the U.S. and Europe have encouraged the inclusion of risk perception throughout the process.

The landmark CARACAS study[13] in the European Union developed a model of the risk analysis process that begins and ends with perceptions. The study identified the fact that a risk analysis process is only begun when someone perceives the potential for risk. In essence, experts do not analyse a situation unless a manager or expert feels some risk is inherent in the situation. These perceptions dictate the kinds of questions experts ask concerning the risk (What kind of risk? How much? When? Where? To Whom?) and hence the data they gather and methods they use to study it. Perceptions, however, must also guide how they communicate results to both managers and stakeholders.

In the U.S., the National Research Council sponsored a similar landmark study[21], which came to a similar conclusion. Risk assessment and risk management cannot be divorced; one informs the other. Risk perceptions must be factored through the process—in the choice of questions, in the data to be gathered, in the analysis methods, in interpreting the results, and finally, in communicating the results and deciding on a choice of action. Only when all three legs of the risk analysis stool—assessment, perception, and management—are included is the analysis successful in developing a lasting, useful solution to an environmental problem.

Risk perception has also been identified in models for how the East and West manage risks. The European Union was the first to act upon the philosophy of the Precautionary Principle[22]. This principle, at its simplest, holds that no action should be undertaken unless it can be proven that no increased risk will result. If the lack of increased risk cannot be proven at levels satisfactory to stakeholders, then the action may only proceed with caution. Thus, risk perceptions heavily drive decisions on technology development and application of scientific breakthroughs, which, by nature, do not have solid proof as to their viability. The U.S. has shied away from following such a stringent principle, although international trade is beginning to be influenced by such practices[23].

The European Union has also developed a method of stakeholder involvement predicated on perceptions. Public trust in government action is based on the notion of transparency. If everything the government does is open and readily accessible to the public, confidence and trust in government will increase, and the need for more visible public involvement will decrease. The U.S., on the other hand, is functioning under a cloud of years of government secrecy, if for national security reasons at the time. Transparent government is viewed as only the beginning steps in public involvement, with stakeholders expecting to have a seat at the decision-making table rather than viewing it from outside. The U.S. model, then, is built upon the principle of debate rather than openness.

2.5.   Complementary Risk Management

The Cold War left a serious legacy across vast areas of land in both the East and the West. Facilities that created, assembled, or stored the nuclear arsenal and its waste products are but the start of the problem. Beyond them are contaminated facilities belonging to the military and acres of land surrounding both types of facilities, some of it heavily contaminated. The social cost, for both the families of those who participated in this work as well as those populations living on or near the contaminated land, is staggering. Managing risks that cross geographical, political, and cultural boundaries is nothing short of challenging.

In this book, we suggest an approach that seeks to balance competing societal demands. A variety of detailed approaches are examined that can be integrated into a single, balanced vision—Complementary Risk Management.

2.6.   References

1. Magee, R.S., Drake, E.M., Bley, D.C., Dyer, G.H., Falter, V.E., Gibson, J.R., Greenberg, M.R., Kolb, C.E., Kossen, D.S., May, W.G., Mushkatel, A.H., Niemiec, P.J., Parshall, G.W., Tumas, W., and Wu, J-S. (1997) Risk Assessment and Management at Deseret Chemical Depot and the Tooele Chemical Agent Disposal Facility. Committee on Review and Evaluation of the Army Chemical Stockpile Disposal Program, National Research Council, National Academy Press, Washington, D.C.
2. U.S. Nuclear Regulatory Commission. (1975) Reactor Safety Study, An Assessment of Accident Risks in U.S. Commercial Nuclear Power Plants. WASH-1400 (NUREG-75/014), U.S. Nuclear Regulator Commission, Washington, D.C.
3. Savage, L.J. (1974) The Foundation of Statistics, Second Revised Edition. Dover Publications, New York.
4. deFinetti, B. (1974-75) Theory of Probability: A Critical Introductory Treatment, translated by Antonio Machi and Adrian Smith. Wiley, London and New York, Vol. 1, 1974; Vol. 2, 1975.
5. Hogarth, R.M. (1975) Cognitive processes and the assessment of subjective probability distribution. *Journal of the American Statistical Association*, **70**(350) 271-94.
6. Budnitz, R.J., Apostolakis, G., Boore, D.M., Cluff, L.S., Coppersmith, K.J., Cornell, C.A., and Morris, P.A. (1998) Use of technical expert panels: applications to probabilistic seismic hazard analysis. *Risk Analysis* **18**(4) 463-469.
7. Pickard, Lowe and Garrick, Inc. (1983) Seabrook Station Probabilistic Safety Assessment. PLG-0300, prepared for Public Service Company of New Hampshire and Yankee Atomic Electric Company.

28

8. Kossen, D.S., Kolb, C.E., Archer, D.H., Armenante, P.M., Bley, D.C., Chandler, J.L.R., Crimi, F.P., Drake, E.M., Gibson, J.R., Greenberg, M.R., Kelly, K.E., Lederman, P.B., Magee, R.S., Mathis, J.F., May, W.G., McGinnis, C.I., Mushkatel, A.H., Rigo, H.G., Saito, K., Short, W.L., Stancell, A.F., Tannenbaum, S.R., Tolman, C.A., and Tumas, W. (1999) Tooele Chemical Agent Disposal Facility Update on National Research Council Recommendations. Committee on Review and Evaluation of the Army Chemical Stockpile Disposal Program, National Research Council, National Academy Press, Washington, D.C.

9. National Research Council. (1994) *Science and Judgment in Risk Assessment*. National Academy Press, Washington, D.C.

10. Hodges, M. (1992) How scientists see risk. *Research Horizons* **Summer** 22-24.

11. Crease, R.P. (1999) Conflicting interpretations of risk: the case of Brookhaven's spent fuel rods. *Technology* **6** 495-500.

12. Collins, D.L., and de Carvaiho, A.B. (1993) Chronic stress from the Goiania ^{137}Cs radiation accident. *Behavioral Medicine* **18** 149-157.

13. Ferguson, C. (1998) *Risk Assessment of Contaminated Sites in Europe, Volume 1: Scientific Basis*. LQM Press, Nottingham, England.

14. Morgan, M.G., Fischhoff, B., Bostrom, A., Lave, L., and Atman, C.J. (1992) Environ. Sci. Technology **26** 2048-2066

15. Sandman, P.M. (1992) Hazard versus outrage: responding to public concerns about the risks of industrial gases. *International Oxygen Manufacturer's Association Broadcaster* **Jan-Feb 1992** 6-17.

16. Covello, V.T., P.M. Sandman, and P. Slovic. (1988) *Risk Communication, Risk Statistics, and Risk Comparisons: A Manual for Plant Managers*. Chemical Manufacturers Association, Washington, D.C.

17. Hance, B.J., Chess, C., and Sandman, P.M. (1988) *Improving Dialogue with Communities: A Risk Communication Manual for Government*. New Jersey Department of Environmental Protection, Division of Science and Research, Trenton, New Jersey.

18. Kasperson, R.E. (1986) Six propositions on public participation and their relevance for risk communication. *Risk Analysis* **6** 275-281.

19. Rumyantseva, G., Allen, P., and Pliplina, D. (1995) Coping strategy and risk communication of the population involved in the Chernobyl accident. Proceedings of the Society for Risk Analysis Annual Meeting, Society for Risk Analysis, Washington, D.C., United States.

20. Collins, D.L. (1992) Behavioural differences of irradiated persons associated with the Kryshtym, Chelyabinsk, and Chernobyl nuclear accidents. *Military Medicine* **157** 548-552.

21. National Research Council. (1996) *Understanding Risk: Informing Decisions in a Democratic Society*. National Academy Press, Washington, D.C.

22. Commission of the European Communities. (2000) *Communication from the Commission on the Precautionary Principle*. Brussels, Belgium.

23. Caswell, J.A. (2000) Regulatory standards and international trade, presented at the Harvard School for Public Health Regulatory Analysis Workshop, Boston, United States.

PART II: LEGACIES

*Although hazards of the Cold War legacy are located at 144 sites
in the United States, the majority of contaminants are located at
six sites in Washington, Idaho, Nevada, New Mexico, Tennessee,
and South Carolina .... with 36 million cubic meters of waste
categorized. By-product materials (e.g., uranium mill tailings)
account for 88% of the volume, while high-level radioactive
wastes account for only 1% of the total volume.*
Dr. Alvin Young, Director, U.S. Department of Energy
Center for Risk Excellence

*A significant feature of the nuclear weapons complex of the former
USSR was the concentration of industrial facilities at special
industrial sites which are widely known nowadays as The
Industrial Association (IA) Mayak in Chelyabinsk Province, The
Siberian Chemical Combine (SCC) in Tomsk Province, and The
Industrial Association for Mining and Chemical Combine (MCC)
in the Krasnoyarsk Territory . . . About 500 million cubic meters
of radioactive waste with an aggregate radioactivity about 1.7
billion curies were accumulated.*
Yuri Gorlinsky, Director RTC Systems Analysis, Russian
Research Centre, Kurchatov Institute

These paragraphs only begin to describe the vast legacies of waste
remaining from nuclear materials production in the U.S. and the former
Soviet Union. The numbers cannot describe the costs in lasting
contamination or the dedication of those seeking to remediate these lost lands.

3.   Radiation Legacy of the Soviet Nuclear Complex

The radiation legacy left from nuclear weapons production is one of the forces compelling countries of the former Soviet Union to undertake risk assessment and risk management. This legacy is also compelling them to understand and manage risk perceptions. Operations of the nuclear production complex of the former Soviet Union resulted in the accumulation of about 500 million cubic meters of radioactive waste with an aggregate radioactivity about 1.7 billion curies. This chapter describes, based on published information, the structure, composition, and arrangement of that production complex; sites of nuclear weapon tests; locations for storage and disposal of radioactive waste; and territories exposed to radioactive contamination as a result of nominal activity and radiation accidents. As the author notes, if large affected areas are considered, then the historical radiation fallout from atmospheric nuclear weapon testing exceeds in magnitude many of the sources discussed in this chapter.

The term "radiation legacy" appeared in scientific publications and the mass media in the U.S. and Russia near the end of the last century practically simultaneously with the end of the Cold War. The first comprehensive publication about a radiation legacy in the former USSR is the book, *Behind the Nuclear Curtain: Radioactive Waste Management in the Former Soviet Union*[1], which was based on widely available materials by Russian and foreign authors.

In the same year as the book was published (1997), Russian agencies published official data in the form of an analytical review of the radiation legacy of the former Soviet Union (FSU)[2]. These data were collected within the framework of the ISTC Project "Radleg" with involvement by the International Institute for Applied Systems Analysis (IIASA) and based on official data published by the Ministry of the Russian Federation for Atomic Energy (Minatom) and other Russian organizations. In 2000, based on the material of this analysis, the English publishing house EARTHSCAN together with IIASA issued the book *The Radiation Legacy of the Soviet Nuclear Complex (An Analytical Overview)*[3].

This chapter uses those resources to describe the structure, composition, and arrangement of the nuclear weapons production complex of the FSU; sites of nuclear weapon tests; locations for storage and disposal of radioactive waste; and territories exposed to radioactive contamination as a result of nominal activity and radiation accidents.

3.1. Background

Historically, the nuclear complex in the Soviet Union produced nuclear and thermo-nuclear weapons. It included
- reactors to produce weapons-grade plutonium and tritium
- industrial facilities to produce nuclear fuel for these reactors
- industrial plants to produce highly-enriched metallic uranium
- facilities to process spent nuclear fuel and separate weapons-grade plutonium
- industrial facilities to manufacture components of nuclear weapons from highly-enriched metallic uranium and plutonium
- industrial plants and organizations to develop and produce nuclear charges and associated components.

Industrial facilities to manufacture nuclear fuel for military ship nuclear reactors and installations to process spent nuclear fuel (SNF) from these reactors can also be included in this complex.

These facilities were housed on unified industrial sites where appropriate technological installations existed. These sites included production of initial nuclear materials and processing of spent materials for military and civilian purposes. These purposes included extraction and enrichment of uranium ore, conversion of uranium hexafluoride and enrichment by an isotope of uranium-235, production of nuclear fuel and radiochemical processing of spent fissile materials, and radioactive waste handling.

A significant feature of the nuclear weapons complex of the FSU was the concentration of industrial facilities at special industrial sites chosen for their geographical position and availability of power supply, water resources, and work force. Such plants created in this manner are widely known nowadays as The Industrial Association (IA) Mayak in Chelyabinsk Province, The Siberian Chemical Combine in Tomsk Province, and The Industrial Association for Mining and Chemical Combine in the Krasnoyarsk Territory (Figure 3.1). As a rule, these industrial complexes

Russian Weapons Complex

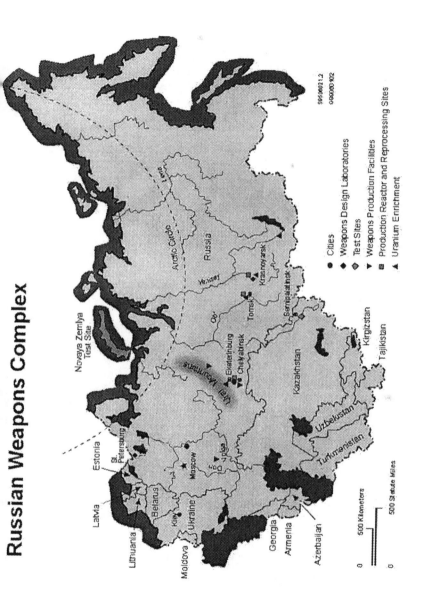

Cities
Weapons Design Laboratories
Test Sites
Weapons Production Facilities
Production Reactor and Reprocessing Sites
Uranium Enrichment

Figure 3.1 Nuclear weapons production complex of the former Soviet Union

have a complete work cycle and include production nuclear reactors, installations for processing fissile material and manufacture of weapons-grade uranium and plutonium, and facilities for processing and handling radioactive waste. In addition, the industrial sites host a full set of servicing activities.

One of most significant peculiarities of this complex was its independent and self-sufficient character. Everything that was required for its proper functioning was produced by the enterprises and organizations. Though these enterprises and organizations were located in territories of different republics of the FSU, they were part of a unified functional system of the former Ministry of Medium Machine-Building Industry of the USSR. The current assignee of this Ministry in Russia is Minatom. Other activities are conducted in cooperation with the enterprises and organizations of other ministries of the FSU.

Although activities were far flung, more than 80% of the nuclear industrial potential of the former USSR remained in the territory of Russia when the USSR broke up. Removal of nuclear weapons from Ukraine, Belarus, and Kazakhstan was completed in 1996, and now all nuclear weapons of the FSU are placed only in Russia. All fission materials produced for military purposes are also disposed of in Russia.

These activities are the main source of radioactive waste. During operations of the nuclear production complex of the FSU, about 500 million m^3 of radioactive waste with an aggregate radioactivity about 1.7 billion Ci were accumulated in Russia. All radioactive waste and spent nuclear fuel are placed in special structures (steel containers, reinforced-concrete and concrete storehouses, underground storage, etc.), thoroughly secured, and permanently monitored.

3.2. Extraction, Enrichment, and Processing of Uranium Ores

The complex for these activities consists of nine mining and processing plants. They are located in six independent states of the Commonwealth (the Russian Federation, Republic of Ukraine, Republic of Uzbekistan, Republic of Kazakhstan, Republic of Tajikistan, and Republic of Kyrgyzstan). The complex includes the following enterprises:

- Argun Industrial Mining and Chemical Association
- Lermontov industrial complex "Almaz"
- Navoi Mining and Metallurgical Combine
- KASKOR Joint-Stock Company (before 1992 known as the Caspian Mining and Metallurgical Combine)
- Industrial Association "Tselinnyi" Mining and Chemical Combine (IA Tselinnyi MCC)

- Industrial complex "The Eastern Combine for Rare-Earth Metals" (IA "Vostokredmet," before 1992 known as The Leninabad Mining and Chemical Combine)
- Scientific and Industrial Association, "Eastern Mining and Enrichment Combine" (SIA Eastern MEC; the association was created on the basis of "The Eastern mining and ore-dressing combine.")
- Industrial Association "Southern Combine for Polymetals" (IA Yuzhpolimetall; the Combine was created on the basis of the "The Kyrgyz Mining Combine.")
- Industrial Association "Dnieper Chemical Plant."

At the initial stage of a nuclear fuel cycle, when uranium ores are being extracted, dressed, and processed, the environment is often contaminated by solid, liquid, and gaseous radioactive waste. The levels of radioactivity in this waste are insignificant when compared to levels in waste generated at other stages in the cycle. Nevertheless, this waste may create a local increase in radiation for a long time (hundreds and thousands of years).

Waste from ore extraction and processing contains long-lived radionuclides. Therefore, for those working throughout their careers in underground uranium mines and associated industrial sites, and those living in territories near these facilities, radiation levels are sometimes higher than background because of the radioactive decay products mentioned above. The main source of contamination in these cases is the waste being generated while the ores are being processed. This waste is accumulated in tailings dumps. Long-term operation of uranium extraction and processing facilities resulted in most of the low-activity waste. This waste includes uranium-238 and thorium-232 in waste material dumps, in tailings dumps from hydro-metallurgical enterprises, and in basins of mine waters.

The activity accrued in soil and bottom sediments in nearby rivers reached 10 to 15 Bq/L while the norm was 0.111 Bq/L. The land area of associated with this mining and extraction includes 130 km^2 (Table 3.1).

3.3.   Production of Uranium Hexafluoride and Isotopic Enrichment of Uranium

Initially, the complex to produce enriched uranium was created only to solve a problem of national defense. The production of highly enriched uranium for nuclear weapons terminated in 1988; now the industrial complex developed for that purpose only provides fuel for nuclear electric power plants. The complex includes plants for deriving uranium hexafluoride and plants for isotopic enrichment of uranium. The infrastructure also includes industrial subdivisions engaged in recycling, processing, and storage of liquid and solid radioactive uranium waste.

TABLE 3.1 Summary data on radioactive mass and activity accumulated at uranium ore mining and processing industry enterprises of the former Soviet Union, as of 1 January 1990

Enterprise	Dumps of Unamenable Ores		Tailings Dumps	
	Mass, 1,000 tons	Activity, Bq 10^{14}	Mass, 1,000 tons	Activity, Bq 10^{14}
Argun Industrial Mining and Chemical Association[a]	211,260	16.7	69,170	29.0
IA Almaz	8,403	1.4	14,047	16.9
Navoi MCC	166	0.034	52,800	74.0
KASKOR Joint-Stock Company	NA	NA	68,145	0.629
IA Tselinnyi MCC	16,200	4.5	56,600	20.0
IA Vostokredmet	1,847	0.07	33,684	2.5
SIA Eastern MEC	3,770	0.34	37,750	21.0
IA Yuzhpolimetall	110,873	6.0	34,461	32.0
IA Dnieper Chemical Plant	NA	NA	52,017	27.0
Total	352,519	29.0	418,674	223.0

Notes: a = As of 1 January 1993; IA = Industrial Association or complex; MCC = Mining and Metallurgical Company; NA = not available; SIA = Scientific and Industrial Association; MEC = Mining and Enrichment Combine.

By 1 January 1997, Russia operated two plants for deriving uranium hexafluoride and four plants for producing enriched uranium:

- Urals Electrochemical Combine, in Novouralsk, Sverdlovsk Province (enrichment plant)
- Electrochemical Plant, in Zelenogorsk, the Krasnoyarsk Territory (enrichment plant)
- Siberian Chemical Combine, in Seversk, Tomsk Province (enrichment plant and plant for deriving uranium hexafluoride)
- The Angarsk Electrolysis Chemical Combine, in Angarsk, Irkutsk Province (enrichment plant and plant for deriving uranium hexafluoride).

All enrichment plants initially used gas-diffusion technology. In 1962, however, gas centrifuges were introduced. By 1992 the operation and maintenance phase of gas-diffusion technology completely ceased. Now only gas centrifuges are used at all enrichment plants.

In operations, enrichment plants offer the same level of risk to the environment as the plants and structures for extraction of uranium and thorium ores. The gases used in production undergo special cleaning before they leave the plant as exhaust. The solutions containing uranium and fluorine are transferred to the liquid waste processing area used for extraction of uranium. In addition, taking into consideration the peculiarities of manufacturing processes to produce uranium hexafluoride and enriched uranium, accidents accompanied by a release of uranium hexafluoride or its compounds cannot have catastrophic consequences, because such accidents would be restricted by the framework of the industrial rooms.

On the other hand, when it comes to waste, plants for deriving uranium hexafluoride and producing enriched uranium are ecologically the cleanest productions in the nuclear fuel cycle. Waste consists of "tailings" in the form of uranium hexafluoride. It is stored in pressure-tight steel containers in special areas and continuously monitored for radiation.

Plants also produce solid and liquid radioactive waste as well as a negligible amount of gaseous release of radionuclides. The treatment technology for fluid waste precludes their accumulation. Therefore, the waste is either buried or discharged into an open hydrologic system (if the radiation concentration in sewer water is lower than permissible). Concentration of radionuclides in sewage water is generally two orders of magnitude lower than specifications regulated for potable water.

At plants being remediated, deactivation and dismantling of equipment may require partial remelting. The resultant sludge containing residual activity is directed to solid waste storage. Any water leaking from this storage is collected in subdivisions in special containers. The water is then periodically pumped to transport containers and transferred for processing (extraction of uranium) to specialized subdivisions of the plants at the same industrial sites. When the material remaining in storage contains only the solid phase, the storage areas undergo isolation by covering their surface with soil.

There are no territories polluted by radionuclides as a result of activity at the given plants (Tables 3.2 to 3.5).

TABLE 3.2 Amounts and radioactivity of accumulated solid radioactive waste, as of 1 January 1997

Enterprise	Production Type	Amount of Radioactive Waste, metric tons	Total Activity, Bq
Siberian Chemical Combine	Uranium hexafluoride production	25,490	9.06×10^{12}
	Uranium enrichment	10,610	1.40×10^{11}
Angarsk Electrolysis Chemical Combine[a]	Uranium hexafluoride production	1,500	6.43×10^{13}
Urals Electrochemical Combine	Uranium enrichment	29,070	1.08×10^{12}
Electrochemical Plant	Uranium enrichment	6,680	8.51×10^{10}

Notes: a = only medium-specific radioactive wastes.

TABLE 3.3 Amounts and radioactivity of accumulated radioactive sludges formed as a result of waste decontamination, as of 1 January 1997

Enterprise	Production Type	Amount of Sludges, m^3	Activity, Bq
Urals Electrochemical Combine	Uranium enrichment	30,194	3.33×10^{11}
Electrochemical Plant	Uranium enrichment	7,140	1.48×10^{10}

TABLE 3.4 Radionuclide[a] releases into the atmosphere in 1996

Enterprise	Release, Bq	% of MPR
Uranium hexafluoride production plants		
1. Siberian Chemical Combine	7.25×10^8	0.6
2. Angarsk Electrolysis Chemical Combine	1.76×10^8	0.05
Total	$\mathbf{9.01 \times 10^8}$	-
Uranium enrichment plants		
1. Siberian Chemical Combine	1.48×10^7	0.006
2. Angarsk Electrolysis Chemical Combine	1.81×10^8	0.05
3. Urals Electrochemical Combine	4.00×10^9	1.0
4. Electrochemical Plant	7.40×10^7	0.1
Total	$\mathbf{4.23 \times 10^9}$	-

Notes: a = The sum of α-active isotopes.

TABLE 3.5 Radionuclide releases into open water reservoirs in 1996

Enterprise	Water Reservoir	Volume of Sewage Water, 1,000 m^3	Radionuclide Release, Bq	% of MPR
Angarsk Electrolysis Chemical Combine	Angara River	27,953	3.70×10^7	0.0001
Urals Electrochemical Combine	Neivo-Rudyansk water storage	13,718	2.58×10^{10}	16.7
Siberian Chemical Combine[a]	Tom River	3,532	--	--
Electrochemical Plant[b]	--	--	--	--

Notes: a = The technological scheme of the sewage water collector makes it difficult to distinguish releases of individual enterprises of the nuclear complex; b = The design does not stipulate any release into open water reservoirs.

3.4. Manufacture of Nuclear Fuel

The industrial manufacture of nuclear fuel was centered at plants in the Russian Federation. Now these plants are included in a structure of the joint-stock company "TVEL" (fuel element). They are as follows:

- Joint-stock company "Machinery Plant," in Electrostal, Moscow Province
- Joint-stock company "Tchepetsky Mechanical Plant," in Glazov, Udmurt Republic
- Joint-stock company "The Novosibirsk plant for concentrated chemical products," in Novosibirsk, Novosibirsk Province
- State Enterprise "The Moscow plant of polymetals," in Moscow.

There is also an Industrial Association in the republic of Kazakhstan named "The PO Ulbinsky Metallurgical Plant."

Initial materials for manufacture of nuclear fuel include uranium ores and concentrates, oxides, hexafluoride of natural uranium, or uranium enriched by an isotope of uranium-235. As a result of chemical and metallurgical processes, metallic uranium, its alloys, and fuel are obtained based on dioxides of uranium, enrichment by uranium-235, and composition mixtures. The finished product includes fuel elements, assemblies, and cassettes intended for nuclear reactors of various assignments.

These plants generate radioactive waste. After relevant processing, this waste it is directed to tailings dumps as pulps or solid waste. Minimal discharge also occurs to air and water. These discharges have radionuclide concentrations lower than established standards. There were no radiation accidents associated with these plants that were accompanied by environmental contamination.

Wastes include low and medium levels of activity. Solid radioactive waste consists of 5,650,000 tons at Russian plants and 1,352,000 tons at the Ulbinsky plant in Kazakhstan (by 1990). At all plants, environmental contamination is caused mainly by nuclides of uranium. The area of contaminated land in Russia, as of 1 January 1996, consists of 1.7 km^2 (Table 3.6).

TABLE 3.6 Characteristics of radioactive waste accumulated at nuclear fuel production enterprises, as of 1 January 1996

Enterprise	Activity, Bq	Main Contaminating Radionuclides
Russian Federation		
1. Machinery Plant	4.2 x 10^{13}	Uranium radionuclides, radium-226
2. Chepetsky Mechanical Plant	7.0 x 10^{13}	Uranium radionuclides, radium-226
3. Novosibirsk Plant of Chemical Concentrates	3.0 x 10^{13}	Uranium radionuclides, radium-226
4. State Enterprise Moscow Plant of Polymetals	NA	NA
Total	1.42 x 10^{14}	--
Republic of Kazakhstan		
1. PO Ulbinsky Metallurgical Plant[a]	3.8 x 10^{13}	Uranium radionuclides, americium-241, strontium-90, plutonium-239

Notes: NA = not applicable; a = As of January 1, 1990.

3.5. Production of Plutonium and Radiochemical Processing of Spent Nuclear Fuel

Work to derive weapons-grade plutonium was centered at the following enterprises:

- Industrial Association Mayak, in Ozersk, Chelyabinsk Province
- Siberian Chemical Combine, in Seversk, Tomsk Province
- Mining and Chemical Combine, in Zheleznogorsk, the Krasnoyarsk Territory.

The goal of extracting plutonium is first to separate the metals and, second, to clean the plutonium and uranium from fission products. The finished product is an article made of metallic plutonium.

This work generates radioactive waste. After relevant processing, the waste is directed to tailings dumps as pulps or solid waste. Liquid and solid radioactive waste with high, medium, and low levels of activity are currently stored at the plants engaged in processing irradiated uranium. These wastes may contain fission products, nuclides of uranium, and transuranic elements. The following subsections provide additional details for each of the industrial associations involved in the production of plutonium or the processing of spent nuclear fuel.

3.5.1. The Industrial Association Mayak

The industrial complex occupies a territory about 160 km^2. This territory is surrounded by a protective zone is of 250 km^2, which is in turn surrounded by a watch zone is 1800 km^2. The enterprise consists of the following basic activities:

- Production reactors
- Radiochemical processing
- Radioisotopic production
- Chemical-metallurgical processing
- Chemical production.

The first industrial uranium-graphite reactor with a power of 100 MW operated from 1948 to 1987. After 1948, four other uranium-graphite reactors were put into operation at the combine. The first heavy-water-moderated reactor with a power of 100 MW began operating in 1951. Early in its operation, this reactor, which was fueled by natural uranium, was used mainly for the production of plutonium. It was shut down in 1965 and then dismantled. A second heavy-water-moderated reactor began operations in December 1955. This reactor operated for 10 years before being shut down in 1965. Later, this reactor was dismantled. Another reactor was built in the same place and began operating in April 1966. That reactor operated for 20 years before its shutdown in 1986.

A radiochemical plant for separating plutonium produced in the reactors was also built at the combine. It began operations at the end of 1948. In that same year, construction of the chemical-metallurgical plant began. This work was carried out in two stages. During the first stage, a facility was built to transform the final solutions of the radiochemical plant into metal and to obtain articles made of metallic plutonium. Trial production at this facility began in early 1949. In the second stage, operations were put in place to obtain highly-enriched uranium-235.

The total amount of solid radioactive waste accumulated from Mayak operations is approximately 451,000 m^3 with an activity of 1.42 x 10^{19} Bq. The amount and activity of liquid waste is 82,500 m^3 and 5.87 x 10^{19} Bq, respectively. Solid waste is stored in reinforced-concrete storehouses equipped with a water-proof cover .

Waters with low levels of activity from the reactors were discharged into the natural lake Kyzyl-Tash. Before 1953, waters of this lake were free flowing. From 1953 until 1956, floods drew down the waters through the basin of the lake. Since 1957, however, the lake waters have only drained internally. It is used as a basin-cooler as part of the recycling water supply for the nuclear reactors (Figure 3.2).

Radionuclides, which were released by the radiochemical plant as well as the production uranium-graphite reactors, also percolated into the Techa River, whose source was Lake Kyzyl-Tash. Since 1951, however, most discharges have been directed to Lake Karachai.

From 1949 to 1951, waste with medium levels of activity from the radiochemical plant was discharged to the Techa River. These discharges caused contamination of bottom sediments and flood-lands in the upper course of the river. The tandem reservoir system (a cascade of storage ponds) was created to prevent further spreading of radionuclides down the river. This system is intended to store liquid waste with low levels of activity. Since 1951, the natural lake of Karachai has been used as a storage reservoir of waste with medium levels of activity. This lake is 0.45 km^2, shallow, and swampy. Since 1988, this basin has been under remediation.

High-level liquid waste is stored in cooled containers made of stainless steel. For high-level radioactive waste from the radiochemical plant, waste is first concentrated by evaporation and then solidified by vitrification. Medium-activity waste is processed by bituminisation. After using ion-exchange resins to remediate to the maximum permissible concentration, low-activity waste is discharged into an open hydrologic system.

The operation of the Industrial Association Mayak was a source of intense radiation contamination in the Ural region, mostly in northern Chelyabinsk Province. There are several reasons for this contamination. First, nuclear production was an imperfect science in the early stages of

42

Techa River and Villages in 1951

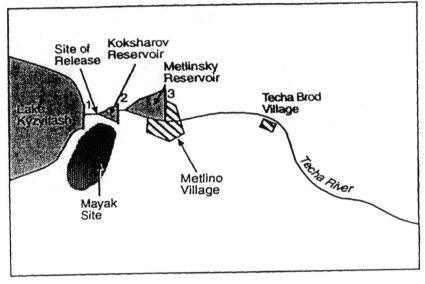

Techa River and Reservoirs in 1961

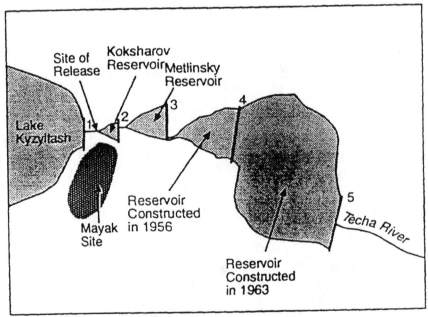

Note: Numbers correspond to dams.

Figure 3.2 Evolution of reservoirs at the Mayak Site

operation of the industrial complex. Second, early workers lacked knowledge about how radioactive materials could impact the environment and the possible consequences of such contamination.

The most intense radioactive contamination of the environment and irradiation of the population caused by it took place during the first half of complex operations. Because the facilities lacked proper technologies to treat liquid waste and because of a simplified approach to discharging such wastes into natural river systems, during 1949 to 1956 wastes with an aggregate activity of 1.0×10^{17} Bq were discharged into the Techa and Iset rivers. As a result, 124,000 people living near the riverbanks in Chelyabinsk and Kurgan provinces have been exposed to radiation.

Besides these intentional releases, construction imperfections of the first storage containers for high-level liquid waste caused an accidental release. In autumn 1957, overheating from radiation resulted in an explosion of nitrate-acetate salts stored in one container. The area contaminated by this explosion, afterward called the East-Ural radioactive trace, was about 20,000 km^2 (Figure 3.3 and Table 3.7). The contaminated territory was inhabited by 272,000 people during this period.

In spring 1967, because of extremely arid conditions, a coastal strip of Lake Karachai's bottom was exposed to air. Radioactive materials accumulated on the sediments were carried over adjacent territory by gusty winds for 2 weeks. About 1,800 km^2 of land has been contaminated. This contamination, though at much lower levels than the accidental release of 1957, was spread predominantly over Chelyabinsk Province. About 40,000 people were exposed to additional radiation.

Another factor in the contamination of the environment and irradiation of the population is the routine release of radioactive substances into the free air from ventilation ducts and stacks of the industrial complex. The greatest releases, caused by buildup of production capacity and imperfection in the gas purification system, occurred during the first 10 years of operation. As of 1 January 1996, the total area of the contaminated land was of 2,736 km^2.

3.5.2. The Siberian Chemical Combine (SCC)

Construction of the Siberian Chemical Combine began in March 1949. The production unit is located on the right bank of the Tom River 12 to 15 km north of the city of Tomsk. The combine includes the following production branches:

44

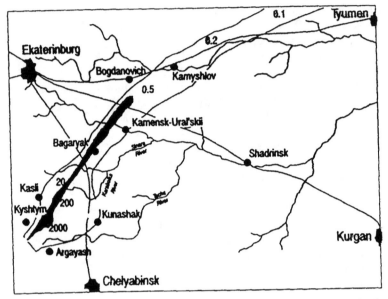

Figure 3.3 Area of the East-Urals radioactive trace. Numbers correspond to the initial density of strontium-90 contamination in Ci/km².

TABLE 3.7 Contaminated lands at the Industrial Association Mayak

Exposure Level, µR/hr	Area of Contaminated Land, km²			
	Production Site	Protective Zone	Watch Zone	Total
Up to 60	--	--	1,055	1,055
60-120	--	--	888	888
120-240	21	31	--	52
240-1,000	15	97	390	502
>1,000	21	71	147	239
Total	57	199	2,480	2,736

- Reactor production (producing plutonium, electrical power, and heat)
- Radiochemical production (processing of irradiated materials to derive and clean salts of uranium and plutonium)
- Chemical-metallurgical production (deriving metallic uranium and plutonium)
- Sublimation production (deriving protoxide-oxides of uranium and uranium hexafluoride)
- Separating production (deriving enriched uranium)
- Warehousing (storing fissionable materials)

- Providing infrastructure (processing, storing, and burying radioactive waste).

Construction of a plant to derive enriched uranium was started in 1951. Facilities from that the first stage of construction were put into operation in 1953. Before 1973, the uranium isotopes were separated by a gas-diffusion method. In subsequent years, separation used a more advanced high-performance and economic technology with ultra-high-speed centrifuges.

In 1952, construction of the first uranium-graphite reactor began. In 1955, this reactor was put into operation. The reactor operated for 35 years, during which time it under went several stages of modernisation. It was shut down in connection with cutting the volume of production of weapons-grade plutonium. In 1958, the combine resolved to build a chemical-metallurgical plant. This plant delivered its first product in August 1961.

Processing irradiated uranium formed liquid and solid industrial wastes with high, medium, and low levels of activity. The total amount of solid radioactive waste accumulated at the Siberian Chemical Combine is 131,153 tons; the activity of this waste is 1.1×10^{15} Bq. The total amount and activity of liquid waste is 5,961,750 tons and 2.23×10^{19} Bq, respectively.

Solid wastes are treated depending on their level of activity. They may be buried in earthen or concrete burial facilities or piled in organized storage in specially chosen rooms.

Most liquid waste is generated at the radiochemical plant. The main treatment method accepted at the Siberian Chemical Combine is underground burial in the form of pumping. High-activity wastes were directed to temporary storage in containers (tanks) made of stainless steel. After preparation, the wastes were directed to injection wells and injected at a depth of 315 to 340 m. These wastes are not longer being buried in this manner.

Wastes with medium levels of activity are also directed to underground burial after relevant preparations. For intermediate storage, liquid radioactive wastes are kept in open land storages and special closed storages. Wastes with medium activity from chemical-metallurgical production are directed to the open basin for storing. Wastes from separation activities are added to other wastes coming into the underground burial site. Liquid radioactive waste from sublimation production is directed to two storages for pulp. For all areas, radiation levels are monitored in the watch zone, which has an area of 1560 km^2 and radius of 15 to 20 km.

During operation of the Siberian Chemical Combine, there have been 36 radiation accidents and incidents of differing scales. The most severe accident took place on 6 April 1993 at the radiochemical plant. During routine operations to prepare uranium solution for extraction, a rapid pressure increase destroyed the preparation device. The following explosion and release of an aerosolic mixture broke through the building roof, causing a fire on part of the roof and partial release of radioactive substances into the environment. All settlements in a 35-km radius of the center of the explosion were contaminated. By 1996, the area of contaminated land amounted to 10,392 km^2. The contamination of soils was caused by isotopes of cesium-137 and strontium-90 (Table 3.8).

TABLE 3.8 Contaminated lands surrounding the Siberian Chemical Combine

Exposure Level, µR/hr	Area of Contaminated Lands, km^2			
	Production Site	Protective Zone	Watch Zone	Total
Up to 60	3.838	--	--	3.838
60-120	1.558	--	--	1.558
120-240	0.958	0.30	--	1.258
240-1,000	1.697	--	--	1.697
>1,000	2.041	--	--	2.041
Total	10.092	0.30	--	10.392

3.5.3. The Mining and Chemical Combine

The decision to construct a mining and chemical combine was made in 1950. The Mining and Chemical Combine occupies about 360 km^2 along the right bank of the Yenisei River, 60 km from the city of Krasnoyarsk.
The combine produces weapons-grade plutonium using uranium-graphite reactors. The irradiated uranium is processed at a radiochemical plant for separating uranium, plutonium, and fission products. The combine delivers plutonium dioxide and uranylnitrate alloy to other plants in the Minatom system.

Three reactors form the center of the work. The first was put into operation in 1958, the second one in 1961. These two reactors are thermal, uranium-graphite, water-cooled types. They discharge cooling water into the Yenisei River. These reactors were decommissioned in 1992. The third reactor was put into operation in 1964 and is used to generate electrical power and heat water for operations.

The radiochemical plant was put into operation in 1964. The plant derives plutonium from natural uranium irradiated in the rectors. Operations have ceased in connection with sharp reductions in the production of weapons-grade plutonium.

All facilities are located underground at a depth of 250 to 300 m. Like IA Mayak and the Siberian Chemical Combine, they have reliable biological protection. The complex is equipped with a system of ventilation with filters that prevent radioactivity from coming in from the outside.

Operations have resulted in liquid and solid industrial wastes of high, medium, and low levels of activity. The total accumulated solid radioactive waste is 105,170 tons. The total liquid waste amounts to 5,622,000 tons, with an activity of 1.46×10^{19} Bq. Gaseous and aerosol releases undergo multiple-step cleaning before they are discharged into the atmosphere.

Solid radioactive waste is stored in deep reinforced-concrete storage in the Combine. Depending on the level of radioactivity, liquid radioactive wastes are transferred to sewage treatment facilities and collected in special tanks made of stainless steel or in open reinforced-concrete storage. After relevant preparation and cleaning, they are shipped to the northern portion of the facility for deep burial. Treated waters are discharged into the Yenisei River.

Deep storage in the northern portion of the facility is used to bury low-activity waste in the second sand horizon in volumes up to 800 m^3/d. The same area allows burial of medium-activity waste in the first sand horizon in volumes up to 500 m^3/d. The deep storage area is located 12 km from the main production facilities in the protective zone of the plant. The aggregate area is 45 km^2; the volume of underground space is 11,000 m^3.

The first and the second sand horizons used for waste burial are bedded in the intervals, at depths of 180 to 280 m and 355 to 500 m, respectively. The horizons are spread under, divided, and overlapped by clay horizons, which isolate horizons containing waste from the surface and shallow-bedded underground waters. Natural velocity of ground water is 5 to 6 m/year in the first horizon and 10 to 15 m/year in the second. Spread of the radionuclides is thus slowed by the soils. This burial of liquid radioactive waste has essentially eliminated huge quantities of radionuclides from reaching the population and the environment.

Releases of radionuclides into the atmosphere in 1993 did not exceed the established norms by all components. In addition, total radionuclide discharges into the Yenisei River after shutdown of reactors using once-through cooling water did not exceed the established norms and lay within the limits of 0.3% to 6.0% of calculated maximum permissible values.

As a whole, after shutdown of once-through reactors, a dose rate from surface water exposure (and the volumetric activity of all radionuclides contained in water) does not exceed permissible values established by the Norms of Radiation Safety[4]. Individual islands and sections of flood-lands 15 to 250 km downstream from the discharge locations hold some "spots" of contamination. This contamination is caused by strong, high water along 300 km of the Yenisei River in 1966 and 1988. During these surges, water flow reached 21,000 m^3/sec, which exported part of the contaminated bottom sediments onto islands and sections of the flood-lands (Table 3.9).

TABLE 3.9 Contaminated lands at the Mining and Chemical Combine

Exposure Level, μR/hr	Area of Contaminated Lands, km²			
	Production Site	Protective Zone	Watch Zone	Total
Up to 60	0.005	0.666	0.106	0.777
60-120	--	0.149	--	0.149
120-240	3.297	0.060	3.394	6.751
240-1,000	--	0.050	--	0.050
>1,000	--	0.062	--	0.062
Total	3.302	0.987	3.500	7.789

3.6. Production of Nuclear Weapons

In the Russian Federation, Minatom is charged with the production of nuclear weapons. The main enterprises are as follows:
- Combine "Electrokhimpribor," in Lesnoi, Sverdlovsk Province
- Industrial Association "Start," in Zarechny, Penza Province
- Instrument Engineering Plant, in Tryokhgorny, Chelyabinsk Province
- Electromechanical plant "Avanguard," in Sarov, Nizhniy Novgorod Province.

As of 1 January 1997, the total solid radioactive waste accumulated at these plants amounted to 4,301 m³ with an activity of 4.04×10^6 Bq. The total amount of sewage water and its activity was 2,546 m³ and 1.9×10^6 Bq, respectively. The basic type of storage is reinforced-concrete tanks.

The discharges and releases of radionuclides into the environment by these plants are extremely insignificant (amounting to part of a percentage from maximum permissible values). Throughout plant operation, no emergency situations resulted in environmental contamination.

3.7. Ship Nuclear Propulsion Plants and Their Infrastructure

Cold War nuclear activities associated with the Russian Navy and the Russian Agency for shipbuilding include ship propulsion plants, plants for their technical support and maintenance, waste storage activities, and objects that sunk or were dumped at sea.

3.7.1. Vessels, Plants, and Waste Storage

The main sources of radiation danger and environmental radioactive contamination are the following objects of the Russian Navy and Russian Agency for shipbuilding:
- Submarines and surface ships with nuclear propulsion plants (NPP)

- Vessels for technical support and maintenance of nuclear ships
- Bases of the ships with NPP
- Places to temporarily store afloat decommissioned ships with NPP and places to carry out recycling of their materials
- Places to temporarily store fresh and spent nuclear fuel
- Ship-repairing and shipbuilding yards that conduct activities on ships with NPP
- Radioisotopic power sources.

By the beginning of 2000, about 156 nuclear submarines (95 at the Northern fleet and 61 at the Pacific) had been withdrawn from active forces of the Navy for decommissioning. This accumulation awaits first defueling and then removal of their reactor compartments to prepare the latter for long-term storage. Only compartment removal and natural radioactive decay reduce radiation danger to a level that allows subsequent disassembly of reactor installations. Long-term storage locations for reactor compartments have not yet been chosen.

Congestion of retired, floating nuclear submarines both with defueled reactors and with reactors containing SNF, combined with the growing number of removed reactor compartments, create a radiation safety problem. Currently, two-thirds of the nuclear submarines undergoing decommissioning are stored afloat with reactors containing SNF.

The Navy has the following volumes for storing and processing radioactive liquid waste (Table 3.10):

- The Northern fleet:
  --Shore storages with bulk volume of 5,300 m^3
  --Floating storages with bulk volume of 3,700 m^3
  --Stationary facility for processing of radioactive waters.
- The Pacific fleet:
  --Shore storages with bulk volume of 3,500 m^3
  --Floating storages with bulk volume of 3,000 m^3
  --Stationary facility for processing of radioactive waters.

SNF from the Northern and Pacific fleets is shipped to Minatom plants for processing.

3.7.2. Emergency Situations

An accident occurred on 10 August 1985 at a Pacific fleet nuclear submarine berthed at the naval base in the bay of Chazhma (the settlement of Shkotovo-22, the Primorye Territory). Personnel broke the requirements of nuclear safety and separated the head from the pressure vessel of the reactor while carrying out refueling. This separation created excessive nuclear reactivity, which caused a spontaneous chain reaction in the reactor on a port side, accompanied by an explosion. Immediately after the explosion, a fire sprang up in the reactor compartment. The fire took 4 hours to extinguish.

TABLE 3.10 Characteristics of radioactive waste stored by the Russian Navy and the Ministry of Economy, as of 1 January 1994

Enterprise	Liquid Radioactive Waste		Solid Radioactive Waste	
	Quantity, 1,000 m^3	Total Activity, Bq	Quantity, 1,000 m^3	Total Activity, Bq
Russian Navy				
Northern Fleet	8.695	3.05×10^{12}	5.863	5.41×10^{13}
Pacific Feet	5.767	1.36×10^{12}	1.703	6.29×10^{14}
Ministry of Economy				
JC Amur Shipbuilding Enterprise	1.0	0.5×10^{9}	0.012	NA
SRZ Nerpa	0.105	1.11×10^{10}	0.5	NA
DVZ Zvezda	0.95	1.78×10^{11}	1.99	4.7×10^{12}
IA Sever	1.033	NA	5.76	5.44×10^{12}
IA Sevmashpred-priyatie	0.123	NA	1.8	NA
Ship Equipment Enterprise	0.02	3.7×10^{5}	0.0015	0.67×10^{10}
Total	17.693	4.6^{12}	17.63	6.91×10^{14}

During this time, radiation fell around the submarine in a radius of 50 to 100 m. This fallout was caused by burning of fission and activation products and release of coarse-grain particles of fuel and slag formed by the explosion. The cloud of gaseous radioactive substances that arose moved to the northwest and crossed the peninsula of Dunai, nearing the sea at the coast of Usury gulf. Full-scale examination of seawater and bottom sediments showed that further movement of the cloud above the Usury gulf (that is 28 to 30 km wide) decreased fallout down to background levels and did not influence radiation levels in the city of Vladivostok. However, as a result of the accident, a center of long-lived radioactive contamination (0.1 km^2) of bottom sediments formed in the bay of Chazhma.

Results of additional in situ observations and numerous radioecological surveys show that this accident did not provide a measurable radiation impact upon Vladivostok, its beach zone, or the settlement of Shkotovo-22. The residual long-lived radioactive contamination of terrain and bottom sediments in the bay of Chazhma is reliably localized and should not cause severe ecological repercussions.

The Navy now has four damaged nuclear submarines: three in the Far East and one in the north.

3.7.3. Dumping Radioactive Waste at Sea and the Sunken Nuclear Ships

The long-lived radionuclides dumped in the Arctic Sea dominate all radioactive waste dumped at seas surrounding Russian territories. The aggregate activity from these radionuclides is 2.4 MCi.

The analysis of information describing solid radioactive waste that was dumped in containers or discharged to near-surface water layers of the Barents and Kara seas has shown that the maximal potential radiation danger can be represented by long-lived radionuclides in SNF and ship NPPs sunk in bays of the eastern coast and near the archipelago of Novaya Zemlya. The doses at these places are insignificant, however, and concentration of radionuclides outside of these areas does not differ from that in the open waters of the Kara Sea (Table 3.11).

Since 1949, their aggregate activity has decreased to two times its original level. It is now four times less than aggregate activity contained in the Atlantic burial sites.

In the Far Eastern region, solid radioactive waste was dumped in several regions of the Sea of Japan, the Okhotsk Sea, near the eastern coast of Sakhalin, in the northwestern part of the Pacific Ocean, and near the eastern coast of Kamchatka. Distribution of an aggregate activity of the dumped radioactive waste in the Arctic, Northern Atlantic, and Far East by 1999 acknowledges assessments comparable to those shown above.

TABLE 3.11 Activity of long-lived radionuclides in reactors sunk near Novaya Zemlya as of late 1994

Submarine or Ship Inventory Number	Fission Products, 1,000 Bq	Activation Products, 1,000 Bq	Actinides, 1,000 Bq	Total, 1,000 Bq
Sunken Reactors with Fuel				
285	634	12.80	8.13	654.93
901	718	5.96	3.44	727.40
421	287	2.88	2.84	292.72
601	375	239.00	1.25	615.25
Total	2,014	260.64	15.66	2,290.30
Sunken Reactors Without Fuel				
254	--	9.47	--	9.47
260	--	5.07	--	5.07
538	--	4.51	--	4.51
Total	--	19.05	--	19.05

In 1989, the nuclear submarine Komsomoletz caught on fire and sank in the Norwegian Sea at the depth of about 1700 m. One nuclear reactor and two torpedoes with nuclear warheads were onboard. However, in comparison with other sources of long-lived radionuclides in the Northern Atlantic, the Komsomoletz does not represent any significant radiation danger for the ambient marine environment (Table 3.12).

TABLE 3.12 Long-lived artificial radionuclides in the North Atlantic, 10^{16} Bq

Radionuclides	Global Fallout	Radiochemical Plants	Komsomolets Nuclear Submarine
Cesium-137	7.6	3.0	0.31
Strontium-90	5.1	4.3	0.28
Plutonium-239/240	0.13	0.06	0.0021

3.8. Nuclear Explosions

Nuclear devices and bombs have been exploded in the USSR from 29 August 1949 (the first nuclear charge) until 24 October 1990 (last nuclear-weapon-related test). During this period, the USSR conducted 559 nuclear-weapon-related tests, with 796 nuclear charges and nuclear explosive devices exploded.

The USSR had two nuclear test ranges for nuclear weapon trials:

- Semipalatinsk test range, put into operation in 1948, at which one the first nuclear device was tested
- Northern test range on the islands of the archipelago of Novaya Zemlya, put into operation by order of the government of the USSR on 31 July 1954. The first nuclear device was exploded at this range was on 21 September 1955.

In addition to tests at these two specially created test ranges, nuclear weapons trials were also being conducted at the following sites:

- At the Missiles Testing Range, settlement of Kapustin Yar, Astrakhan Province, rocket missiles with warheads equipped with nuclear and thermonuclear charges were launched to conduct tests in high layers of the atmosphere and space
- At the training grounds of the Ministry of Defense near Totsk (Orenburg Province), a 40-Kt nuclear weapon was tested in the air on 14 September 1954 during a combined-arms exercise
- Near Aralsk (Kazakhstan), a 0.3-Kt surface nuclear explosion occurred on 2 February 1956.

The environment was contaminated by global radioactive fallout from nuclear weapons tests from the end of the 1940s until the middle of the 1960s. By the beginning of 1986, the mean level of cesium-137 contamination in the eastern European plains amounted to 0.08 Ci/km². At higher elevations, contamination levels were up to 0.35 Ci/km². Today, excluding contamination resulting from the accident at the Chernobyl Atomic Power

Station, this level should have decreased on the average of 20%, with a range of 0.05-0.06 Ci/km^2.

3.8.1. Semipalatinsk Nuclear Test Range

According to the requirements introduced in 1948, the test range for nuclear weapon trials was to be located in a wilderness area with a diameter of about 200 km adjacent to a railway station and aerodrome. The site was chosen 160 km from Semipalatinsk, in an area naturally bordered by the Shagan River (a tributary of the Irtysh) and by the mountains of Deguelen and Kalyastan, which are 100 km apart. The initial area of the site was approximately 5,200 km^2. The geographical position together with predominantly eastern (to the east, southeast, and northeast) movement of air masses (within the framework of an overall atmospheric circulation) predetermined the most likely regions to be contaminated in the USSR (now Russia and the Republic of Kazakhstan). These areas are the Altai Territory, the Republic of Altai (Russia), as well as Semipalatinsk and the East Kazakhstan and Karaganda provinces (Republic of Kazakhstan).

The main biologically hazardous radionuclides in the areas contaminated by radioactive fallout from atmospheric nuclear weapons tests are strontium-90, cerium-137, and plutonium (Table 3.13).

TABLE 3.13 Tentative data on external exposure doses of the population (until the complete decay of radionuclides) in the area influenced by nuclear tests at the Semipalatinsk Test Range

Region	Distance from the Site, 1,000 km	Population, 1,000 persons	Maximum External Dose, CSv	Mean External Population Dose, CSv	Collective External Dose, 1,000 person-Sv
Altai land	0.14-0.7	2,514	52.00	0.50	13.50
Republic of Altai	0.40-0.8	174	0.50	0.20	0.30
Republic of Khakassia	0.70-1.0	508	0.20	0.15	0.76
Novosibirsk region	0.50-0.7	2.657	1.00	0.05	1.44
Kemerovo region	0.70-1.0	2,990	1.00	0.06	1.64
Krasnoyarsk land	0.90-2.2	600	0.12	0.04	0.24
Irkutsk region	1.30-2.7	1,340	0.10	0.04	0.47
Chita region	2.00-3.0	1,258	0.05	0.04	0.44
Tomsk region	0.70-1.3	887	0.15	0.04	0.35
Total		15,928			19.14

3.8.2. Northern Nuclear Test Range (the Archipelago of Novaya Zemlya)

Nuclear weapons were tested at this range at three locations (Figure 3.4):

Zone A (near the Chornaya Fjords and Cape Bashmachny). At this site, three underwater and two surface water tests (from 1955 to 1962), one

54

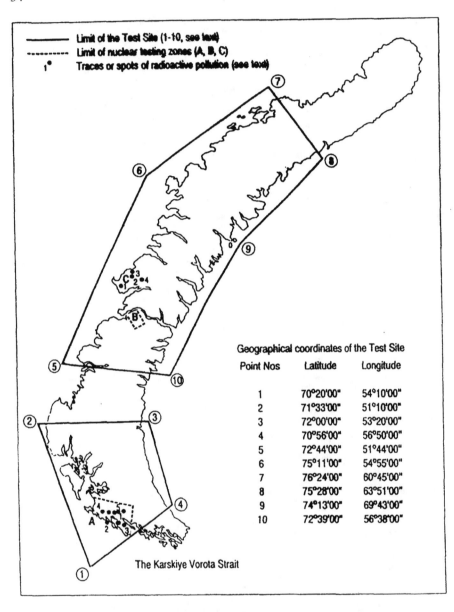

Figure 3.4 Novaya Zemlya northern test site

- land surface explosion (on 7 September 1957), and six underground (in wells) nuclear explosions were made (from 1972 to 1975).
- **Zone B** (an area in the western part of the Strait of Matochkin Shar). At this site, 36 nuclear explosions were made.
- **Zone C** (near the peninsula of Sukhoi Nos). At this site, atmospheric nuclear explosions were conducted at four spots at different altitudes before 1962.

The density of contamination on the test range is nearly identical to the density of contamination and background radiation everywhere at mean latitudes of the northern hemisphere. The highest density characterizes the site of the 1957 land nuclear explosion on the coast of the fjord of Chornaya. The area of contamination is about 1 km^2 (Table 3.14)

3.8.3. Missile Testing Range

At this range, rockets equipped with nuclear and thermonuclear charges were fired for testing in space and the upper atmosphere. The work did not cause contamination of the range and regions adjacent to it because all tests were conducted at a high altitude.

3.8.4. Area of the Totsk Combined-Arms Exercises of 1954

A 40-Kt atomic bomb was exploded at an altitude of 350 m. The fiery ball did not touch the underlying surface; therefore, fission products and residual plutonium were deposited across a wide area. At the explosion epicenter, however, increased activity was observed as a result of absorption of neutrons by the soil stratum. The radionuclides were characterized by cobalt-60, europium-152, and europium-154. A column of dust containing these radionuclides rose above the epicenter and fell out in a trail extending 210 km. The maximum accumulated dose reached about 1 Roentgen up to 70 km from the epicenter (Table 3.15).

3.8.5. Total Impact of Global Fallout

The world-wide nuclear-weapons testing that occurred from the end of 1940s until the middle of the 1960s left a global legacy from the Cold War. Locally in eastern Europe at the beginning of 1986, the total radioactive contamination of land surfaces by global fallout of cesium-127 amounted 0.08 Ci/km^2. In mountainous areas, the contamination level was up to 0.35 Ci/km^2. It is estimated that currently (year 2000), these surface concentrations should have been reduced by 20% by natural processes.

TABLE 3.14 Tentative data on external exposure dose to the population (until the complete decay of radionuclides) of various regions of Russia in the area influenced by nuclear tests at the Northern test range

Region	Distance from the Site, 1,000 km	Population, 1,000 persons	Maximum External Dose, CSv	Mean External Population Dose, CSv	Collective External Dose, 1,000 person-Sv
Krasnoyarsk land (ADs not included)	1.3-3.0	2,693	0.7	0.10	3.0
Taimyr (Dolgano-Nenets) AD	0.9-2.2	48	2	1.0	0.5
Evenk AD	1.6-2.4	17	1.5	0.7	0.12
Republic of Sakha (Yakutia)	2.0-3.7	883	1	0.8	7.0
Tyumen region (ADs not included)	1.8-2.2	1,165	0.3	0.15	1.8
Yamalo-Nenets AD	0.5-1.8	193	0.4	0.13	0.25
Khanty-Mansi AD	0.9-1.9	673	0.3	0.17	4.9
Perm region (ADs not included)	1.3-2.0	2,830	0.3	0.17	4.9
Magadan region (together with the Chukotsk AD)	3.8-4.5	490	0.6	0.25	1.2
Republic of Komi	0.8-1.6	1,147	0.4	0.17	2.0
Khabarovsk land	3.6-4.5	1,610	0.6	0.2	3.2
Nenets AD of Arkhangelsk region	0.4-0.8	50	0.3	0.10	0.5
Republic of Udmurt	1.7-2.0	1,516	0.2	0.11	1.6
Sverdlovsk region	1.4-2.0	4,500	0.3	0.20	9.5
Kurgan region	2.0-2.2	1,085	0.2	0.14	1.5
Chelyabinsk region	2.0-2.4	3,480	0.2	0.14	4.8
Republic of Bashkortostan	2.0-2.4	3,865	0.2	0.10	4.0
Omsk region	1.9-2.4	1,963	0.15	0.10	2.0
Republic of Tatarstan	1.9-2.2	3,453	0.15	0.06	2.4
Irkutsk region	2.6-3.4	2,616	0.3	0.005	0.8
Chita region	3.4-3.9	1,258	0.2	0.001	0.15
Total	--	35,535	--	0.15	52.27

Note: AD = autonomous district.

TABLE 3.15 Preliminary data on external exposure dose to the population (until the complete decay of radionuclides) in the area influenced by the nuclear explosion at the 1954 Totsk military exercises

Region	Distance from Firing Ground, 1,000 km	Population, 1,000 persons	Maximum External Dose, CSv	Mean Population External Dose, CSv	Collective External Dose, 1,000 person-Sv
Near zone (Orenburg region)	0-0.2	20	1	0.3	0.065
Remote zone (Krasnoyarsk land)	1.5-2.1	150	0.12	0.05	0.081
Total area	--	170	--	0.175	0.146

3.9.   Conclusion

The main objective of this chapter is to provide an overview of the nuclear weapons production complex of the FSU and its radiation legacy. To gain a greater understanding of the technological processes of the complex, isotopic structure of radioactive releases, and their impact on the population and environment, consult the literature on which this chapter has been based[1,2,3,5,6].

3.10.   References

1.   Bradley, D. (1997) Behind the Nuclear Curtain: Radioactive Waste Management in the Former Soviet Union. Battelle Press, Columbus, Ohio.
2.   RADLEG. (1997) Radiation Legacy of the Former USSR: The Available Data on the Main Directions of Examinations. ISTC, Project 245, RADLEG, Moscow.
3.   Egorov, N.N., Novikov, V.M., Parker, F.L., Popov, V.K. (2000) The Radiation Legacy of the Soviet Nuclear Complex (An Analytical Overview). IIASA, Earthscan Publication Ltd, London.
4.   State Committee of Russia. (1996) Norms of Radiation Safety in Russia, 1996. HPБ-96, Hygienic Standards GS 2.6.1.054-96, State Committee of Russia, Moscow.
5.   Israel, Y.A. (1999) Radioactive contamination measurements of the environment. IGKE of the Rosgidromet, and RAS, International Symposium HISAP'99, IIASA, Laksenburg, Austria.
6.   U.S. Department of Energy. (1997) Linking Legacies: Connecting the Cold War Nuclear Weapons Production Processes to their Environmental Consequences. U.S. Department of Energy, Washington, D.C.

4. Status and Challenges of Managing Risks in the U.S. Department of Energy Environmental Management Program

Even in the United States, where various methods of risk assessment, risk management, and risk communication have been attempted over the years, these disciplines continue to evolve to meet the needs of decision makers faced with legacy wastes. This chapter provides an overview of the Cold War legacy challenges as currently understood by the U.S. Department of Energy (DOE), as well as the risk-based methodologies currently being applied to assist the DOE in managing those challenges. In the past decade (1990s), the DOE created a single organization within their waste management structure to coordinate their risk activities: the Office of Environmental Management's Center for Risk Excellence. The chapter describes the formation, operation, and contributions of that organization, which was created to encourage the use of risk-based approaches to DOE site management and to provide consistency in the use of such approaches across the DOE complex. Of particular interest are the effective communication concepts developed by this organization for summarizing site risk and risk-related information as risk profiles.

In the U.S., the Department of Energy (DOE) bears the responsibility for stabilizing, treating, storing, and disposing of hazardous and radioactive wastes, materials, and facilities from more than 50 years of research, development, testing, and production of nuclear weapons and civilian research activities. The DOE nuclear complex included uranium mining, nuclear reactors, chemical processing, metal machining plants, laboratories, and maintenance facilities. This complex manufactured tens of thousands of nuclear warheads and conducted more than 1,000 nuclear explosion tests. Weapons production stopped in the late 1980s, initially to correct environmental and safety problems, but it was later discontinued indefinitely because of the Cold War's cessation[1,2]. Simultaneously, during this 50-year period, the federal government funded and conducted research in support of civilian applications of nuclear technology, which also resulted in some legacy waste.

Residual materials and contaminated facilities pose a risk[1] to workers, the environment, and members of the public. Eliminating and managing urgent risks is one of the primary goals of DOE's Office of Environmental Management (EM). Risk management, unlike the past production of defence material and research, must be accomplished in a social and legal setting that considers site-specific conditions, is highly visible, has external oversight, and meets regulatory standards for safety. Risk management is also an intangible product subject to diverse interpretation from risk professionals and the lay public. Thus, a primary challenge in risk management is completing credible technical work and communicating it in a public forum.

This chapter provides an overview of the Cold War legacy challenges as currently understood by DOE as well as the risk-based methodologies that are currently being applied to assist DOE in managing those challenges. In the past decade (1990s), DOE created a single organization within their waste management structure to coordinate their risk activities. This chapter provides background information on the formation, operation, and contributions of that organization. The concept in creating such an organization is to encourage the use of risk-based approaches to help manage particular sites and to provide consistency in the use of such approaches across DOE sites. Of particular interest are the effective communication concepts developed by this organization for summarizing site risk and risk-related information as risk profiles.

The organization that DOE-EM established is the Center for Risk Excellence. The Center has the assignment to address the difficult questions concerning the management of risks. The Center's mission is to provide leadership, expertise, and integration of risk activities through strategic partnerships and to be a catalyst for improved environmental decisions through sound risk management. One of the initial charges to the Center was to assist remediation sites in the DOE complex to develop "site risk profiles."

---

[1] Risk is defined as "the probability that a substance or situation will produce harm under specified conditions. Risk is a combination of two factors: the **probability** that an adverse event will occur and the **consequences** of the adverse event. Risk encompasses impacts on human health and the environment and arises from **exposure** and **hazard**. Risk does not exist if exposure to a harmful substance or situation does not or will not occur. Hazard is determined by whether a particular substance or situation has the potential to cause harmful effects." [3]

This chapter discusses some of the Cold War legacy challenges being studied by the Center, presents example hazard/risk profiles for managing and communicating the risks and corrective actions associated with some these challenges, and provides a potential methodology for developing these hazard/risk profiles.

4.1. Cold War Legacy Challenges

The EM scope of work is one of the most technically challenging and complex of any environmental program in the world[4,5,6]. Although the DOE complex comprises almost 9,710 km^2, the majority of this land is uncontaminated (more than 85%). However, the 75 million m^3 of contaminated soil present difficult technical challenges because of the presence of radionuclides. In addition, there are currently no effective technical solutions for remediating much of the 1.8 billion m^3 of contaminated ground water. Millions of cubic meters of radioactive and mixed waste (waste that is both hazardous and radioactive) also need to be disposed. Disagreement among experts regarding how the waste should be disposed, the enormity of the task, and a shortage of disposal capacity mean that final disposition of the wastes, and management of residual risks, will require many decades of commitment from the federal government.

Characterization of risks is difficult for many reasons, the lesser ones being the size, diversity, and functions of the sites involved. For example, although hazards of the Cold War legacy are located at 144 sites in the United States, the majority of contaminants are located at six sites: the Hanford Site in Washington State, the Idaho National Engineering and Environmental Laboratory, the Nevada Test Site, Los Alamos National Laboratory in New Mexico, Oak Ridge National Laboratory in Tennessee, and the Savannah River Site in Georgia (Figure 4.1). Even among these six sites, the level of contamination is not uniform. Figure 4.2 shows the relative amount of contamination in waste at the largest sites.

Curies already released to the environment are substantial at some sites and not included in Figure 4.2 because comprehensive information is not available. The Nevada Operations Office, for example, estimates that an additional 310 million curies are in the soil and water there. In the early history (1950s) of the Hanford Site, radioactive liquid was disposed of in trenches or directly on dry soil. Contaminated equipment and drums of waste were disposed of similarly.

62

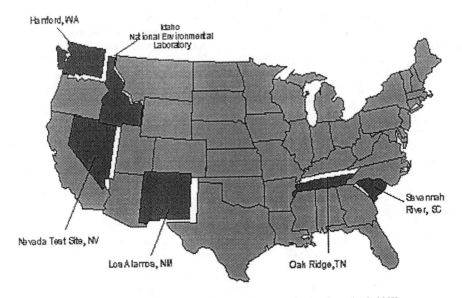

Hanford, WA

Idaho
National Environmental
Laboratory

Savannah
River, SC

Nevada Test Site, NV

Los Alamos, NM

Oak Ridge, TN

Figure 4.1 Six states with the majority of environmental contamination from the Cold War

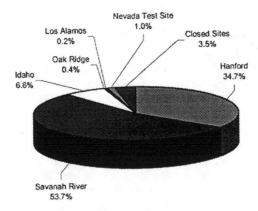

Nevada Test Site
1.0%

Los Alamos
0.2%

Closed Sites
3.5%

Oak Ridge
0.4%

Hanford
34.7%

Idaho
6.6%

Savanah River
53.7%

Figure 4.2 Relative radioactive waste curie inventories for major U.S. Department of Energy sites

Material volumes are often used to describe progress in risk management for projects removing, disposing, or treating of material. Figure 4.3 shows how 36 million m^3 of waste are categorized. By-product materials (e.g., uranium mill tailings) account for 88% of the volume, while high-level radioactive wastes account for only 1% of the total volume.

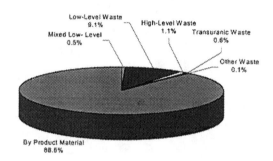

Figure 4.3 Volumes of waste at U.S. Department of Energy facilities

The large volume of by-product material is not located at the six DOE sites shown in Figure 4.1. However, the six sites are important to risk management because, as shown in Figure 4.4, the small volume of high-level waste at Hanford, Idaho, and Savannah River contains 95% of the more than one billion curies of radioactivity. In addition, much of the high-level waste is in liquid form awaiting conversion to a stable solid, suitable for disposal. After conversion to a solid, high-level waste requires a deep geologic disposal facility for long-term risk management. No disposal facility for high-level waste is currently operating.

While this complex set of risks poses a significant challenge, more problematic is the definition of risk itself. This chapter focuses on the potential for impacts from contaminants; however, other factors, such as cultural and socio-economic risks, are important at some locations. Also, there are significant uncertainties in the specifics of the source term, potential future accessibility of the hazard, and potential future receptors. Finally, the wide number of approaches to risk management varies greatly in complexity, comprehensiveness, and clarity of communication. All of these factors have limited the use of risk management in the EM program at the national level.

Several attempts have been made over the last 5 years to develop an integrated program of risk management, but all have been abandoned as

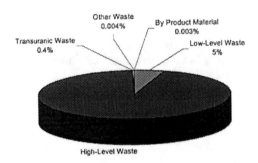

Figure 4.4 Percentage of curies in U.S. Department of Energy waste (total 1.01 billion)

either burdensome or not scientifically credible.[7] Lack of specific measures is not necessarily bad or contributing to excessive risks or hazards at DOE sites, but it does constrain the discussion of progress by the program in meeting their objectives. This constraint may be undermining confidence in the DOE, confidence necessary to carry out the long-term program.

Hazards and risks under control of the EM program have varied greatly in time, extent, longevity, and remedy. Some sites are still operating, requiring active risk management (people and equipment) to minimize or eliminate the potential for releases to the environment, public, or workers. In some cases, however, active management is being used even at sites that are no longer producing nuclear materials. With the cessation of production, the costs of active risk management are more visible and unsustainable for the life of the hazard. Active management also places workers in a higher-risk environment than is acceptable. Even with no limitations on resources, this situation could lead to increased risk on and off site from using equipment in ways that it is not designed (e.g., using temporary waste storage tanks well beyond their design life).

Over time, DOE's strategy is to move many materials and sites to a state in which the cost of risk management and the risk levels themselves have been reduced. This risk reduction is being accomplished through the use of barriers to control releases, stabilization of materials to reduce their mobility and reactivity, and treatment to place materials in a long-term stable condition for passive storage and/or disposal.

On the other hand, some materials present a persistent and significant hazard for periods of time that exceed human experience. For these materials, isolation in the earth is planned where no human or active equipment would be required for safe management. The Waste Isolation Pilot Plant in New Mexico is an example of this type of facility for the disposal of transuranic waste. However, all facilities necessary for long-term-passive risk management are not currently available. For this reason, facilities are being constructed for supervised surface storage until the disposal facilities are available. Many of the disposal facilities are at different sites than those used for waste treatment and storage. This diffusion will require management of transportation risks as part of waste disposal operations.

One hundred and eleven sites will have residual hazards that are not planned for removal and further treatment[8]. Many of these sites will require DOE responsibility for residual contamination in perpetuity. The DOE[9] has not yet completed its strategy for these types of sties, as many of these sites will continue to operate for many years. Long-term risk management strategies will be site specific to allow integration with enduring DOE responsibilities for natural resource management, adaptive reuse of federal assets, and management of long-term exposures to contaminants.

4.2. New Approach to Risk Management– Risk/Hazard Profiles

Regardless of the many complexities of managing risks within the DOE complex, the public expects effective management and communication. The Center for Risk Excellence has developed a new approach to risk management that uses semi-quantitative methods to describe reductions in hazards and risk at major DOE sites[10]. These methods consider the physical form, management, and environmental behaviour of these materials in addition to their volume and radioactive and toxic components. The methods provide a balance between the complexity of a full risk assessment and the desire to clearly show progress in the EM program. Preliminary results for two sites were discussed in a non-technical focus group setting and were judged to be more comprehensive and clear in their communication of program objectives and progress than either the curie information or volume information.

This new approach involves the development of risk/hazard profiles for some selected sites and processes. The profiles include graphic illustrations to provide the reader with a high-level mental picture to associate with all the qualitative risk management information presented. The methodology presented later in this chapter was developed to provide a means of calculating the risk values to use in developing these graphic illustrations.

The relative hazard (RH) equation, as presented in this methodology, is primarily a collection of key factors that are relevant to understanding the hazards and risks associated with projected risk management activities. The

RH equation has the potential for much broader application than was used in generating the risk profiles. For example, it can be used to compare one risk management activity with another, instead of just comparing it to a fixed baseline as was done for the risk profiles. If the appropriate source term data are available, it could be used in its non-ratio form to estimate absolute values of the associated hazards. These estimated values of hazard could then be examined to help understand which risk management activities are addressing the higher hazard conditions at a site. Graphics could be generated from these absolute hazard values to pictorially show and compare these high-hazard conditions. If the RH equation is used in this manner, however, care must be taken to specifically define and qualify (e.g., identify which factors were considered and which ones tended to drive the hazard estimation) the estimated absolute hazard values.

Another component of the methodology is the risk measure (RM), which was developed to extend the RH analysis to a measure of the potential risk from the hazardous material. The RM value includes the likelihood of a release to the environment based on the facility conditions and material packaging configurations. As the material is processed for safety improvements or waste treatment, the likelihood of the release event will usually be reduced and the risk measure will also be reduced. The RM and RH values are both normalized to the same quantity (i.e., denominator of the RH equation) so the parameter can be plotted on the same graph.

The risk/hazard profiles are intended to provide a brief narrative summary of risk-related activities. They are tailored to each site so that the most informative story can be told, yet a standard format is maintained to facilitate combining the documents into a cohesive national story. Flexibility is critical because some sites want to emphasize the importance of certain hazards at their site, others want to point out the risk avoidance activities at their sites, and still others wish to point out the lack of hazards and/or risks.

The following sections describe the initial design and construction of the profiles as well as their limitations.

4.2.1. Initial Design and Construction of Risk/Hazard Profiles

Information developed for the profiles represents a significant departure from previous efforts by DOE-EM to collect and communicate risk information. The profiles focus on how EM program activities result in *hazard reduction* through remediation activities, describing hazards in physical terms. Previous discussions of risk focused on the *potential risk from hazards if they were not managed*. The difference is significant in that the former discusses the realities of program progress, while the latter requires hypothetical evaluations of a non-action scenario. This new approach has three major advantages:

1. Hazard reductions can be summarized at the site level by material/waste type (i.e., hazard type, location, etc.) and avoids differences in site management structures and overlap with regulatory compliance.
2. Hazard reductions can be directly linked to intermediate and final milestones and described without the complexity of speculating what the risk would be at each stage of completion.
3. A more accurate and commendable risk picture can be presented by focusing on efforts to control hazards to ensure that risks are low.

The technical approach is to focus on current site hazards. The profiles include a brief overview of the field office, site histories, and other general information (see Figure 4.5 for example), in addition to a brief description of the public hazards and planned actions to address them. Next, the potential pathways for the release of the hazards are discussed. This is followed by a look at the control, storage, treatment, disposal, characterization, and other actions that limit the risks posed by the hazards. The hazard-pathway relationship is discussed in the context of the potential "receptors" which are described in the introductory site description. A description of the hazard-pathway-receptor relationship is displayed in a table or series of tables (see example in Table 4.1).

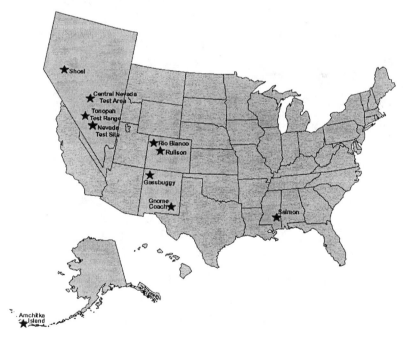

Figure 4.5 Example map from profile

Table 4.1 Example hazards and risks table from profile

Material Category	Nature of Hazard	Nature of Potential Risk	Status of Current Risk Management	Planned Risk Reduction and Control Measures	Anticipated Risk Reduction Progress	End-State Disposition and Risk
Contaminated ground-water	Ground water offsite is contaminated with Volatile Organic Compounds (VOCs) at levels above the drinking water standards. Onsite ground water is contaminated with VOCs, tritium and strontium-90 above drinking water standards.	Contaminated ground water is migrating south-southeast. Risk is consumption of contaminated water.	Hydraulic control and treatment currently manage migration and spread of onsite VOC-contaminated ground water. Risks offsite are mitigated by the provision of alternate water supplies (i.e., public water).	Additional VOC-contaminated ground-water will be treated by in-well air sparging. Strontium-90-contaminated ground water associated with chemical holes will also be treated. Extensive ground-water monitoring will also be performed.	Contaminants will be removed or will attenuate and decay.	End-state water in the aquifer will have contamination levels below drinking water standards.
Contaminated soils, debris, and sediments	Contaminated materials onsite include 39,000 m³ of radiologically contaminated soils and 7,000 m³ of contaminated river sediments. Primary contaminants are radionuclides (cesium-137 and strontium-90) and VOCs.	The primary exposure pathway is direct exposure to cesium-137-contaminated soils. Contaminated sediments pose an ecological risk.	Risks associated with direct exposure to cesium-137-contaminated soils are currently managed by restricted access (i.e., fences, security guards, etc.)	Remediation of soils and sediments will be performed to reduce risks and a variety of technologies (including excavation and off-site disposal) will be used. Soil remediation goals will consider ground-water protection.	Removal of the contaminated soil and sediment will reduce the risk of direct exposure and the potential for additional ground-water contamination.	Contaminated materials will be excavated and disposed of at licensed commercial facilities. Planned land use will be restricted.

After the narrative description of the relationship, an approximation of the relative hazard reduction over time is presented (Figure 4.6). The method used to derive this example is a quantitative evaluation; however, the evaluation stops short of computing absolute risk and makes no attempt to define difficult terms such as "high" and "low" hazard. Although these graphs were later removed, they were a first attempt at revealing a high-level picture to associate with all the qualitative information presented. The methodology used to create these illustrations consisted of using site-specific information and applying factors from applicable site-specific risk assessment results or look-up tables to generate relative hazard ratio values by waste/material type.[10] This methodology has since been expanded and updated, and is presented in more detail in Section 4.3.

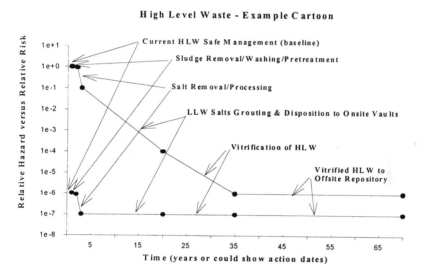

Figure 4.6 Example relative hazard profile

.2.2. Limitations of Initial Profiles

During the course of development and review of the initial risk profiles, several limitations were identified. For example, in a review of the draft risk profile produced for the Savannah River Site, their Citizen Advisory Board expressed a concern that the risk stories told by their draft risk profile were not complete. They agreed that it was important to tell the hazard reduction story, because that was the focus of the actual clean-up activities. However, they believed that just presenting the clean-up actions and details for risk reduction and showing how the hazards were reduced over time left out a very import aspect to the overall risk story. The missing component of the story is the ongoing work and resource commitment dedicated to maintaining acceptable risks to the public.

In addition to this concern, limitations of the initial profiles are generally that they are too narrow in focus. They only included public health (not worker, ecological, project, cultural, or other risks). They did not include all activities at a site; rather, they focused only on EM program activities and/or hazards. To portray the major risks at the site(s), hazards that pose minimal or no risk were omitted for simplification. The profiles also excluded discussion of regulatory issues, detailed site risk assessments, and safety assessments of specific projects conducted for regulatory compliance or to establish safety bases for specific facilities.

4.3. Relative Hazard and Risk Measure Methodology

The profiles could not have been developed without a credible scientific underpinning. The formula is based on state-of-the-art risk assessment techniques and methodologies. Key to these assessments is the evaluations of relative hazard values and the risk measure. These assessments are described in the following sections.

Note that the term "controlling constituent" is used often in this discussion. Controlling constituents are defined as those radionuclides and/or hazardous chemicals in a particular waste type that tend to control the impact or hazardousness of the consequences associated with the waste material. That is, they are the radionuclides and/or hazardous chemicals that tend to most heavily influence the concern over the need to control the waste material. In the analysis methods discussed, it is advantageous to limit the number of controlling constituents to as few as possible while still adequately representing the hazards of the waste material. In most risk assessments, usually just one or two constituents tend to most influence the risk.

4.3.1. Relative Hazard (RH) Calculation

The methodology to calculate RH consists of using site-specific information (e.g., information from site disposition maps, site-specific project information, and other site documents that address elements of the overall risk story for a site) and applying factors from applicable site-specific risk assessment results or look-up tables to generate RH ratio values by waste type.

RH was calculated using the following relationship of key risk-related parameters that can be extracted from the information provided for the risk profiles:

$$RH = \frac{\sum_{cc=1}^{n} Q_{cct} RF_{ccct} HM_{cct} HC_{cct}}{\sum_{cc=1}^{n} Q_{cct0} RF_{ccct0} HM_{cct0} HC_{cct0}}$$

(4.1)

where

Q_{cct} = quantity of the controlling constituents (radionuclides, in curies and hazardous chemicals, in kilograms) at time t (i.e., time when specified risk management action is completed)

Q_{cct0} = quantity of the controlling constituents (radionuclides, in curies and hazardous chemicals, in kilograms) at time t0 (i.e., the original baseline or starting time)

RF_{cct} = fraction of controlling constituent quantity that is releasable to the controlling pathway at time t

RF_{cct0} = fraction of the controlling constituent quantity that is releasable to the controlling pathway at time t0

HM_{cct} = hazard measure factor for controlling constituent and controlling pathway at time t (hazard measure factors from look-up tables)

HM_{cct0} = hazard measure factor for controlling constituent and controlling pathway at time t0 (hazard measure factors from look-up tables)

HC_{cct} = hazard control factor for risk management control action specific at time t (hazard control factors may be estimated from site risk data or approximated using supplied look-up tables)

HC_{cct0} = hazard control factor for risk management control action specific at time t0 (hazard control factors may be estimated from site risk data or approximated using supplied look-up tables)

N = number of controlling constituents

Note: If only one controlling constituent is identified, the equation will not need to be summed over the number of controlling constituents.

The RH equation calculates a relative ratio representative of the hazard reduction associated with a specified risk management action compared to a baseline. It does *not* calculate an absolute hazard value. For most DOE sites, the level of data available in the disposition maps and other site information is not detailed enough to support the calculation of absolute hazard values. In the risk profiles, the current state is assumed as the baseline for which to compare each risk management action (i.e., each factor$_{cct}$ is compared to its corresponding baseline factor$_{cct0}$). If it is desired to compare each risk management action step with the previous risk management action time step, the baseline factors (i.e., factor$_{cct0}$) can simply be replaced with the corresponding previous time factor (i.e., factor$_{cct-1}$). For additional information on RH calculations and graphing situations, consult Stenner et al.[11]

4.3.2. Risk Measure (RM) Calculation

The RH factor tracks the change in hazard over time. Another important consideration is the change in risk for the facility. As mentioned previously, the relative risk (RR) is related to the RH by the frequency of a release event for the facility. A risk measure (RM) can be calculated in a manner similar to the RH factor by addition of the hazard likelihood (HL) to the RH equation. The HL is represented as the expected frequency of the event that results in release of a contaminant to the environment. This can be represented by the following equation.

$$RR = \frac{\displaystyle\sum_{s=1}^{S} \sum_{cc=1}^{N} HL_{ccts} Q_{ccts} RF_{ccts} HM_{cct} HC_{ccts}}{\displaystyle\sum_{s=1}^{S} \sum_{cc=1}^{N} Q_{cct0s} RF_{cct0s} HM_{cct0} HC_{cct0s}}$$

(4.2)

where

RM	=	risk measure at time t (per year)
HL_{ccts}	=	likelihood that a release will occur for the controlling constituents at time t for scenario s (i.e., time when specified risk management action is completed)
Q_{ccts}	=	quantity of the controlling constituents (radionuclides, in curies and hazardous chemicals, in kilograms) at time t for scenario s
Q_{cct0s}	=	quantity of the controlling constituents (radionuclides, in curies and hazardous chemicals, in kilograms) at time t0 for scenario s
RF_{ccts}	=	fraction of controlling constituent quantity that is releasable to the controlling pathway at time t for scenario s
RF_{cct0s}	=	fraction of the controlling constituent quantity that is releasable to the controlling pathway at time t0 for scenario s
HM_{cct}	=	hazard measure factor for controlling constituent and controlling pathway at time t (hazard measure factors from look-up tables)
HM_{cct0}	=	hazard measure factor for controlling constituent and controlling pathway at time t0 (hazard measure factors from look-up tables)
HC_{ccts}	=	hazard control factor for risk management control action specific at time t for scenario s (hazard control factors may be estimated from site risk data or approximated using supplied look-up tables)
HC_{cct0s}	=	hazard control factor for risk management control action specific at time t0 for scenario s (hazard control factors may be estimated from site risk data or approximated using supplied look-up tables)
N	=	number of controlling constituents
S	=	number of controlling events (accident scenarios) for the analysis.

Note that the HL is only added to the numerator of the RH equation. This simplification allows the RM to be compared to the RH values and plotted as a function of time.

The equation includes the number of controlling events (scenarios) that may result in significant releases of the controlling constituents. For some facilities, there may be only one controlling event, while others may have multiple events that need to be included. As facilities and operations are improved, the likelihood of an event is expected to decrease. As the likelihood of one event is reduced, there may be a corresponding change in the likelihood of another event. This reduction could lead to one event dominating the RM initially, and a second event dominating at a later time.

The summation over controlling events may be used to represent multiple events for processing of one hazardous material, or to represent more than one hazardous material. In the latter case, each hazardous material would have one or more events defined for evaluation of the RM.

In using these formulas to develop risk profiles, consultation with appropriate site representatives is critical to ensure that the controlling scenarios adequately represent the hazard and risk management activities. It is also critical to consider the life cycle of the waste material being analysed.

4.4. Conclusions

The first generation of the risk/hazard profiles provided the Center for Risk Excellence with a number of insights into the challenges of assessing, managing, and communicating risks posed by the Cold War legacy. Some of these insights include the following:

- **EM's significant hazards pose little risk to the public.** Significant hazards exist at DOE sites. However, these hazards are currently managed in such a way as to minimize risk to the public, workers, and environment.

- **Risk management practices will require change.** Current risk management approaches are not viable for the long term. Issues such as cost, effectiveness, and legal requirements preclude maintaining old approaches. For these reasons, EM activities are focused on improved storage and remediation to alter site hazards as a method of long-term risk reduction. The lack of disposal facilities is a daunting problem to completing the EM mission and achieving long-term risk management objectives.

Communicating risk is problematic. For a variety of reasons, EM has not been as successful as it might have been in communicating risks to those outside the agency. Many existing communications methods have not achieved credible results, leaving the program vulnerable to external reviews.

A desire to communicate sites risk stories exists. The original idea behind the risk profiles came as a result of a series of meetings hosted by the Center for Risk Excellence with other DOE offices around the country. Most of these offices felt that the problems with past efforts to communicate risk at the national level could be overcome by simply letting the sites tell their stories. The profiles have received positive comments from focus groups and stakeholders and represent a good start toward improved communication of the site risk stories.

- **Tools exist for describing worker, ecological, and cultural risk communication.** The Center has demonstrated some of the possible methods that can be used to go beyond addressing public risk to describe worker, ecological, and cultural risks.
- **Risk assessments must become more comprehensive.** DOE's large, complex sites are increasingly being challenged to consider all potential types of risks and impacts over a range of spatial and temporal scales in long-term decisions. The Center is positioned to support sites in this effort and in fact, continues to play a supporting role in some sites' efforts to assess risks in a more integrated fashion.

The role of the Center is to continue to communicate and debate the lessons learned that were discussed above and to prepare methods that will be broadly accepted for responding to EM's needs.

4.5. References

1. U.S. Department of Energy (1997) Linking Legacies: Connecting the Cold War Nuclear Weapons Production Processes to their Environmental Consequences. DOE/EM-0319, U.S. Department of Energy, Washington, D.C.
2. Tyborowski, T.A. (1996) Putting a price tag on the environmental consequences of the Cold War. Federal Facilities Environmental Journal 7(3) 35-45.
3. The Presidential /Congressional Commission on Risk Assessment and Risk Management (1997) Framework for Environmental Health Management. Presidential/Congressional Commission on Risk Assessment and Risk Management, Washington, D.C.
4. U.S. Department of Energy (1996) The 1996 Baseline Environmental Management Report. DOE/EM-0290, Office of Environmental Management, U.S. Department of Energy, Washington D.C.
5. U.S. Department of Energy (1998) Accelerating Cleanup: Paths to Closure. DOE/EM-0362, U.S. Department of Energy, Washington, D.C.
6. Probst, K.N. and Lowe, A.I. (2000) Cleaning Up the Nuclear Weapons Complex: Does Anybody Care? Report 1001, Resources for the Future, Washington, D.C.
7. Consortium for Risk Evaluation with Stakeholder Participation (1999) Peer Review of the U.S. Department of Energy's Use of Risk in its Prioritization Process. Consortium for Risk Evaluation with Stakeholder Participation, New Brunswick, NJ.
8. U.S. Department of Energy (1999) From Cleanup to Stewardship. DOE/EM-0466, U.S. Department of Energy, Washington, D.C.
9. U.S. Congress (1999) FY2000 Defense Authorization Act Conference Report, "Long-term stewardship plan." Congressional Record, August 5, 1999, page H7855.
10. Stenner, R.D., White, M.K., Strenge, D.L., and Andrews, W.B. (1999) Relative Hazard Calculation Methodology. PNNL-12008, Pacific Northwest National Laboratory, Richland.
11. Stenner, R.D., White, M.K., Strenge, D.L., Aaberg, R.L., and Andrews, W.B. (2000) Relative Hazard and Risk Measure Calculation Methodology. PNNL-12008, Rev. 1., Pacific Northwest National Laboratory, Richland.

5. Perception of Risk, Health, and Inequality

While risk assessment and management are becoming more common place activities in the United States and countries of the former Soviet Union, understanding risk perceptions, particularly perceptions of the lay public, is still a far from perfect science. This chapter maps out the legacy of public opinion. This survey of perceptions of the lay public (taken from a random sample of individuals from all walks of life in Bulgaria) can help decision makers, risk managers, and risk analysts understand why the public responds as it does to their overtures. The survey examines the views of the lay public on the risks they face, their ability to control such risks, and their perceptions of those who are charged with risk management.

Today's societies create risks, which are the subject of analysis by experts, managers, and researchers. The current industrial development is connected not only with reducing the number of life and health risks but also with creating new risks and sometimes also with renewing old ones. This connection is particularly true for Bulgaria, where the transition to a market economy and the change of the old administrative system made the real environmental risks obvious for people. The pauperisation of the population, the increase in the cost of living, unemployment, fear of being involved in war, and crime are only a few examples of risk issues that attract attention and show increasing contradiction between the societal demand to see these risks reduced and the real activities of the institutions that manage these risks. Societal awareness of risks demands objective information about them and adequate standards for their prevention and reduction.

This chapter describes a study that explored the connection among the perception of different societal risks, health concerns, and attitudes of people who perceive their social security threatened. This research overcomes one common weakness of risk analysis studies: to treat people like "lay people" as opposed to "experts," and thus miss looking at differences connected with social status, material state, and health situation. The research findings showed that the socio-economic conditions are very powerful and significant factors when people judge the level of risks in the society today. The following sections describe the general setting of the study, risk perceptions identified, health and environmental issues identified, and how perceptions are related to socio-economic status. This information can assist experts in the social policy area in making decisions by providing information about how people perceive the social risks today.

5.1.   General Setting of the Study

Starting in the mid-1980s, several studies of risk perception have been conducted in different countries but most of the research has been based on small convenient samples, often only on samples of students. This narrow vision is obviously unsatisfactory, especially when the results are interpreted in terms of "public risk perception" and "cross-national comparison." Another unsatisfactory aspect of previous studies is that they often have been concerned with well-educated individuals, perhaps because these experts were easy to reach. The study described in this chapter used a random sampling methodology, which guarantees that the respondents represent different parts of the population.

The use of qualitative data collection also provided possibilities to more deeply understand people's concerns in four study areas: risk characteristics, perceptions of health issues, perceptions of environmental issues, and the relationship of risk perception to socio-economic status. Using semi-structured interviews and focus groups discussions made it possible to touch the "hot spots" in people's lives. This connection contributes to the current risk perception analysis and provides considerations for the scientist. The traditional risk perception surveys framed respondent judgements according to the interests of the researchers and little attention was paid to the real and often complicated nature of everyday concerns.

The study of risk perception was carried out during the late 1990s and included 748 individuals from four industrial towns in Bulgaria–Sofia, Pernik, Varna, and Devnja. The sites were chosen to observe the social and psychological price of the structural changes in industry, its effects such as unemployment, and its reflection on the household and individual social attitudes. Most of the respondents were employed. The highest percentage of unemployment was found in Devnja, which appeared to have serious economic problems. The highest proportion of low-income people was found in Devnja, although a high number of respondents also belong to the low-income group.

After preliminary qualitative research to identify relevant issues, a questionnaire was designed and a random sample was surveyed. The content of the questionnaire included a section for judging risk with regard to society and some sections asking for judgements of the subset of risks with regard to two targets: the respondents themselves (personal risk) or people in general (societal risk). The respondents rated several dimensions, such as demand for risk mitigation, perceived control over risks, probability of harm, severity of consequences, and trust in institutions and media. The survey also included questions about the living standard of the family, where several indicators of material deprivation were used. These indicators included unemployment, overcrowding, households not owner-occupied, household with no car, and household with incomes lower than the official existence-minimum for the country. The surveys also included questions about health state (permanent sickness or disabilities, long-standing illness, etc.) and demographic characteristics.

5.2. Risk Perception

In general, a high level of risk sensitivity and health concerns was found to be a basic feature of public risk perception when people felt threatened by lack of social and economic security. Respondents perceived the average life risk to be relatively high; this was especially true for those living in small industrial regions that suffered from a high rate of unemployment following the privatisation of the existing plants and their reconstruction. Comparing against some of the outrage factors described in Section 5.4, the respondents judged the risk of being harmed by the socio-economical situation to have severe consequences and as being unfairly distributed, unacceptable, and involuntary (Table 5.1).

TABLE 5.1 Characteristics of perceived risk to life because of the difficult socio-economic situation in Bulgaria

Extreme 1 (N1 in the scale)	Mean Magnitude	Extreme 2 (N7 in the scale)
Consequences are not severe	6.00	Consequences are severe
Fairly distributed	5.78	Unfairly distributed
Acceptable	5.60	Unacceptable
Voluntary	5.55	Involuntary
Can be tolerated	5.50	Cannot be tolerated
Ethically right	5.40	Ethically wrong
Under individual control	5.37	Out of individual control
Familiar	4.20	Unfamiliar
New	4.20	Old
Can be sensed	4.00	Cannot be sensed

Note: The judgment for each risk was made in a seven-step scale, beginning at 1 and ending at 7.

For the specific risks mentioned in the survey (Table 5.2), the perception of personal risk was found to be very high in the cases like being assaulted (crime), having poor treatment when ill (medical care), being injured by exhaust from motor vehicles, becoming ill because of stress (illness), being injured by corruption of power (distrust in authorities), and being unemployed (lack of job). When questioned about society as a whole, the main concerns where connected with unemployment, crime, inability of people to have adequate housing and sufficient nutritious and tasty food, bad medical treatment, and diseases caused by a stressful life.

Generally, there was a tendency among the respondents to perceive the risk to themselves as lower than the risk to people in society. Respondents also considered they had the greatest opportunity to protect themselves in the cases of being injured by

TABLE 5.2 Mean magnitude of perceived personal and social risk (whole group)

N	Risks	Personal Risk	Societal Risk
1	To be injured by smoking	2.43	4.39
2	To be injured by alcohol consumption	1.74	4.50
3	To be injured by exhaust from motor vehicles	3.57	4.30
4	To be injured from industrial pollution	4.00	4.45
5	To become ill because of stress	4.00	4.74
6	To have the children's health become worse	3.85	4.59
7	To be assaulted	4.47	4.96
8	To be injured by corruption of power	4.00	4.62
9	To be injured by depletion of the ozone layer	3.68	3.89
10	To be injured by a nuclear power accident	3.52	3.72
11	To be lonely	3.79	4.05
12	To have a serious road traffic accident	3.82	4.29
13	To be unemployed	3.98	4.96
14	To be unable to afford adequate housing	3.73	4.76
15	To be unable to afford sufficient nutritious and tasty food	3.69	4.68
16	To have poor treatment when ill	4.08	4.74
17	To be injured by a water shortage	2.43	3.21
18	To be injured by dirty public places	3.65	4.00
19	To be injured by domestic civil turmoil	2.79	3.44
20	To have inadequate education for self or family	3.53	4.50
21	To be poor	3.99	4.75
22	To be injured by wild dogs in the streets	3.75	4.17

Note: The judgment for each risk was made in a seven-step scale, beginning at 0–"no existing risk" and ending at 6–"extremely high risk."

alcohol consumption and being injured by smoking. These harmful activities were perceived to be controllable, familiar, and less likely to cause injury. Respondents expressed an inability to protect themselves from risks like being injured by a nuclear power accident, depletion of the ozone layer, industrial pollution, and corruption of power. They related the highest probability of harm to corruption of power, crime, bad medical treatment, and poverty. The most trusted sources of reliable information about the various risks and dangers of life were considered to be friends, physicians, and teachers. Least trusted sources were unions, municipality authorities, and government. Generally, respondents considered that they are powerless to influence any state policy.

5.3.   Health and Environmental Concerns

Overall, individuals considered themselves personally healthier than people in general (Table 5.3). The overall perception based on the respondents' answers was that the health situation is worsening. Respondents were most pessimistic about their health and that of their families in relation to life expectancy, respiratory diseases in adults, allergies, and heart disease. When asked about society, respondents considered the

TABLE 5.3 Perceived changes in health aspects

Health Aspects	For Me and My Family (Mean score)	For People in the Country as a Whole (Mean score)
Life expectancy	4.05	4.48
Respiratory diseases in adults	3.77	4.24
Allergies	3.76	4.35
Heart diseases	3.73	4.45
Children's health in general	3.62	4.39
Cancer	3.60	4.30
Other respiratory diseases in children	3.47	4.21
Traffic accidents	3.45	4.26
Rheumatism	3.38	3.90
Diabetes	3.28	4.20
Children's asthma	3.23	4.14
Alcoholism	3.22	4.40
Inborn defects	3.21	4.18

Note: Respondents' answers was estimated according to a five-step scale starting at 1 (becomes better) and ending at 5 (becomes worse).

biggest concerns to be life expectancy, heart disease, alcoholism, and children's health in general.

Most respondents (84% of all) estimated their health condition as good or satisfying. According to respondents, the people's complex socio-economical situation causes or complicates diseases of the nervous system, heart diseases, injuries or death caused by violence, tuberculosis, cancer, and accidents at work, as can be seen in Table 5.4. On the other hand, respondents considered that environmental pollution causes or complicates lead poisoning, bronchitis, asthma, a damaged immune system, cancer, inborn defects, and tuberculosis.

Respondents indicated willingness to participate in activities to improve the environment by not smoking in non-smoking areas, maintaining green areas, encouraging the formation of pedestrian zones in the town, not using pesticides around homes or gardens, not smoking in homes, and separating garbage into glass, plastics, and paper for recycling. Generally, most respondents were willing to participate in ecological activities, but only if this participation was not related to any personal inconvenience. The lowest degree of willingness to participate involved limiting the usage of electrical appliances and demanding that the government decrease traffic. The damages/costs of the air pollution were considered to exceed to some extent the economic benefits.

TABLE 5.4 Perceived causes of health problems

Diseases	"Does the complex socio-economical situation cause or complicate the following diseases?" (% respondents)			"Does environmental pollution cause or complicate the following diseases?" (% respondents)		
	Yes	No	I Don't Know	Yes	No	I Don't Know
Inborn defects	32.4	28.9	25.7	66.3	8.7	14.4
Diseases of the nerve system	88.9	4.4	3.7	32.9	27.3	23.1
Damaged immune system	45.7	19.0	21.5	69.9	5.6	15.0
Poisoning with lead	21.3	39.4	23.0	78.7	3.7	10.6
Asthma	36.8	27.7	20.6	76.1	4.1	11.6
Tuberculosis	65.4	11.4	13.4	57.4	10.3	19.7
Bronchitis	41.0	29.0	16.0	73.8	4.7	13.0
Cancer	62.0	14.8	13.1	69.9	8.7	13.8
Heart diseases	75.8	7.5	11.1	45.6	20.2	21.5
Traffic accidents	47.9	23.8	19.5	21.0	38.2	25.0
Accidents at work	59.9	14.3	17.2	30.9	30.9	23.3
Injuries or death caused by assault	72.1	8.6	12.6	13.6	46.0	24.5

Note: The percentage is not 100 in all cases because some respondents did not answer.

5.4. Risk Perception and Socio-Economic Factors

Another aspect of this study was the influence of the standard of living on people's risk perceptions. The results strongly supported the idea that low-income groups have higher sensitivity to risks created by society. The following tendencies were found:

1. People defining themselves as poor (M=75.1, SD=22.3, N=147) perceived *risk created by the difficult socio-economic situation in the country* as higher than did people defining themselves as wealthy or better than average (M=56.87, SD=21.89, N=80). The analysis of variance showed a significant difference for the perceived level of risk to life between the "poor" and "wealthy" group (F ratio=19.9, F probability=0.000).

2. Respondents from the "poor" group (M=3.08, SD=22.3, N=147) were more likely to perceive their *health state* as worse than were respondents from "wealthy" and "better than average material state" groups (M=2.26, SD=0.57, N=80). The analysis of variance showed significant differences of the perceived health state between these groups (F ratio=32.21, F probability=0.000).

3. The lower-income group perceived a higher level of *personal and societal risk* than the higher income group. There were statistically significant differences between these two groups in the perception of almost all of the 22 personal and societal risks judged by the respondents. See Tables 5.5, 5.6, 5.7, and 5.8.

4. Judging the same 22 risks in relation to their perceived *personal controllability*, the lower-income respondents perceive many risks to be less controllable than does the

group with higher incomes. These risks include, among others, to be unemployed; to be ill or have children become ill; to be assaulted; to be unable to afford adequate housing, good medical treatment, education, or nutritious food; and to be poor. It was obvious that low-income people more often express a sense of helplessness than do high-income people.

5. The same 22 risks were judged in relation to perceived *personal probability* of harm to occur. Compared to the high-income group, the low-income group had a significant tendency toward higher perceived probability for all risks.

6. The results also showed that the low-income group found their top risk (to be injured by the socio-economic situation in the country) to be *more involuntary, uncontrollable, unfairly distributed, unacceptable, familiar, new, and containing dreaded consequences* than did the high-income group. These differences were found to be significant according the T-test for comparison of means.

7. In comparison with the high-income group, the low-income group judged the *possibility to influence public policy* as lower connected with unemployment; prices of housing, goods, electricity, and food; the agricultural policy; education; and other aspects of social life. This tendency definitely influences the ability to change attitudes among people who are mostly influenced by the economic crisis.

TABLE 5.5 Rank order of perceived personal risk for the higher income group

Rank	Risks	M	SD	N
1	To be assaulted	3.89	1.43	79
2	To have a serious road traffic accident	3.44	1.33	79
3	To be injured by corruption of power	3.41	1.74	79
4	To be injured by industrial pollution	3.38	1.84	80
5	To be injured by wild dogs in the streets	3.30	1.70	79
6	To have the children's health become worse	3.16	1.86	50
7	To become ill because of stress	3.15	1.81	79
8	To be injured by a nuclear power accident	3.05	1.78	78
9	To be injured by depletion of the ozone layer	3.01	1.58	79
10	To be injured by dirty public places	2.91	1.78	80
11	To be injured by exhaust from motor vehicles	2.89	1.66	79
12	To be injured by domestic civil turmoil	2.35	1.57	79
13	To be unemployed	2.20	1.73	80
14	To have poor treatment when ill	2.16	1.73	79
15	To be lonely	2.16	1.81	79
16	To be injured by a water shortage	1.96	1.74	79
17	To be unable to afford adequate housing	1.92	1.77	79
18	To be injured by smoking	1.87	1.85	79
19	To be poor	1.82	1.50	79
20	To have inadequate education for self or family	1.77	1.70	79
21	To be unable to afford sufficient nutritious and tasty food	1.59	1.54	79
22	To be injured by alcohol consumption	1.46	1.71	79

TABLE 5.6 Rank order of perceived personal risks for the lower-income group

Risks	M	SD	N	T-test (prob.)
1. To be poor	5.44	1.07	142	18.91; (.000)
2. To be unable to afford sufficient nutritious and tasty food	5.21	1.19	137	15.00; (.000)
3. To have poor treatment when ill	5.17	1.18	138	13.76; (.000)
4. To be unemployed	4.93	1.73	139	11.24; (.000)
5. To be assaulted	4.92	1.55	133	4.95; (.000)
6. To be unable to afford adequate housing	4.70	1.89	129	10.66; (.000)
7. To become ill because of stress	4.68	1.49	130	6.31; (.000)
8. To be injured by corruption of power	4.63	1.60	126	5.08; (.000)
9. To have the children's health become worse	4.58	1.60	116	4.70; (.000)
10. To have inadequate education for self or family	4.48	2.04	128	10.32; (.000)
11. To be injured by industrial pollution	4.48	1.55	131	4.49; (.000)
12. To be injured by wild dogs in the streets	4.44	1.77	133	4.61; (.000)
13. To have a serious road traffic accident	4.38	1.46	121	4.69; (.000)
14. To be injured by depletion of the ozone layer	4.22	1.82	124	4.98; (.000)
15. To be injured by dirty public places	4.11	1.74	126	4.79; (.000)
16. To be injured by exhaust from motor vehicles	4.09	1.63	127	5.09; (.000)
17. To be injured by a nuclear power accident	4.04	1.94	119	3.69; (.000)
18. To be lonely	3.80	2.09	124	5.91; (.000)
19. To be injured by domestic civil turmoil	3.35	1.76	126	4.21; (.000)
20. To be injured by a water shortage	3.17	2.06	124	4.48; (.000)
21. To be injured by smoking	2.72	1.28	131	2.71; (<.004)
22. To be injured by alcohol consumption	1.88	2.27	127	-

TABLE 5.7 Rank order of perceived societal risk for higher-income group

Rank	Risks	M	SD	N
1	To be unemployed	4.59	1.13	78
2	To be assaulted	4.53	1.28	77
3	To be unable to afford adequate housing	4.35	1.17	77
4	To be injured by alcohol consumption	4.31	1.20	78
5	To be poor	4.29	1.25	78
6	To become ill because of stress	4.27	1.21	78
7	To be injured by wild dogs in the streets	4.19	1.66	78
8	To be unable to afford sufficient nutritious and tasty food	4.17	1.21	78
9	To be injured by corruption of power	4.15	1.27	78
10	To have poor treatment when ill	4.14	1.12	78
11	To be injured by smoking	4.09	1.16	78
12	To have the children's health become worse	4.09	1.11	78
13	To be injured by industrial pollution	3.88	1.56	78
14	To have a serious road traffic accident	3.87	1.37	78
15	To have inadequate education for self or family	3.86	1.40	78
16	To be injured by exhaust from motor vehicles	3.82	1.38	78
17	To be injured by dirty public places	3.46	1.58	78
18	To be lonely	3.37	1.51	78
19	To be injured by depletion of the ozone layer	3.26	1.43	78
20	To be injured by a nuclear power accident	3.21	1.66	78
21	To be injured by domestic civil turmoil	3.03	1.42	78
22	To be injured by a water shortage	2.92	1.52	78

TABLE 5.8 Rank order of perceived societal risk for the lower-income group

Risks	M	SD	N	T-test (prob.)
1. To be poor	5.24	1.01	141	5.73;(.000)
2. To be unemployed	5.23	1.18	140	3.92; (.000)
3. To be assaulted	5.18	1.17	134	3.64; (.000)
4. To have poor treatment when ill	5.11	1.13	138	6.06; (.000)
5. To be unable to afford sufficient nutritious and tasty food	5.02	1.25	132	4.85; (.000)
6. To be unable to afford adequate housing	5.01	1.25	132	3.82; (.000)
7. To become ill because of stress	5.01	1.26	134	4.21; (.000)
8. To have the children's health become worse	4.91	1.37	132	4.73; (.000)
9. To be injured by corruption of power	4.91	1.24	133	4.20; (.000)
10. To have inadequate education for self or family	4.89	1.34	129	5.26; (.000)
11. To be injured by industrial pollution	4.84	1.23	131	4.63; (.000)
12. To be injured by exhaust from motor vehicles	4.68	1.27	131	4.48; (.000)
13. To be injured by alcohol consumption	4.60	1.57	130	-
14. To be injured by wild dogs in the streets	4.46	1.53	128	-
15. To have a serious road traffic accident	4.45	1.46	126	2.87; ($<$.005)
16. To be injured by depletion of the ozone layer	4.41	1.63	128	5.31; (.000)
17. To be injured by smoking	4.39	1.59	130	-
18. To be injured by dirty public places	4.39	1.45	126	4.20; (.000)
19. To be lonely	4.38	1.38	128	4.81; (.000)
20. To be injured by a nuclear power accident	4.27	1.79	126	4.32; (.000)
21. To be injured by domestic civil turmoil	4.10	2.06	126	4.96; (.000)
22. To be injured by a water shortage	3.58	1.28	129	2.79;($<$ .006)

5.5. Discussion and Conclusions

The main concern of people from the study areas is the socio-economic situation of the region and the question of everyday survival. Even in a region like Devnja (which traditionally has had a bad ecological image in public opinion), the environmental risks were judged to be less important than social problems like unemployment, income, and health. These concerns influence all the judgments connected with social and environmental health policy. The concerns were caused by the reorganization of the economy, which includes privatisation and subsequent structural changes. For example, the new managers of companies attracted a small group of people with higher qualifications and discharged hundreds of employees with low or medium qualifications. The local government programs for temporary jobs (mainly connected with environmental activities like cleaning or laying out lawns) are usually unsuccessful because they cannot satisfy the main need of having sufficient incomes.

The analysis of the interviews showed that respondents perceived high levels of risk associated with different societal issues. Risks like environmental pollution and related health problems were perceived lower than the economic risks but this perception is still high in the risk scale. According to the assimilation/contrast model of risk perception[1], the presence of a great threat makes other risks less significant in the consciousness of harmed people. This difference could be the reason that issues such as environmental health problems in Devnja and Pernik are not thought to be so important as the problems of everyday survival, like family security, health, and income.

Many people think that the cost of industrialisation is too high for people living near industrial enterprises. Nevertheless the attention of these people is directed mostly to other problems. All the respondents mentioned that people with low material status are involved mainly in survival problems and are not interested in anything else. In some sense the lack of interest in other social problems helps them to cope with the situation. As one of them said: "We have to become more sensitive to environmental problems for example, but the problem is that the present situation demands reversed behaviour: if we want to survive we have to be insensitive and neglectful." According to Korte[2,3], such non-involvement helps individuals to adapt.

A number of conclusions can be made from this study. First, social and material differences of the people influence significantly the perception of level of risk, controllability of risks, and the possibilities for protection. People with low income perceive high probability that harm will occur to them. At the same time, their low confidence in authorities and institutions and their sense of helplessness to influence the public policy make this group unwilling to start any actions for change. To counter these tendencies, authorities must work to increase trust through open dialogue with people about reasons for economic structural changes and measures taken to improve the situation.

Second, all results agree that, for the public, the socio-economic situation in Bulgaria determines all other concerns including health and ecology. In the public consciousness, the main problems with the present unstable situation are the lack of security and an unclear perspective for the future. The pauperisation of the population and other real risks in the environment are very obvious. A rise in the cost of living (food, medical service, medicines, housing, etc.), the unemployment rates (about 18%), continuing inflation even with a currency board, and increasing crime rates are some of the main concerns of the common people. It could therefore be said that life in the country today is not as safe as before. People at risk of material deprivation constitute a majority of the population.

Third, some of the risk characteristics influence the risk sensitivity of people with low incomes. For example, lay people (the participants in the interviews and in the focus groups who were teachers, employees, workers, journalists, representatives of non-government organisations, public health doctors, etc.) felt pessimistic about being able to control the societal risks themselves. For many years they have been put in the situation where authorities completely ignored public opinion. The previous communist policy neither considered the cost to society of industrialisation nor put the interests of common people in the centre. Under such conditions, it is not surprising that people do not believe that they are able to influence industrial and other policies today. The observed hopelessness and helplessness are the negative effects of the lack of sense of control.[4] Often the expressed pessimism also reflects a lack of knowledge and skills for communicating with responsible agents.

Fourth, lay people and experts considered the question of taking measures in different ways. The first group expressed strong demands for risk reduction by responsible agents and at the same time feelings of helplessness for personal influence over the processes. The experts and other decision makers realized their responsibility for the situation and suggested different measures against unemployment, environmental pollution, and health problems of the population. Unfortunately, lay people showed low confidence in institutions that are responsible for control of social and economic problems. At the same time the measures that were pointed out as necessary were connected with the institutions (e.g., the government).

The main recommendation of this work to policy makers is connected with motivating socially responsible behaviour. This motivation means that decision makers must:

1. Help the public develop an attitude for active problem-solving behaviour (through education, media policy, etc.)
2. Ensure support for social projects (e.g., laws)
3. Stimulate the public to become self-responsible and less dependent on the behaviour of institutions
4. Take measures against poverty and provide opportunities for active involvement of people in decision making when these decisions influence their lives.

The results of this study represent one part of the risk management circle–namely consideration of public opinion. The work showed that public awareness of risks calls for valid information and adequate safety standards. The present research could also serve as a basis for a risk communication process and could facilitate a better understanding of risk and improvements to risk management practices.

5.6. References

1. Sjoberg, L., and Drottz-Sjoberg, B.M. (1991) Knowledge and perception of risk among nuclear power plant employees, *Risk Analysis*, **11**, 607-618.
2. Korte, C. (1980) Urban-nonurban differences in social behavior and social psychological models of urban impact, *Journal of Social Issues*, **36**, 29-51.
3. Korte, C. (1981) Constraints of helping in an urban environment. In J.P. Rushton and R.M. Sorrentino (eds.), *Altruism and Helping Behaviour*, Erlbaum, Hillsdale, New Jersey.
4. Schaeffer, M.H., Street, S.W., Singer, J.E., and Baum, A. (1988). Effects of control on the stress reactions of commuters, *Journal of Applied Psychology*, **18**, 944-95.

6. Risk-Based Ranking Experiences for Cold War Legacy Facilities in the United States

Over the past two decades, a number of government agencies in the United States have faced increasing public scrutiny for their efforts to address the wide range of potential environmental issues related to Cold War legacies. Risk-based ranking was selected as a means of defining the relative importance of issues. Ambitious facility-wide risk-based ranking applications were undertaken. However, although facility-wide risk-based ranking efforts can build invaluable understanding of the potential issues related to Cold War legacies, conducting such efforts is difficult because of the potentially enormous scope and the potentially strong institutional barriers. The U.S. experience is that such efforts are worth undertaking to start building a knowledge base and infrastructure that are based on a thorough understanding of risk.

In both the East and the West, the legacy of the Cold War includes a wide range of potential environmental issues associated with large industrial complexes of weapon production facilities. The responsible agencies or ministries are required to make decisions that could benefit greatly from information on the relative importance of these potential issues. Facility-wide risk-based ranking of potential health and environmental issues is one means to help these decision makers. The initial U.S. risk-based ranking applications described in this chapter were "ground-breaking" in that they defined new methodologies and approaches to meet the challenges. Many of these approaches fit the designation of a population-centred risk assessment. These U.S. activities parallel efforts that are just beginning for similar facilities in the countries of the former Soviet Union. As described below, conducting a facility-wide risk-based ranking has special challenges and potential pitfalls. Little guidance exists to conduct major risk-based rankings. For those considering undertaking such efforts, the material contained in this chapter should be useful background information.

6.1. Background

The Cold War participants face legacy issues related to potential environmental contamination and releases at major military and military-support facilities. These include nuclear production operations (mining, milling, enrichment, fabrication, and waste storage/disposal), research and development centres, missile bases, airfields, and remote observation installations. Potential environmental issues include risks from both radiation and chemicals though air, water, and soil exposure routes.

In the United States, a range of regulatory requirements drives the assessments for specific sites. Major actions often require environmental impact statements to meet requirements of the National Environmental Protection Act (NEPA). Operating facilities are covered under the Resource Conservation and Recovery Act (RCRA), and activities at inactive sites are covered under the Comprehensive Environmental Response, Compensation, and Liability Act (CERCLA). The guidance for the required regulatory evaluations varies from state-to-state, from region-to-region, and agency-to-agency. In addition, the products of required assessments are often incompatible in terms of comparisons and rankings with the different end-points, time-scales, locations, etc.

Consequently, risk-based ranking often does not use endpoints based directly on local norms, standards, and regulations. To do relative risk comparisons, equivalent risk-based endpoints are required that allow a consistent comparison of various potential environmental issues over a wide range of environmental conditions and contaminants. It is, however, possible to select and compute the risk-based endpoints in a manner that is consistent with local norms, standards, and regulations.

Over the past two decades, a number of government agencies in United States have faced increasing pubic scrutiny for their efforts to address the wide range of potential environmental issues related to these Cold War legacies. Risk-based ranking was selected as a means of defining the relative importance of issues. Ambitious facility-wide risk-based ranking applications were undertaken. Several of these applications and lessons learned are described below. Also described are risk estimation support tools that were developed and continue to be used in ongoing efforts to understand the risks associated with legacies from the Cold War.

6.2. Risk-Based Ranking Approaches

Major complex-wide risk-based ranking applications were undertaken by the U.S. Environmental Protection Agency (EPA), the U.S. Department of Energy (DOE), and the U.S. Department of Defense (DOD) to help set remediation priorities. These applications used different approaches to accomplish their objectives. The emphasis of this chapter is health and environmental risk estimation approaches based on characterizing the fate and transport of potential contaminants in the environment.

The DOE and DOD faced similar challenges in terms of having many large Cold War facilities facing a massive number of environmental remediation issues with relatively little information for prioritisation of those issues. DOE facility-wide applications in the mid-1980s attempted to directly assess individual and population risks as endpoints[1]. Risk-comparison applications employed a quantitative approach using the outputs of fate and transport models. DOD facility-wide risk evaluations took a different approach. The DOD applications employed a qualitative approach based on using indicators of risk. The DOD effort was an attempt to extend the hazard ranking system approach to more detailed relative risk-based ranking applications. These applications, with their successes and failures, have progressively led to the development of more effective approaches for conducting facility-wide risk-based ranking.

Applications conducted in the 1990s for Cold War legacies benefited from these earlier experiences. Buck et al.[2] developed and applied an integrated risk assessment approach for considering potential public health impacts from the underground tanks containing high-level radioactive wastes at DOE's Hanford Site. Buck et al.[3] also conducted a risk-based analysis of the long-term potential impacts of transuranic waste. A highly effective and compelling risk-based approach to long-term stewardship was proposed and demonstrated by Jarvis et al.[4]. The premise of this approach is that the only meaningful measures of remediation and containment effectiveness is a consideration of resulting risks to human health and the environment--as calculated and summed over at least ten half-lives for each radionuclide (i.e., for some materials, hundreds and thousands of years).

The following subsections describe three of the early risk-based approaches: the EPA qualitative approach for sorting sites, the DOE qualitative risk-ranking approach, and the DOD qualitative risk-ranking approach. These risk-based ranking approaches do have a common link in that each is based on measures of potential environmental and public health risks. These ranking efforts have been unique applications largely outside the area of media-specific U.S. standards and regulations. Instead, risk-based holistic approaches were used to assess impacts from potential water, soil, and air pathways. Recent risk evaluation efforts for DOE have shifted to a risk indices approach (see Chapter 4). Some of the more recent site-based risk estimation efforts by DOD are also covered in other chapters of this book (see for example, Chapter 8).

"Qualitative risk ranking" means that the risk potential is characterized based on a rule-based system. The idea is to use available surrogate site and regional parameters to define the risk potential. The rule-based process attempts to account for the major physical, chemical, and toxicological processes that result in exposures and impacts. The strength of the approach is based on known, or readily available, information. A limitation is that the surrogate parameters may not adequately represent the risk.

"Quantitative risk ranking" means that the risks are estimated based computations that account for the major physical, chemical, and toxicological processes determining potential exposures and impacts. The critical estimate of exposures may use monitoring or modelling inputs. A key effort is the definition of conceptual models for the source-to-exposure routes. The strength of the approach is that it can incorporate our best understanding of the pathways for potential risks. The weaknesses include related limitations both in our understanding of the potential risk processes and the resources that are reasonably available to estimate the risks.

The U.S. National Research Council conducted a review of these approaches. Their review included hands-on use of the software systems used as part of these approaches. Their 1994 report, *Ranking Hazardous Waste Sites*[5], details the various approaches, associated ranking systems, and applications, and is a valuable reference for additional information on the approaches. Laniak et al.[6] and Mills et al.[7] document comparisons of performance among DOE and EPA models. This comparison was performed to provide information to decision makers to help them understand their options when selecting a model for a specific application.

6.2.1. Qualitative Risk Scoring: Hazard Ranking System (HRS)

The EPA, as part of the Superfund program to clean up sites contaminated with hazardous wastes, needed to define which of a wide variety of contaminated sites located across the United States should be included in a National Priorities List (NPL). These sites include a mixture of activities, some of which are related to Cold War legacy issues and others that are strictly commercial sites. The EPA developed the Hazard Ranking System (HRS), a qualitative system for making that initial screening decision[8,9].

The HRS is a numerically based screening system that uses information from initial, limited investigations—the preliminary assessment and the site inspection—to assess the relative potential of sites to pose a threat to human health or the environment. Based on a score generated from qualitative site information, the HRS generates a score from 0 to 100 for the site. Only if the site scored above a certain number was that site listed on the NPL. The national review of this system resulted in an update (Revised HRS) in which the score generation is more directly tied to underlying physical processes.

The HRS uses a structured analysis to score sites; this analysis emulates the factors determining the potential hazard. This approach assigns numerical values to factors that relate to risk, based on conditions at the site. The factors are grouped into three categories: 1) likelihood that a site has released or has the potential to release hazardous substances into the environment; 2) characteristics of the waste (e.g., toxicity and waste quantity); and 3) people or sensitive environments (targets) affected by the release.

The HRS represents one of the early successful attempts to create a multimedia risk scoring system. Four pathways can be scored under the HRS: 1) ground water migration (drinking water); 2) surface water migration (drinking water, human food chain, and sensitive environments); 3) soil exposure (resident population, nearby population, and sensitive environments); and 4) air migration (population and sensitive environments). After scores are calculated for one or more pathways, they are combined using a root-mean-square equation to determine the overall site score. If all pathway scores are low, the site score is low. However, the site score can be relatively high even if only one pathway score is high. This factor is an important requirement for HRS scoring, because some extremely dangerous sites pose threats through only one pathway.

The EPA makes a point in their documentation that HRS scores do not determine the priority in funding EPA remedial response actions. They explicitly state that the information collected to develop HRS scores is not sufficient to determine either the extent of contamination or the appropriate response for a particular site. They rely on more detailed studies conducted subsequent to the listing to define the appropriate actions. However, despite these strong statements, some attempts were made to extend and apply the HRS as a risk ranking system. These later efforts generally were not well received because of difficulties defining importance to the relative values of HRS scores that had not been created for this use.

The HRS has fulfilled its objective of defining whether or not a candidate site is placed on the NPL. Indeed, the HRS is the principal mechanism EPA uses to place uncontrolled waste sites on the NPL.

Qualitative rule-based systems, such as the EPA's revised HRS, are an alternative approach for facility-wide applications that generally require less input data but have more uncertainty in the output parameters. Such systems are designed for identifying potential problems on minimal site data.

5.2.2. Qualitative Risk Ranking: Defense Priority Model (DPM)

The DOD efforts resulted in the development of a computer-based risk-ranking system called the Defense Priority Model (DPM). Using approximate physical relationships as a basis, the DPM allowed use of general site information to generate risk indices. A system that required less time and cost was certainly needed to make the efforts feasible; conducting risk assessments for facility-wide ranking at complex sites was potentially much too expensive and time consuming.

Despite the good design intentions, the applications of the model in the early 1990s did not go as planned. Because prioritisation of major cleanup budgets depended on the results, the data collection efforts invested by the sites were much larger than had been expected. In the end, DPM was phased out because it did not provide a defensible risk-based ranking system for a reasonable expenditure of resources; that is qualitative estimates were hard to defend and the actual costs of application were high.

Subsequently the DOD has continued to use risk estimates to plan cleanup and closure activities. In general, they now rely on site-specific risk information as opposed to the earlier broader nation-wide relative ranking approach.

5.2.3. Quantitative Risk-Based Ranking

Environmental and public health impacts have been, and are being, considered as part of the environmental restoration and waste management activities by the DOE. The DOE conducted a number of facility-wide environmental and public health evaluations based on a qualitative risk-ranking approach. These efforts provided DOE with a much better understanding of the nature of risks associated with the Cold War legacies.

The DOE approach uses a source-to-receptor analysis to estimate potential risk indices. As with the DOD efforts, a system that required less time and cost was required to make the efforts feasible; conducting these site-specific risk assessments for facility-wide ranking at complex sites was potentially much too expensive and time consuming.

The initial nation-wide risk-ranking efforts by the DOE in the late 1980s and early 1990s took the approach to reduce required resources by developing better computer tools. The Multimedia Environmental Pollutant Assessment System (MEPAS) was developed for the DOE by the Pacific Northwest National Laboratory as a unique computer-based system to consider potential health risks from chemical carcinogens and non-carcinogens as well as ionising radiation. MEPAS integrates impact computations of radioactive and hazardous contaminants for major air, soil, and water exposure routes via air, surface water, ground water, and overland flow. MEPAS allows for active operations (such as stack and vent releases) as well as inactive storage site and environmental contamination. By putting all computation models in a single, integrated, linked system, the complexity of using different, often incompatible, systems was avoided.

A number of multimedia models have been developed for various applications. DOE collaborated with the EPA to compare MEPAS and two other multimedia models being used in the United States[6,7]. The international BIOMOVS effort also included comparisons of applications of multimedia models[10].

The DOE efforts have maintained and continued to develop MEPAS risk modelling support systems. In subsequent DOE applications as described below, a modular risk computation approach was created that greatly reduced the time and resources required to conduct a facility-wide assessment. The combination of an integrated system such as MEPAS and the modular risk approach has been used in a number of DOE risk-based ranking applications related to Cold War legacies[3,4,11].

An important advance is the development of an open architecture system, Framework for Risk Assessment of Multimedia Environmental Systems (FRAMES), for linking assessment modules[12]. FRAMES is a Windows™-based operating system that retains the advantages of the integrated MEPAS while allowing flexibility in the selection of specific modules and models. The DOE, EPA, and DOD are currently jointly sponsoring FRAMES development to provide a common platform for conducting multimedia-based risk analyses. The expansion to the FRAMES concept is greatly broadening the usefulness and applicably of this risk modelling support system such that recent efforts have been conducted for major rule-making actions such as EPA's Hazardous Waste Identification Rule (see Chapter 14 and [13,14,15,16]).

6.2.4. Risk Ranking Versus Risk Modelling Support Systems

It is critical that the planner of major risk-based ranking applications makes a clear distinction between tools used to conduct the risk ranking and modelling tools used to support the ranking effort by generating risk information. A pitfall has been adoption of the risk modelling support tools before defining the risk-ranking system. The correct procedure is to start with the definition of what measures and methods are needed for ranking and then define what risk modelling tools will be required. A unique risk-ranking system normally needs to be developed for each application whereby there often are risk-modelling systems that can be used directly, or with some updates.

6.3.   Quantitative Health and Environmental Risk Estimation

A major challenge in conducting qualitative risk rankings is the estimation of potential impacts from Cold War legacy facilities with complex environmental issues. As illustrated in Figure 6.1, the analyses must address environmental releases to, and linkages between, air, ground water, surface water, and soil. For each of these media, the major pathways of interest for exposure must be defined and evaluated. Also the effects of either, or both, radioactive contaminants and hazardous wastes must be considered as well as carcinogenic and non-carcinogenic impacts.

Figure 6.1 illustrates a site with potential multimedia transport of contaminants. The major elements of a multimedia analysis required to address such a site are shown in Figure 6.2. A multimedia analysis starts with the contaminants at the "source" and their potential release in the various media by vitalization, suspension, infiltration, overland flow, and direct contact. "Transport" of the contaminants occurs within and between the air, surface water, and ground-water transport media. When that transport moves the contaminants to a receptor, an "Exposure" results by inhalation, ingestion, external dose, or dermal contact. Exposures also can result from proximity to the source though external dose. The "Impact" endpoints for an analysis are risks related to human health and environmental effects.

The planner of risk-ranking efforts must understand that using detailed site-specific models for each potential problem is generally not feasible. Although detailed analysis has the appeal of doing the "best" analysis for each potential impact end-point, the logistics will often be prohibitive and thus result in failure if attempted. The expanded data and extended analysis time required for each case, multiplied by the large number of cases, leads to very high costs and long implementation times. These factors will normally preclude selecting this approach as a practical method to address a large number of potential impacts.

The planner of risk-ranking efforts should also select a suite of "approved" and/or "accepted" models to conduct a multimedia assessment. The FRAMES development effort (see Chapter 14 and [12]) facilitates the use of this approach. The suite-of-models approach for prioritisation and ranking applications has the advantage of allowing the use of models specifically designed for each of the various issues. This approach has the limitation of having to deal with the logistics of running and combining the outputs from disparate models. Also the question of model output comparability needs to be addressed when different codes are used for different types of impacts.

The best modelling approach depends mainly on the objectives of an application. The selected model(s) must be able to address the range of potential problems associated with that particular application. In practice, for facility-wide applications, more than one model is normally required. Using a linked multimedia model such as MEPAS, or a

Figure 6.1 Pathways for multimedia risk assessment

model-linking computer system such as FRAMES, can greatly reduce auxiliary efforts required by coupling codes into an integrated system. No one model will fill all the modelling needs, and some combination of models is normally used.

Facility-wide applications of MEPAS that consider major pathways for human exposures are described below. In these evaluations, the main endpoints are mainly for human impacts, either as a risk of cancer or as a ratio representing proximity to a "safe" level. Although not addressed in this chapter, several of the applications did consider ecological endpoints. The lessons and experiences for these facility-wide applications discussed below should be appropriate for other applications of similar scale.

6.4. Facility-Wide Application Experience

While a wide variety of models address specific site characteristics, transport media, and impact type, only a few models have been developed to address the broad range of long-term public health issues. As noted above, MEPAS integrates risk computations for radioactive and hazardous materials across major exposure routes via air, surface water, ground water, and overland flow transport[17]. This section details experiences in applying MEPAS to DOE facility-wide applications.

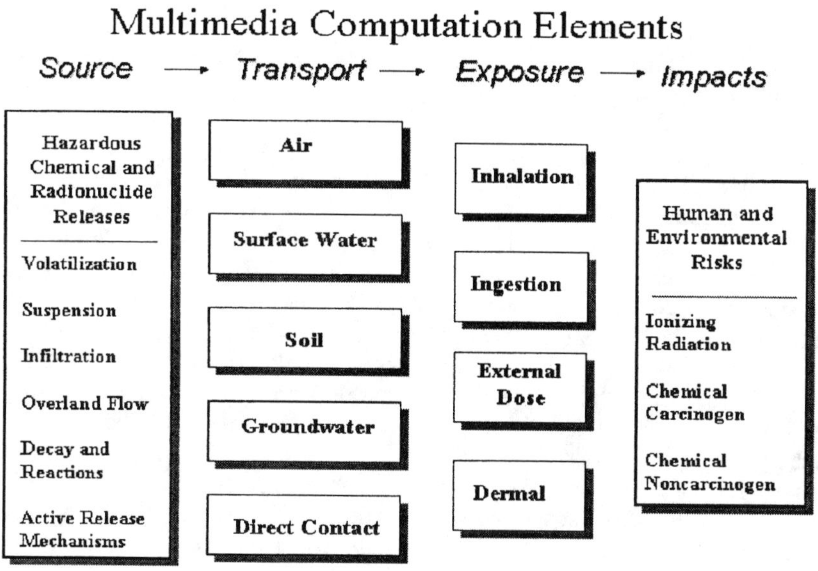

Multimedia Computation Elements

Source → **Transport** → **Exposure** → **Impacts**

Hazardous Chemical and Radionuclide Releases	Air	Inhalation	Human and Environmental Risks
Volatilization	Surface Water	Ingestion	Ionizing Radiation
Suspension	Soil		
Infiltration		External Dose	Chemical Carcinogen
Overland Flow	Groundwater		
Decay and Reactions		Dermal	Chemical Noncarcinogen
Active Release Mechanisms	Direct Contact		

Figure 6.2 Elements of Multimedia Environmental Pollutant Assessment System source to risk analyses

5.4.1. Early DOE Risk-Ranking Efforts

Although the concept for the underlying computer software has remained essentially the same, the overall approach of using the MEPAS software has undergone an evolution since its first application. The major challenge has been to reduce the resources required to evaluate risks for a large number of potential problems.

The first major facility-wide application of MEPAS was in DOE's Environmental Survey. That effort involved a nation-wide comparison of potential environmental problems at 36 DOE facilities[18]. To provide consistency in such a broad application, detailed instructions were generated for creating a conceptual site model[19] and defining model inputs[20]. Also to assure consistency for non-site specific modelling values, a constituent database was published with values for chemical, physical, uptake, and toxicity parameters[21].

The effort started with the plan to apply MEPAS to all potential sources. However, it very quickly became evident that, with thousands of potential sources, the projected time and costs for completion of the effort were not acceptable. The solution was to group the potential sources into a workable number of composite release sites. That solution made the study feasible--but in the end resulted in concerns over the validity of the computed risk numbers based on these composite release sites. The aggregated approach did allow DOE to generate preliminary Environmental Survey risk estimates for 16 major DOE facilities involving about 500 potential problems with about 1,000 transport pathways[18,22].

As an example of the preliminary Environmental Survey results, Figure 6.3 shows a scatter plot of the survey's population-based "maximum risk" ranking parameter (x-axis) versus a similarly derived "sum of risk" ranking parameter (y-axis)[22]. The former is based on the maximum risk value computed for any of the contaminants modelled for that source. The latter is based on the sum of risks for all the chemical and radionuclides modelled for that source. A ten-point change in plotted ranking parameters represents an order of magnitude change in risk. The wide range of values for the "maximum risk" in Figure 6.3 allowed DOE to effectively divide the sources from these 16 facilities into broad ranges of categories of concern from a potential risk standpoint[18]. The plot in Figure 6.3 also illustrates that, if one assumes that risks are simply additive, the process of adding or not adding the risks from a source is not an important issue for the overall risk ranking.

The preliminary Environmental Survey results had a profound influence on the DOE. There was considerable controversy over the preliminary results, and the final survey results were never released. Although these risk estimates were generated in a relatively short period (about 2 years), the results held up over time as being valid representations of risks.

A subsequent effort was to develop a DOE "Priority System" for application to environmental cleanup efforts[23]. The concept was to optimise risk reduction in DOE's environmental remediation efforts. The design required input of site-specific risk information. Initial applications used health and environmental risk data from the Environmental Survey. Complex-wide multimedia risks were not estimated for the DOE Priority System. The estimation of risks at a detailed site-by-site level was judged too large of an effort. This system was used for several years with site-generated estimates of risk reduction that included some of the data from the DOE survey. The DOE is not currently using the Priority System.

6.4.2. Modular Risk Estimation Efforts

The next major facility-wide application that considered health and environmental risks was for the environmental restoration portion of the DOE Programmatic Environmental Impact Statement (PEIS). This effort was nation-wide in scope and was to consider potential risks to all sites of the Environmental Survey plus a number of additional smaller sites. The implementers of the PEIS effort were faced with a major logistics challenge. Estimating risks using conventional approaches (even with MEPAS) would be prohibitively expensive and would take longer than their schedule allowed. The scientists conducting the assessment jointly proposed a solution: a unit-factor risk computation approach, later renamed the Modular Risk Approach. The concept is to use unit-factors representing the component factors used in a risk computation. The approach allows the extensive reuse of large portions of the risk computation and the

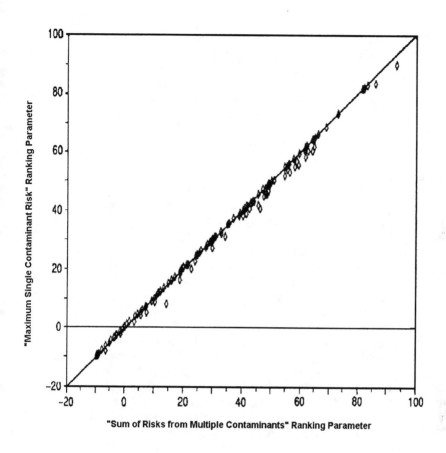

Figure 6.3 Preliminary environmental survey of potential sources at 16 U.S. Department of Energy facilities (comparison of risk rankings based on maximum single contaminant risk versus summed risks for multiple contaminants)

imultaneous computation of component unit-factors. The result is that facility-wide isks can be computed with less effort (i.e., cost) in a much shorter time. The unit-factor pproach allows roll-up of risks computed for each potential source and avoids the need ɔ use the large aggregated release sites of early efforts.

In the PEIS application, unit factors were generated for each type of potential nvironmental problem at eight of DOE's major facilities. Several models were used to enerate the unit factors: MEPAS for the multimedia transport and other models for ertain on-site exposures not addressed by MEPAS. The overall potential impacts for a ite were estimated by combining the unit factors and the contaminant inventories for ach potential problem.

The data from the earlier DOE survey provided a starting point to develop an updated and expanded database for the PEIS effort. Although risk computations were conducted for all the major DOE sites, a decision to limit the PEIS to waste management resulted in the data for environmental restoration not appearing in the PEIS final report.

The application for the Hanford Remedial Action Environmental Impact Statement[24] for waste sites at DOE's Hanford Site resulted in important expansions and enhancements to the renamed "Modular Risk Approach." This approach considered separate unit factors for the transport (unit-transfer factor [UTF]) and impact (unit risk factor [URF]; [25]). Gelston et al.[26] documents the URFs for air and soil computed for this application.

Unit risk factors such as these were used to compute risk estimates across a large complex DOE facility (Hanford) for very conservative land use at different time periods[27]. Figure 6.4 shows the resulting spatial distribution of risk for "current baseline conditions." The risks in Figure 6.4, based on monitored soil, air, and water concentrations, represent the risks for unrestricted residential use by some very uninformed people that do everything to maximize their risk. The result is that there are some areas where the risk will be fatal–and there are areas were there is no, or very little, risk. The very high risk levels are associated with assumptions of direct use and contact with the contaminated media. Similar plots provided a visual representation for a range of future land-use options at time periods 50, 100, 1,000, and 10,000 years into the future. In Figure 6.5, the 10,000-year potential risk plot corresponding to Figure 6.4 shows that, even with the movement, dispersion, and decay processes, the site will have the potential to cause harm for hypothetical residential access for a long time.

The PEIS and Hanford efforts laid the foundation for risk-estimation efforts for the preparation of the DOE Baseline Environmental Management Report (BEMR)[28] for the U.S. Congress. The PEIS information on sites and their potential impacts provided a starting point to develop risk-based cost-drivers for the BEMR effort. The unit factor approach was revised to consider a series of unit factors that started with the source and progressed to the risk computation.

An important aspect of the unit-factor approaches that merits some discussion here is the "anchoring" of the computations to available site concentration and risk data. By demonstrating that the estimated risk values are consistent with site data, the analysis provides a level of validation for the site. The anchoring requires that the estimated risks be the same when differences in assumptions are accounted for. The effort typically involves re-computing the modular risk assessment risks for specific problems using the same assumptions as a previous detailed risk analysis. This anchoring effort has the advantage of formally and clearly explaining any differences between applications in the risks estimated for a site[26].

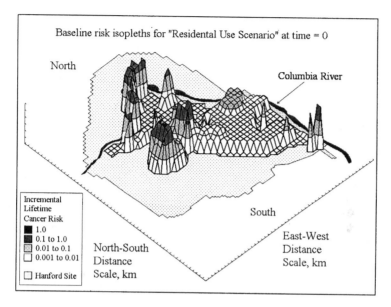

Figure 6.4 Hanford risk isopleths for assumed residential use at time=0 years.
Computed risks are based on monitored environmental concentration data.

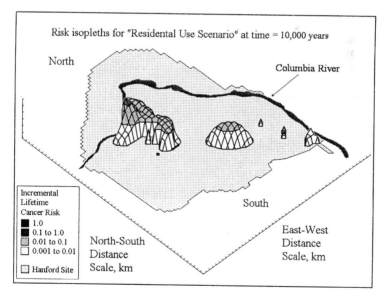

Figure 6.5 Hanford risk isopleths for far future residential use.
Computed risks are based on MEPAS outputs.

The modular risk assessment approach can be applied much faster and for less cost than a detailed site-by-site analysis. For example, all the risk computations for the Hanford Site were completed in about a 3-month period and for a fraction of the cost of conducting detailed risk analyses of all potential sources. Because the modular risk approach allows a rapid re-computation of overall risk for alternative assumptions (involving changes in factors such as the strength of the source, the land-use scenario, and the exposure/uptake/toxicity factors), the risk-estimation database from this Hanford effort was used in several subsequent Hanford risk estimation efforts conducted for other purposes.

6.5. Lessons Learned: Factors to be Considered and Managed

These facility-wide risk evaluation efforts had varying degrees of success. All were successful in generating important information that better defined the risk issues related to Cold War legacies. The lessons learned include a better understanding of the challenges of conducting such efforts along with some solutions and suggestions for future similar efforts.

The undertaking and completion of a facility-wide risk characterization effort is not an easy task. Although such efforts have the highly commendable goal of providing information to better manage risks, these efforts in the U.S. have encountered considerable opposition. The sources of opposition discussed below are factors that need to be considered and managed as part of the facility-wide risk characterization effort.

6.5.1. Internal and External Resistance

Interactions with stakeholders and the public are important to direct efforts and thus facilitate the acceptance of the results. The DOE had various levels of input and review from stakeholders and the public in these efforts. However, each of the efforts suffered in significant ways from lack of support from stakeholders and the public. Early in the efforts, very unfavourable and damaging testimony was given directly to the U.S. Congress. A common experience was that the stakeholders and the public did not trust the risk-ranking approach and/or the motives behind using it.

In each of the above facility-wide applications, there were also varying degrees of institutional resistance. The strongest in the initial efforts came from the view that the DOE Headquarters staff was trying to shift from locally to nationally managed programs. The initial nation-wide environmental analyses did provide DOE Headquarters with an overall knowledge and understanding of the DOE complex that was practically nonexistent. However, many of the individual sites completely refused to accept the initial results and embarked on efforts to disprove them. The original estimates of risk proved to be relatively robust and held up relatively well under the subsequent scrutiny. Whether this type of resistance will occur in other countries will depend mainly on their institutional structures and cultures.

Another source of opposition was the fear that the risk-based rankings would change the priorities for specific environmental management programs. People wanted to protect their projects and programs. Moreover, their fear was well founded: problems that are important from a regulatory standpoint, or other framework, may not be as important from an overall risk standpoint. Several such major shifts in DOE funding priorities occurred as the result of use of relative risk data from the efforts described above.

A science-based concern was that the risk-ranking parameters might not be consistent with locally computed risk values. The risk parameter anchoring efforts described above were invaluable in addressing concerns related to the comparability of risk estimates. Checking and affirming that any difference in risk values is the result of assumptions (and not the result of using different data or models) is important in getting local acceptance of the risk rankings.

Another component of the resistance is subtler and from scientists involved in detailed site characterization and modelling. For some, it was the reluctance to shift from traditional separate analyses of media to a holistic multimedia approach. Some of the more vocal critics felt that their detailed site-specific efforts were being inappropriately trivialized, seeing facility-wide risk estimation as overly simplified and thus incorrect. Typically no merit is seen in using approximate values to guide decision makers. It is important to understand the origin of these concerns and to keep such opposition, often by very senior and respected scientists, from stopping the risk evaluation efforts.

The planner of risk-ranking efforts should not let the factors discussed above deter the undertakings. The process of having characterized and estimated risks is an important part of the product--which results in a more risk-aware infrastructure. The fact that several of the above efforts stopped short of producing a final report did not greatly diminish the influence of the efforts. The resultant better knowledge of the risks related to legacies became an important part of the DOE infrastructure as well as provided a strong base for subsequent risk evaluation efforts.

5.5.2. Multimedia Analysis Experience

The above efforts have identified a number of advantages specifically related to using an integrated multimedia system. The consistency of analysis provides the ability to compare impact endpoints between sites that are not normally possible with separate analyses for those sites. Although no single computer model is appropriate for all situations, multimedia computer models such as MEPAS have the advantage of covering a wide range of potential problems. These models are particularly useful in the facility-wide, programmatic, risk-based, and multiple-issue applications.

In the applications described above, the multimedia analysis always resulted in some surprises in terms of defining important pathways. By being able to easily evaluate the range of possible impacts from waterborne and airborne pathways, new information was routinely obtained on the relative importance of pathways. One such new piece of information was that, at some sites, chemical carcinogens turned out to have potential impacts on the same order as, or greater than, radioactive contaminants. Such results were unexpected because of a historical concern only for radioactive contaminants.

At one site, DOE Headquarters staff felt the ground-water pathway was most important, while local DOE site officials felt the air pathway was most important. The multimedia results led to a mutual understanding that the water and air pathways were both important. At other sites, the analysis identified previously unsuspected pathways. At several of these sites, operations changed to better protect the public because of new insights from the multimedia analysis.

When ranking or comparing different sites in facility-wide applications using computed human health endpoints, it is essential to consider the inherent uncertainty in the computed values. Although a single-value deterministic approach can rank problems in broad groups separated by many orders of magnitude, a ranking of risks closer in magnitude requires consideration of the inherent uncertainty[22,29]. For such applications, a sensitivity-uncertainty module was developed for MEPAS.

There are many aspects to risk, and decision makers need information that covers all those aspects. DOE[18] initially based their rankings on a single risk parameter, a measure of population impact discounted for future impacts. In the pubic review, the use of a single risk ranking parameter received strong criticism. DOE then changed to a multifaceted definition of discounted and undiscounted risk-ranking parameters that included population, individual, and environment risk measures. This larger view of risk as having many aspects and dimensions has continued though the other risk-ranking and risk-characterization efforts described above.

An important aspect of the facility-wide applications is the generation of a database of information in the process of doing the analysis. The environmental site, regional characterization, and contaminant data are collected for the potential problems. The database of model inputs and outputs can be an invaluable resource for other analyses. This aspect has been an important factor in allowing the facility-wide applications described in this paper to build on the data collected by the preceding effort. Typically, the database provided a starting point for data review and collection efforts on subsequent projects.

It is important that these applications have external and internal reviews of proposed risk-based approaches. In the case of MEPAS, a formal scientific peer review was conducted during its development[30,31]. Subsequently the EPA conducted two independent reviews in terms of potentially using MEPAS: for possible listing sites on the NPL[32] and for analyses of hazardous, mixed, and radioactive waste sites[33]. EPA also reviewed MEPAS as part of their review of DOE's Priority System development effort[23]. MEPAS also was the subject of three scientific reviews by the U.S. National Research Council. Health and Welfare, Canada, also commissioned an independent review of multimedia models[34].

The facility-wide applications have also shown that no one model can apply to all situations. For each application, although the majority of cases were covered, certain

ιew special situations arose that were not covered by the current model formulations.
;ome cases were covered with model updates and others by using an alternative model.

;.6. Conclusion

'acility-wide risk-based ranking efforts can build invaluable understanding of the
•otential issues related to Cold War legacies. Conducting such efforts is difficult both
rom the standpoint of the potentially enormous scope of such an effort and the
tandpoint of potentially strong institutional barriers. The U.S. experience is that such
fforts are worth undertaking to start building a stronger risk-based knowledge and
nfrastructure.

Multimedia models have proven to be an effective tool in facility-wide applications.
'hese models integrate waterborne and airborne pathways into a single system for
stimating various impacts. An open architecture system such as FRAMES retains the
dvantages of the integrated multimedia assessment system such as MEPAS, while
llowing flexibility in the selection of specific modules.

Despite the improved efficiencies, the application of the multimedia software to
pplications with very large numbers of potential sources can still be prohibitive. The
nodular risk-estimation approach, based on unit factors, makes it feasible to conduct
acility-wide risk-estimation efforts. The Hanford application example discussed above
vas conducted for a reasonable cost and within a relatively short time.

The modular risk-estimation approach has a special advantage for Cold War legacy
ites where a restricted status will preclude access to certain required data. Because the
arious unit factors can be prepared separately, all the "insensitive" factors can be
repared independently of any of the sensitive data. Those with access to the sensitive
ata can generate unit factors and then combine all unit factors to estimate potential
sks. The modular risk-estimation approach will thus allow the responsible parties for a
te to conduct multimedia evaluations without having to reveal or release sensitive data.

.7. References

Droppo, J.G. Jr., Buck, J.W., Strenge, D.L., and Hoopes, B.L. (1993) Risk computation for
environmental restoration activities, Journal of Hazardous Materials, **35** 341-352.
Buck, J.W., Gelston, G.M., and Farris, W.T. (1995) Integrated Risk Assessment Program: Scoring
Methods and Results of Public Health Impacts from the Hanford High-Level Waste Tanks. PNL-10725,
Pacific Northwest Laboratory, Richland, Washington.
Buck, J.W., Bagaasen, L.M., Bergeron, M.P., Streile, G.P., Staven, L.H., Castleton, K.J., Gelston, G.M.,
Strenge, D.L., Krupka, K.M., Serne, R.J., and Ikenberry, T.A. (1997) Analysis of the Long-term
Impacts of TRU Waste Remaining at Generator/Storage Sites for No Action Alternative 2. Support
Information for the Waste Isolation Pilot Plant Disposal-Phase Final Supplemental Environmental Impact
Statement. PNNL-11251, Pacific Northwest National Laboratory, Richland, Washington.
Jarvis, T.T., Andrews, W.B., Buck, J.W., Gelston, G.M., Husties, L.R., Miley, T.B., and Peffers, M.S.
(1998) Risk-Based Requirements for Long-Term Stewardship: A Proof-of-Principle Analysis of an
Analytic Method Tested on Selected Hanford Locations. PNNL-11852, Pacific Northwest National
Laboratory, Richland, Washington.

104

5. U.S. National Research Council. (1994) Ranking Hazardous Waste Sites. National Research Council, Washington, D.C.

6. Laniak, G.F. et al. (1997) An overview of multimedia benchmarking analysis for three risk assessment models: RESRAD, MMSOILS and MEPAS. Risk Analysis, 17(2) 203-214.

7. Mills, W.E. et al. (1997) Multimedia benchmarking analysis for three risk assessment models: RESRAD, MMSOILS, and MEPAS. Risk Analysis, 17(2) 187-202.

8. Federal Register. (1990) Hazard Ranking System; Final Rule (55 FR 51532), December 14, 1990.

9. U.S. Environmental Protection Agency. (1992) The Hazard Ranking System Guidance Manual; Interim Final, November 1992. NTIS PB92-963377, EPA 9345.1-07, U.S. Environmental Protection Agency, Washington, D.C.

10. Camus, H., Little, R., Acton, D., Agro, A., Chambers, D., Chamney, L., Doroussin, J.L., Droppo, J.G., Ferry, C., Gnanapragosam, E., Hallam, C., Horyna, J., Lush, D., Stammose, D., Takahashi, T., Toro, L., and Yu, C. (1999) Long term contaminant migration and impacts from uranium mill tailings, *J. Environmental Radioactivity*, 42, 289-304.

11. Buck, J.W., Whelan, G., Droppo, J.G., Jr., Strenge, D.L., Castleton, K.J., McDonald, J.P., Sato C., and Streile, G.P. (1995) Multimedia Environmental Pollutant Assessment System (MEPAS) Application Guidance, Guidelines for Evaluating MEPAS Input Parameters for Version 3.1. PNL-10395, Pacific Northwest Laboratory, Richland, Washington.

12. Whelan, G., Castleton, K.J., Buck, J.W., Gelston, G.M., Hoopes, B.L., Pelton, M.A., Strenge, D.L., and R.N. Rickert. (1997) Concepts of a Framework for Risk Analysis In Multimedia Environmental Systems (FRAMES). PNNL-11748. Pacific Northwest National Laboratory, Richland, Washington.

13. Buck, J.W., McDonald, J.P., Pelton, M.A., Lundgren, R.E., Whelan, G., Castleton, K.J., Gelston, G.M., Hoopes, B.L., and Taira, R.Y. (1998) Documentation for the FRAMES-HWIR Technology Software System Volume 2: System User Interface. PNNL-11914, Vol. 2, Pacific Northwest National Laboratory, Richland, Washington.

14. Buck, J.W., Castleton, K.J., Pelton, M.A., Hoopes, B.L., Lundgren, R.E., Gelston, G.M., Whelan, G., Taira, R.Y., and McDonald, J.P. (1998) Documentation for the FRAMES-HWIR Technology Software System Volume 11: System User's Guide. PNNL-11914, Vol. 11, Pacific Northwest National Laboratory, Richland, Washington.

15. Whelan, G. and Laniak, G.F. (1998) A risk-based approach for a national assessment, in C.H. Benson, J.N. Meegoda, R.B. Gilbert, and S.P. Clemence (eds.), Risk-Based Corrective Action and Brownfields Restorations, Geotechnical Special Publication Number 82., American Society of Civil Engineers, Reston, Virginia, pp. 55-74.

16. 60 FR 66344-469 (1995) Hazardous waste management system: identification and listing of hazardous waste: Hazardous Waste Identification Rule (HWIR), Federal Register, Thursday, December 21, 1995.

17. Whelan, G., Jr., Buck, J.W., Strenge, D.L., Droppo, J.G., Jr., and Hoopes, B.L. (1992) Overview of the Multimedia Environmental Pollutant Assessment System (MEPAS). Journal of Hazardous Waste and Materials 9(2) 191-208.

18. U.S. Department of Energy. (1988) Environmental Survey Preliminary Summary Report of the Defense Production Facilities. DOE/EH-0072, U.S. Department of Energy, Washington, D.C.

19. Droppo, J.G., Jr., Strenge, D.L., Buck, J.W., Hoopes, B.L., Brockhaus, R.D., Walter, M.B., and Whelan, G. (1989) Multimedia Environmental Pollutant Assessment System (MEPAS) Application Guidance Volume 1 - User's Guide (for MEPAS Version 2). PNL-7216, Pacific Northwest Laboratory, Richland, Washington.

20. Droppo, J.G., Jr., Strenge, D.L., Buck, J.W., Hoopes, B.L., Brockhaus, R.D., Walter, M.B., and Whelan, G. (1989) Multimedia Environmental Pollutant Assessment System (MEPAS) Application Guidance Volume 2 - Guidelines for Evaluating MEPAS Parameters (for MEPAS Version 2). PNL-7216, Pacific Northwest Laboratory, Richland, Washington.

21. Strenge, D.L., and Peterson, S.R. (1989) Chemical Data Bases for the Multimedia Environmental Pollutant Assessment System (MEPAS): Version 1. PNL-7145, Pacific Northwest Laboratory, Richland, Washington.

22. Droppo, J.G., Jr., Buck, J.W., Strenge, D.L., and Siegel, M.R. (1990) Analysis of Health Impact Inputs to the U.S. Department of Energy's Risk Information System. PNL-7432, Pacific Northwest Laboratory Richland, Washington.

23. Longo, T.P., Witfield, R.P., Cotton, T.A., and Merkhofer, M.W. (1990) DOE's formal priority system for funding environmental cleanup. Federal Facilities Environmental Journal **1**(2) 219-231.

24. U.S. Department of Energy. (1994) Hanford Remedial Action Draft Environmental Impact Statement Volumes I and II. U.S. Department of Energy, Washington, D.C.

25. Strenge, D.L., and Chamberlain, P.J., II. (1994) Evaluation of Unit Risk Factors in Support of the Hanford Remedial Action Environmental Impact Statement. PNL-10190, Pacific Northwest Laboratory, Richland, Washington.

26. Gelston, G.M., Jarvis, M.F., Von Berg, R., and Warren, B.R. (1995) Risk Information in Support of Cost Estimates for the Baseline Environmental Management Report (BEMR): Section I, Development and Applications of Unit Risk Factor Methodology: Nevada Test Site. PNL-10608, Pacific Northwest Laboratory, Richland, Washington.

27. Whelan, G., Buck, J.W., and Nazarali, A. (1994) Modular risk analysis for assessing multiple waste sites. In Proceedings of the U.S. DOE Integrated Planning Workshop, U.S. Department of Energy, Idaho National Engineering and Environmental Laboratory, Richland, Washington.

28. U.S. Department of Energy. (1995) Estimating the Cold War Mortgage, The 1995 Baseline Environmental Management Report, Volumes I and II. U.S. Department of Energy Office of Environmental Management, Washington, D.C.

29. Doctor, P.G., Miley, T.B., and Cowan, C.E. (1990) Multimedia Environmental Pollutant Assessment System (MEPAS) Sensitivity Analysis of Computer Codes. PNL-7296, Pacific Northwest Laboratory, Richland, Washington.

30. Whelan, G., Strenge, D.L., Droppo, J.G., Jr., and Steelman, B.L. (1987) The Remedial Action Priority System (RAPS): Mathematical Formulations. PNL-6200, Pacific Northwest Laboratory, Richland, Washington.

31. Droppo, J.G., Jr., Whelan, G., Buck, J.W., Strenge, D.L., Hoopes, B.L., and Walter, M.B. (1989) Supplemental Mathematical Formulations: The Multimedia Environmental Pollutant Assessment System (MEPAS). PNL-7201, Pacific Northwest Laboratory, Richland, Washington.

32. EPA. (1988) Analysis of Alternatives to the Superfund Hazard Ranking System. prepared by Industrial Economics, Incorporated, Cambridge, Massachusetts.

33. Moskowitz, P.D., Pardi, R., Fthenakis, V.M., Holtzman, S., Sun, L.C., and Irla, B. (1996) An evaluation of three representative multimedia models used to support cleanup decision-making at hazardous, mixed, and radioactive waste sites, *Risk Analysis*, **16**(2), 279.

34. Intera Information Technologies Corporation. (1992) Review and Assessment of Two Multimedia Exposure Models: MEPAS and MUITIMED. Intera Information Technologies Corporation, Environmental Sciences Division, Ontario, Canada.

7. Cleanup of Radioactive Floating Refuse at Vromos Bay

The Cold War legacies in the United States and countries of the former Soviet Union are daunting enough in the aggregate. However, even when looking at a single example site, the difficulties of clean up can be staggering. This chapter presents a case study of remediation of the contamination legacy in Bulgaria. Tailings from mine milling operations dumped a total of about 8,000,000 tons of refuse in Vromos Bay on the Black Sea. The heavy iron sulphides and oxides, copper, and uranium minerals remained deposited in the surf area, right on the beach, where they formed a field about 2,300 meters long, up to 150 meters wide, and 2.3 meters thick. In 1995, the Bourgas Copper Mines chose to apply for the PHARE-ECOLOGY Programme to sponsor the restoration project.

From 1954 to 1977, part of the refuse resulting from operations at the Rossen Flotation Mill in Bulgaria was discharged into the Black Sea, to the west of the village of Chernomorets, in Vromos Bay. Vromos Bay is a smaller bay located at the southern end of the Bay of Bourgas (Figure 7.1). To the east and west, Vromos Bay borders on two rocky capes: Atia and Akin (Figure 7.2). A long (2,500-m) and comparatively narrow beach covered with sand stretches between the two capes. The sand is naturally tiny and yellow; detritus prevails to the west.

After 1968, all mill refuse, a total of about 8,000,000 tons, was discharged there. As a result, the beach at Vromos Bay has been covered with flotation refuse between Cape Atia and Chernomorets (Tschernomorez in Figure 7.3), and the coast line has been extended some 150 m into the sea in the area of the discharge. Being a source product, the flotation tailings consist mainly of rock-forming minerals (feldspars, pyroxene, quartz, and chlorite) and gangue minerals (quartz, calcite, dolomite, anchorite, fluorite, and clays), as well as five to six ore minerals (pyrite, chalcopyrite, magnetite, haematite, molybdenite, chalcocite, etc.).

Certain radioactive materials have also been detected–uraninite, nasturane, and uranium resins. Both lighter rock-forming and gangue materials have been carried far

107

into the sea, where they are building a thick layer of slime. The heavy iron sulphides and oxides, copper, and uranium minerals have been chiefly deposited in the surf area and right on the beach area, where they formed a field about 2,300 m long, up to 150 m wide, and 2.3 m thick. This field included about 10% of the total amount of flotation refuse, but with copper, iron, and uranium contents several times higher[1].

As a result of the combined influence of sea waves and other processes, the border between the beach and adjacent areas was encircled by a continuous bank of flotation refuse of height 2 to 3 m. In the central parts where the beach is over 150 m wide, these same processes led to the formation of single dunes up to 5 to 6 m in height.

This chapter presents a case study of how Vromos Bay was restored. It describes conditions before restoration, initial restoration attempts, a major restoration project, and its results.

7.1. Conditions Before Restoration

As early as the 1970s, research sought to determine the concentration of radioactive components on the beach and in bottom sediments of Vromos Bay. In 1970, Ouzounov[2] pointed out an increased value of the exposition power of concentrations in the "black" sand of the beach. The most detailed studies were carried out in 1976, 1978,

Figure 7.1 Location of Vromos Bay

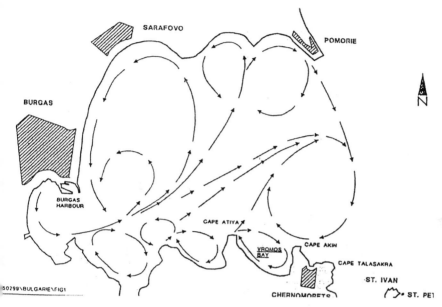

Figure 7.2 Bay of Bourgas and sea currents

983, and 1991. In the 1976 studies, some quite high values of P_x–up to 940 μR/h–were ctected[3]. It was then that a larger area was measured for the first time, and "hot oots" were detected where the radium-226 concentration was as high as 41,400 q/kg[4].

.1.1. Beach Area

he results from three studies carried out by Bourgas Copper Mines Co. pointed out that unique technogenic placer deposit had been formed in the west and central part of the each. This deposit has the component distribution and typical oblique structures of dimentation characteristic of coastal sea placers, as determined by the combined ravity and separating effect of the sea in the surf zone and the accumulative processes the confluence point. The black colour of that beach was not natural, but resulted ainly from the magnetite and haematite brought by the tailings (see Figure 7.1) as well the rock-forming materials.

The copper contents were within the range of 0.06% to 2.20%, the iron content was ithin 8.4% to 49.2%, and concentrations of radionuclides were present. Results from e radioactivity measured as of September 15, 1993, are given in Table 7.1. The same ble also shows the natural radioactivity of some adjacent beaches for the sake of mparison.

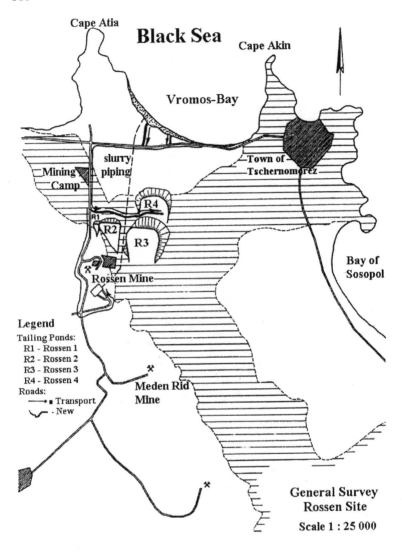

Figure 7.3 Plan of the Rossen Mine and Vromos Bay

TABLE 7.1 Radioactivity measured in September 1993 near Vromos Bay

Location	Power of Exposition, µR/h	Specific Activity, Bq/kg		
		Uranium-235	Radium-226	Thorium-232
Vromos Bay				
1. Exurb I	50	950 + 12%	600 + 20%	30 + 10%
2. Exurb II	110	1800 + 15%	3700 + 20%	30 + 18%
3. Central part	100	5000 + 7%	4900 + 12%	40 + 17%
4. Central part (soil)	35	300 + 25%	200 + 30%	45 + 7%
5. Poplar forest	80	2900 + 10%	2800 + 15%	30 + 19%
Adjacent Beaches				
6. Bourgas	6-7	< MDA *	25 + 5%	20 + 10%
7. Village of Chernomorets	15	< MDA	18 + 10%	20 + 10%
8. Gradina campground	7-8	< MDA	12 + 10%	10 + 10%
9. Sozopol	10-12	< MDA	20 + 5%	15 + 10%

MDA=Minimal Detectable Activity

.1.2. Bay Bottom

Of the 8,000,000 tons of tailings thrown into Vromos Bay, about 1,000,000 tons ave been re-deposited on the beach area, and over 900,000 tons have been dredged and sed in the fundamental construction of the Port of Bourgas-West. The remaining ,000,000 tons cover the greater part of the bay bottom with a layer over 1 m thick. The ailings have been separated by the bottom stream and waves according to the size and ensity of the primary material (Table 7.2). Table 7.3 shows the number of samples ken in the respective year, and the range of radium-226 contents.

TABLE 7.2 Screen analysis of tailings in Vromos Bay

Particle Size, mm	Percent
>0.4	1.4
<0.4 > 0.3	3.1
<0.3 > 0.25	6.9
<0.25 > 0.12	10.1
<0.12 > 0.08	12.3
<0.08	61.3
Other	4.9

TABLE 7.3 Range of radium-226 in samples in Vromos Bay, 1978, 1983, and 1991

Year	No. of Samples	Specific Activity of Radium-226 (Bq/kg)
1978	7	10- 640
1983	37	54-1380
1991	78	10-1150

7.1.3. Requirements for Radiation Protection

Bulgarian main rules for radiation protection (the Committee for Peaceful Use of Nuclear Power, 1992) defined the basic levels of radiation and their control (Table 7.4). Regulation No.7, 1986, the Environmental Ministry, provides instructions and rules to define the allowed pollution of flowing surface water for radium-226 as 150 mBq/L. Pursuant to Regulation No.7, the Committee for Peaceful Use of Nuclear Power/07.01.1992, the residue characterised by a value of over 1 mSv/h of the equivalent dose of gamma radiation at a 0.1-m distance from the surface shall be deemed hard radioactive residue.

TABLE 7.4. Regulations for radiation exposure

Groups	Definition	Basic Limits for Effective External Radiation for 1 year
A	People working temporarily of permanently with sources of ionising radiation; people exposed to such radiation by profession; people involved in extreme lifesaving activities.	50 mSv
B	People or groups of population including both sexes above 18 years of age.	5 mSv
C	The population of the country as a whole.	1 mSv

7.2. Initial Restoration, 1991-1994

Initial restoration efforts sought to clean the beach area and find a way to utilise the useful components of deposit. The Bourgas Copper Mine developed an effective technology for this purpose. They started by studying the mineral and material composition of the tailings. According to their studies, the copper in the tailings is mainly represented by chalcopyrite. Chalcocite and bornite are detected much more rarely. The highest copper contents can be found in classes of +0.08, +0.12, and +0.16 mm. Most of copper minerals are covered with a thin film of copper oxides. The iron is found as magnetite and haematite, mostly in the classes below 0.16 mm. Most radioactive minerals are distributed in the non-magnetic fraction (the classes below 0.16 mm).

The technology developed based on these findings was put in place in early 1991. Tailings are collected and carried to the Rossen Flotation Mills for additional grinding to remove the oxide film from the mineral surface. Tailings are then subject to magnetic separation and flotation in accordance with the operative technological scheme. The result is a copper concentrate with copper contents of 13% to 15% and gold contents of 4 to 5 g/ton, as well as an iron concentrate with 55% iron contents.

These flotation tailings (about 75% of the treated quantity) contain nearly all the radioactive minerals. They are stored in the Rossen-3 tailings dump (noted as R3 on Figure 7.3).

7.3.   PHARE Project, 1995

Despite the technology developed to prevent additional contamination of the beach and bottom sediments, the areas still required restoration from the original deposits. In 1995, the Bourgas Copper Mines chose to apply for the PHARE-ECOLOGY Programme to sponsor the restoration project. This project involved purposeful research work to define the degree of pollution of both beach and sea bottom as well as the way of removing the radioactive flotation tailings.

7.3.1.   Gathering Information

To prepare for the project, all information on the quantity of the ore treated in the Rossen Flotation Mill between 1954 and 1977, the time when flotation tailings were thrown in the sea, was collected and used. The quantity of flotation tailings was determined from the quantity of ore treated. All information related to the study of the beach area and bay (made in 1979, 1982, 1991, and 1993) as well as the partial study of sea sediments in the bay (made in 1978, 1983, and 1991) were collected and analysed.

7.3.2.   Sampling

Additional information was gathered by sampling the beach and bay areas.

Beach Area

The beach area was sampled by profiles located a distance of 125 m apart; two points were sampled for each profile. The gamma radiation background was measured at these points, and a sample quantity of about 2 kg was taken for analysis. That sampling network proved optimal based on the wind and surf effect of many years on the beach area and the information collected from previous studies. (The 1993 study used profiles 50 m apart with three points on each profile.) The network was defined by using the rarefaction method and the information from the 1993 study. This network is denser than the one in the bay because the surf area has a more dynamic sedimentation environment (the formation of oblique structures), and the quantities subject to treatment and direct disposal required a more precise determination.

Bay Bottom

The bay bottom was sampled using a 250-m by 250-m network . A scuba diver took the samples, which weighed about 2 kg, from the upper 15 to 20 cm of the sediment. These samples are characteristic of the whole sediment thickness at the particular point based on a number of constants influencing the process of sedimentation. These constants include composition of flowing tailings, fixed place of flow, and typical and continuous streams in the bay. These factors resulted in the formation in the central part of the bay of an elliptical geological body, whose long axis coincides with the direction of the tailings flow (Figure 7.4, the information refers to radium-226 contents in bottom sediments and comes from a research carried out in 1995).

114

The sampling network was selected following an analysis of the information resulting from the above research and considering the slower sedimentation at depths over 10 m[5]..

The water layer was also sampled by a 500-m by 500-m network. The sample volume was about 3 litres, and samples were analysed for radium-226.

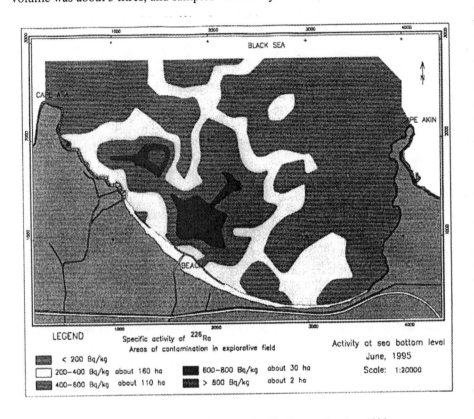

Figure 7.4 Activity at sea bottom level in Vromos Bay, June 1995

The motor boat used by the scuba divers was fixed to a particular point from the shore by an EOT-2000 light-distance-meter geodetic apparatus (made in Germany).

Adjacent Bays
The bays adjacent to Vromos–at Atia, the village of Chernomorets, the Chernomorets camping site, and the Gradina camping site–were sampled by single profiles. A sample was taken from the beach area and from sampling points at a distance of 250 m in the bottom sediments of the respective bays.

7.3.3. Independent Assessment of Project Plans

The project for the Vromos Bay cleanup was reviewed by URANERZBERGBAU, a German consulting company. The company consultants revised the project and researched alternatives. Variants considered included

- Pebbling the beach

    Sucking bottom radioactive deposits out of the Vromos Bay and disposing of them far out to sea

    Disposing the flotation tailings into the Rossen Flotation Mill

    Disposing the flotation tailings into the liquidated mine galleries of adjacent mines in the Rossen mine field.

To assess the variants, consultants developed a qualification system. According to their analysis, the most appropriate variant is the disposal in the mine galleries. Unfortunately, the galleries do not have the capacity required. Because of this, disposal of the tailings in the Rossen Flotation Mill was chosen as the optimal and environmentally friendly alternative.

7.3.4. Removal and Disposal of Tailings from the Beach

Cleanup of the Vromos Bay beach area started as early as September 1997 in compliance with the techniques in the project plans. The radioactive flotation tailings were collected by bulldozers, loaded onto dump trucks by front loaders and excavators (Figure 7.5), and then disposed either to the R-4 tailing dump or the Rossen Flotation Mill (see Figure 7.3). To speed the process, sand was excavated at three or four points simultaneously. This arrangement required 30 to 40 dump trucks of 14- to 16-ton capacity to be simultaneously operating.

The beach area was cleaned mainly in winter and spring, when storms frequent the bay. Thus, black sand from the beach area had to be extracted three times following each new release of new quantities by the sea. Before each new cleanup, the area was sampled. Samples were analysed in the laboratory of the Regional Inspection of Environment and Water, Bourgas. The beach was cleaned to the base layer of hard grey-black clays. Visual inspection show that, following each cleanup, the sands newly released by the sea are of a lighter and lighter colour and smaller radionuclide concentration (detected by the gamma spectrometric analyses of intermediate samples).

To dispose of the tailings, a tailing dump was built on a site owned by the Bourgas Copper Mines. The dump was built in agreement with the Bourgas Regional Inspection of Environment and Water, and was approved by the Ministry of Environment and Water. The soil layer and weathering material were removed in advance and disposed of separately. The south and west part of the tailing dump was walled by soil material. The wall has the following dimensions:

    Base, 8 m wide
    Crown, 4 m wide
    Height, 6 m.

Figure 7.5 Cleaning of the beach at Vromos Bay, 1998

The wall was shaped at two levels at a slope angle of 1:1. It was compacted by a road roller during construction.

7.3.5. Treating Residue in the Rossen Flotation Plant

A part of the tailings was treated in the Rossen Flotation Plant. The quantity was considerably smaller than the one originally suggested because of low prices at the London Metal Exchange. The quantities were characterised by copper content over 0.38%. After being transported to the plant reception bunkers about 7 km away from the beach, tailings were treated by grinding, flotation, and magnetic separation. The total output of both copper and iron concentrate was about 10% to 12%. The remaining quantity (88% to 90%), containing most of the radioactive minerals, was disposed of in the operating Rossen-3 tailing pond (noted as R3 on Figure 7.3).

7.3.6. Monitoring

In compliance with project plans, areas were sampled after cleanup in the same volume and within the same network as the original sampling in 1995. Areas were sampled in July and August 1998 (Table 7.5).

TABLE 7.5 Sampling conducted near Vromos Bay, 1998

Type of Activity	Dimension	Quantity
1. Beach area sampling	No	65
2. Beach area radiometric sampling	No	65
3. Beach area drilling	1 m	33
4. Taking samples from drill holes	No	66
5. Taking samples from bottom sediments	No	158
6. Taking water samples from bottom layer	No	37
7. Referring samples to a geodetic basis	No	251
8. Analysis of radionuclides	No	289

Samples from bottom sediments, the beach area, drill holes, and water taken from the bottom layer were analysed for uranium-238, radium-226, thorium-232, and potassium-40. Samples were analysed using an ORTEC low-background gamma-spectrometric apparatus in the laboratory of the Bourgas Regional Inspection of Environment and Water, and the Dosimetric and Radiation Protection Laboratory at the Physical Department of the Sofia University "St Kliment Ohridski."

7.4. Project Results

Restoration activities resulted in noted improvement to both the beach and bottom sediments.

7.4.1. Beach Area

As a result of the removal of some 800,000 tons of flotation tailings, the beach area of the bay was considerably influenced:

The width of the beach area was reduced in the various profiles from 60 to150 m to 20 to 50 m, which has brought it near its original condition (Figure 7.6.)

The thickness of the flotation tailings on the beach, which ranged from 2 to 6 m, was reduced to about 0.30 m, as detected by the drill holes (Figure 7.6). A number of places, particularly in the east part of the bay, were totally cleaned of their flotation tailings.

Radionuclide concentration in the beach sand decreased considerably since 1979, mostly because of the excavation of flotation residue but partly because of its spreading over a larger surface (Table 7.6)

Sand colour was essentially changed: from dark grey and black before restoration, to light grey and yellow after restoration (Figures 7.7 and 7.8).

118

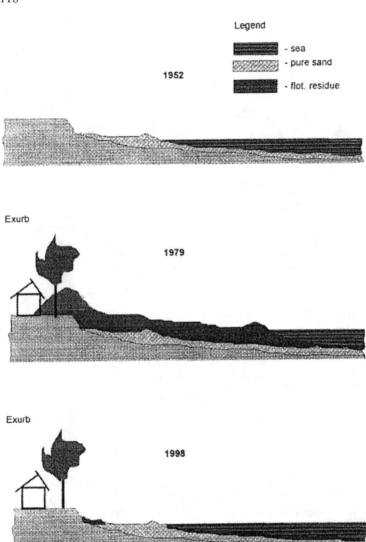

Figure 7.6 Profile No. 4, typical section of Vromos beach

TABLE 7.6 Changes in radionuclide concentrations over time at Vromos Bay

Year	No. of Samples	Exposition Power, μR/h		Specific Activity, Bq/kg	
		Min	Max	Min	Max
1979	15	25	940	74	41400
1993	28	35	110	200	3 700
1995	44	17	56	100	609
1998	44	15	32	83	400

7.4.2. Bottom Sediments

The results from the research work carried out to determine the bay bottom lithodynamics show that, 20 years after the discharge of flotation residue ceased, the newly formed lithobodies are still slowly moving. This dynamic is a result of three main factors, the first and most significant being the removal of flotation residue from the beach. The other two factors involve bottom streams and waves. At the time of sampling, two areas of increased radioactivity of over 200 Bq/kg were located on the sea bottom. One of them was located against the discharge zone, with a centre of about 600 m away from the shore (Figure 7.9). Two comparatively smaller areas were detected within it, characterised by a radium-266 concentration of over 600 Bq/kg. Bottom

Figure 7.7 Beach at Vromos Bay, a view from the east (August 1998)

Figure 7.8 Beach at Vromos Bay, central part (August 1998)

sediments here comprised compact black clay materials, formed during the sedimentation of the pellet fraction of the flotation residue. The sand component, which had moved to that point during the discharge, was pushed back to the shore. The second area was located at about 1,500 m farther into the bay, at a depth of over 15 m below the surface. Sediments in that area comprised black slimes in a semi-liquid state. The area distribution and activity as determined in three successive studies are shown in Table 7.7.

Analyses also showed that the size of high concentration areas in bottom sediments is continually decreasing. The absolute value of specific activity is also decreasing.

7.5. References

1. Bonev, I., Boiadjan, O., Tzonkov, T., and Stanchev, S. (1989) Recreation of the beach area of Vromos Bay by exploitation of the technogenic deposit of the same name formed by the tailings of the Flotation Plant "Rossen." *Sofia, Mining Journal*, **314**-15 (in Bulgarian).

2. Ouzounov, I. (1970) Studying of the health condition of Mine Rossen workers in connection with the silicosis and radioactive regulations and measures for protecting their working activity. A report for NIOTPZ, Sofia (in Bulgarian).

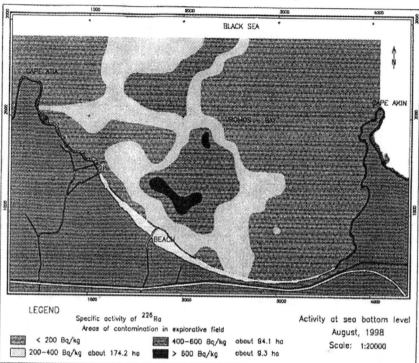

Figure 7.9 Activity at sea bottom level in Vromos Bay, August 1998

TABLE 7.7 Distribution of bottom sediments in Vromos Bay by area and activity

Specific Activity of Radium-226, Bq/kg	Area, m²		
	1991	1995	1998
200-400	1,400,000	1,600,000	1,742,000
400-600	1,050,000	1,100,000	941,000
600-800	470,000	300,000	93,000
800-1000	60,000	20,000	–
> 1000	25,000	–	–

Minev, L., Teofilov, S., Ouzounov, I., Kritidis, P., and Konstantinov, V. (1976) Studying of the concentration of natural radioactive nuclides in bottom sediments at the South Black Sea Coast. Sofia, Second National Conference of Biomedical Physics (in Bulgarian).

Ouzounov, I., et al. (1980) Final Report on Theme 16808, "Studying of the natural radioactivity at the Black Sea Coast – 1978-1980." Sofia University "St. Kliment Ohridski," Department of Physics, Sofia.

Zenkovich, 1960 in Bulgarian

8. Integrated Accident Risk Analysis and Applications for the Disposal of Chemical Agents and Munitions

Although many studies could benefit from the aspects of risk assessment described in this book to clean up Cold War legacies, only a few studies have integrated risk management and risk assessment well. This chapter describes one such study—efforts of the U.S. Army to remediate the legacy of chemical weapons stored in the United States. This effort addressed the human risk associated with that storage, developed and implemented a process to destroy the chemical weapon energetics and agent, analysed the facility and human risk associated with the destruction process, and used a risk management approach to control the process. The risk analysis is an accident analysis. Risk from routine operations and mild accidents is examined in other studies and is small compared with the risk of more severe accidents.

Chemical weapons have played an important role as a United States military deterrent over the past 50 years. Weapons materials are stored at eight sites within the continental United States. Chemical agents included in this stockpile are of two basic types, nerve and blister, and are configured in a variety of munitions and bulk containers. The changing global political climate, however, has led to an elimination of the need for these weapons and the chemical agents used in their manufacture.

In 1985, Congress enacted Public Law (PL) 99-145[1]. This law directed the Department of the Army to establish a program to dispose of the U.S. stockpile of unitary chemical weapons and agents. In 1997, the U.S. ratified the Chemical Weapons Convention. This treaty commits the signatories to destroy all their chemical warfare materiel in an environmentally safe manner by April 2007[2]. The Chemical Stockpile Disposal Program was established to achieve these goals. The U.S. Army Program Manager for Chemical Demilitarization (PMCD) has responsibility for the disposal program. This program is committed to meeting the disposal objectives while protecting the environment and the safety and health of the workers and the people of the surrounding communities.

This chapter introduces the principles the program applies to achieve its risk management goals. The focus then shifts to the integrated accident risk analysis [or Quantitative Risk Assessments (QRAs)] used to assess the greatest hazards associated with chemical agent weapons disposal–agent and explosives. The QRA uses a number of techniques to estimate risk. The chapter discusses each in turn as well as introduces the software code being developed to facilitate the analysis. The chapter also provides an introduction to presenting risk results because this activity poses unique challenges in presentation and communication. Finally, the chapter summarizes applications of integrated facility analysis for Cold War legacy facilities.

8.1. Managing Chemical Demilitarisation

The primary vehicle for managing the Army's chemical demilitarisation efforts is the Risk Management Program. This program is a life-cycle activity that started in the conceptual design phase and will continue until the disposal of chemical warfare material is complete and the associated facilities decommissioned. The primary objective of this program is outlined in the PMCD safety policy. This policy was developed from classical risk management principles[3-8] and includes military regulations[9] and industry standards[10,11].

The Army's chemical demilitarisation efforts are managed by programs for risk assessment, risk management, peer review, and public involvement, all of which support the goals of quantitative risk assessment. The following subsections deal with the various aspects of the program in more detail.

8.1.1. Program Manager for Chemical Demilitarization Risk Assessment Activities

The PMCD has developed a risk management program in keeping with U.S. Army regulations and state and federal laws and to meet the goals of minimizing risks to the worker, environment, and communities. To accomplish these goals, the U.S. Army uses risk assessments to understand and control risks. Several different types of risk assessments are performed and, taken together, they form a complete picture of the risks of storage and disposal.

The following hazards are studied in risk assessments:
- Chemical agent
- Explosives
- Stack emissions
- Occupational hazards, such as lifting injuries or hearing damage
- Industrial hazards involving other chemicals and materials, such as caustic chemicals.

Identifying and understanding hazards through risk assessment is the first step in successfully reducing risks. Several risk assessments, each with a different purpose and scope, are done for each chemical agent disposal facility. Some hazards are examined in more than one assessment. Three main types of risk assessments provide a comprehensive analysis of storage and disposal risk:

1. Health Risk Assessments–examine the risks to the surrounding communities and environment from incineration stack emissions. Assessments for each site include a Human Health Risk Assessment and an Ecological Risk Assessment.
2. Hazard Evaluations–identify and rank potential hazards resulting from disposal operations. Multiple evaluations are performed for each site and cover risks associated with chemical agent and explosives, as well as industrial and occupational hazards.
3. QRAs–evaluate the likelihood and effects of an accidental release of chemical agent during storage and disposal. Risks to the public and workers are both studied.

To help reduce the chance of an accidental release of a chemical agent, the program completes QRAs, which are limited to the greatest hazard, the chemical agents, and any associated explosives. The QRA studies the complete disposal process, as well as munition storage, and considers:

1. Human errors, such as an accident driving a forklift
2. Equipment failures, such as a drain line valve failure
3. Loss of support utilities, such as electrical power
4. External influences, such as aircraft crashes
5. Acts of nature, such as storms and earthquakes.

Hundreds of potential accidents, including very rare events, are studied using models of the facility processes. The QRA is intended to represent, to the maximum extent possible, a best estimate of the frequency of potential accidents, and the magnitude of the consequences (number of people affected). The result of the QRA is a list of events most likely to occur or to cause the greatest harm to human health. The combination of likelihood and health consequence is called risk. The program manager reviews this list to make changes to equipment or procedures to further increase safety. Risks can also be compared, such as the risk of storage to the risk of processing. Risks can also be compared to other risks in life, although that requires great care in understanding what is being compared.

In terms of risk management, the QRA results are translated into the U.S. Army and program's existing system of risk assessment codes. This translation allows QRA risks to be considered within the existing and accepted decision framework, rather than having separate numerical decisions associated with QRAs and all other hazard analyses.

5.1.2. Program Manager for Chemical Demilitarization Risk Management

Assessments are the first step in risk management, the process by which risks are identified, controlled, and reduced. Risk management also includes:

- Establishing requirements to minimize risks
- Monitoring to continuously ensure that safety measures are effective
- Assessing and tracking changes to maintain safety throughout the life of the plant
- Encouraging public participation to ensure that members of the public are informed and involved.

By identifying and managing risks, the Army's program achieves its objective of providing maximum protection to the health and safety of the public, workers, and the environment. The Army's chemical demilitarisation risk management program is summarized in *Guide to Risk Management Policy and Activities*[12] and *Chemical Agent Disposal Facility Risk Management Program Requirements*[13].

8.1.3. National Research Council and the Qualitative Risk Assessment

Ongoing review of the Chemical Stockpile Disposal Program by a standing committee of the National Research Council of the National Academy of Sciences helps ensure that the program is technically sound and uses available technology. To this end, the committee makes recommendations with respect to the implementation of various technologies and takes other steps that have the potential for minimizing adverse impacts of the program.

In a letter to the Assistant Secretary of the U.S. Army (Installations, Logistics, and Environment), dated January 8, 1993[14], the National Research Council Committee for Review and Evaluation of the U.S. Army Chemical Stockpile Disposal Program recommended that a comprehensive plan be developed to manage the risk associated with the disposal of chemical munitions and associated chemical agents. The recommendation indicated that site-specific QRAs be performed before developing a site risk management program. In a 1994 report[15], the council reiterated their recommendation to perform site-specific risk analyses using the most recent information and methods. They recommended that analyses be conducted to compare the relative risk of continued storage and disposal at each stockpile storage site. The principal objectives would be to identify major risk contributors and to use the QRA models in ongoing risk management. The QRA also updates conclusions drawn from the risk analysis developed in 1987[16] to support the Final Programmatic Environmental Impact Statement. This risk analysis compared several programmatic alternatives and concluded that maximum safety dictates prompt disposal.

In response to these recommendations, the program directed that a QRA and a risk management program be developed for the first of eight planned facilities in the continental United States: the Tooele Chemical Agent Disposal Facility. The goal of these activities was to minimize the risk that could be posed to the public, site workforce, and environment by potential agent-related accidents during chemical disposal operations. The Tooele Chemical Agent Disposal Facility QRA was published in 1996[17].

The National Research Council has continued to provide oversight of the program and has consistently reinforced their view of the importance of the QRA as part of the risk management program. In 1996, *Review of Systemization of the Tooele Chemical Agent Disposal Facility*[18] was published. The review recognized and expressed general satisfaction with the ongoing risk management efforts including the QRA and recommended that QRAs be completed before the start of agent operations at the Tooele facility. This report was followed by a more specific report, *Risk Assessment and Management at the Deseret Chemical Depot and the Tooele Chemical Agent Disposal Facility*[19], which included a review of the QRA and other risk management efforts.

The committee found that the Tooele facility QRA met their previous recommendations and offered the following with regard to the QRA:

> The Stockpile Committee has followed the [Deseret Chemical Depot] DCD/TOCDF [Tooele Chemical Agent Disposal Facility] QRA project closely since its inception and has maintained oversight of the Expert Panel independent peer review process. The QRA has achieved the goals set out in the committee's 1993 letter report and the *Recommendations* report (NRC, 1994). The success of the QRA was a direct result of a skilled SAIC technical team, firm support from the U.S. Army and TOCDF personnel, and frequent and positive interactions between the TOCDF field staff and the QRA team. The resulting QRA was significantly improved during the Expert Panel review. The findings of the QRA are consistent with the interim findings in the *Systemization* report (NRC, 1996).[19]

The committee urged some additional work to promote integration of the QRA activities and other endeavours within a complete risk management program. They also reinforced their view that the QRAs should be maintained current and used to evaluate ongoing operations.

Finally, the National Research Council has issued an update to the report[20]. That report urged that the QRAs for facilities under development be performed as soon as feasible to allow risk mitigation measures to be implemented into the design. The committee also recommended formalization of the risk management programs. Activities are currently underway with the individual sites to ensure that risk management efforts meet program goals.

National Research Council reviews continue for the Anniston Chemical Agent Disposal Facility QRA published in the fall of 2000[21].

3.1.4. Objectives of the QRA

The QRA is used to help efficiently manage and minimize the risk associated with facility operations, as part of the program's overall risk management program. A principal goal of this assessment is to identify those systems, components, and activities that govern the risks associated with disposal of chemical munitions and agents.

Risk is quantified for several reasons. Primarily, quantification provides for ranking those items that govern risk. Insights derived from the QRA will be used to identify potential improvements in systems or operations that could further reduce the public and worker health risks during disposal operations. In addition, the quantitative results will allow determination of whether proposed modifications to the facility, operating procedures, or the schedule for disposal would actually avert a significant amount of risk relative to the complexity of the change. The QRA provides the plant-specific inputs for each site's risk management program as documented in the requirements document[13]. Finally, the evaluation of risk will serve as the basis for communicating the risk insights to the operating staff and other interested parties.

3.1.5. Scope of the QRA

The scope of the QRA includes analysing the risk to the public and site workers from accidental releases of chemical agent during chemical munitions and agent storage and disposal activities at stockpile disposal sites. The QRA includes an estimate of the risks associated with the following aspects of chemical storage and disposal activities:

1. Stockpile munitions handling associated with moving munitions to prepare for transport to the facility
2. Transportation of munitions from the stockpile storage area to the chemical disposal facility
3. Disposal processes within the facility.

In addition, the QRA estimates risk associated with storing munitions in the stockpile storage area.

The QRA considers the effects of postulated accidental releases of chemical agent on both the public (the population outside the depot boundary) and workers (within the depot boundary). Only accidental releases of agent large enough to cause adverse health effects to the public or workers are included.

Both public and worker risk are calculated in terms of acute fatality risk, which is the probability of fatality over a specified period of time as a result of a one-time exposure to postulated releases of chemical agent. The risk of exposure-induced cancers is also considered for potential releases of mustard agent (nerve agents have not shown long-term effects such as cancer). Risk is not assessed for accidents involving workers where there is no potential for agent releases (i.e., typical industrial accidents that do not involve handling munitions or agent).

For all operations and storage activities, a full range of potential events that could lead to an agent release is considered. Both releases that result from internal events (originating inside the plant or directly from the activity being performed) and those initiated by external events (such as earthquakes, tornadoes, and aircraft crashes) are modelled.

Walkdowns of systems and structures are performed to support the analyses. In a walkdown, the analyst physically examines systems and structures in the actual facility to determine whether it was built and is being maintained and operated according to design, maintenance, and operations documentation. System walkdowns support development of the system fault tree models. Seismic, lightning protection, tornado, and fire analysis walkdowns are conducted to support the external event analyses. The transportation analysis is based on actual road conditions and traffic patterns. Discussions are held with plant staff regarding munitions handling and disposal operations. This approach is preferred over obtaining information only from design drawings and other reference documents.

Risks lie in a continuum between a definite outcome (for example, a 100% chance that a worker would be injured) and very rare occurrences (for example, one chance in a billion that the person would be injured). The estimated risks are uncertain because of limitations in knowledge concerning both the likelihood and consequences of events. They may also be uncertain because of randomness involved in the risk phenomena (for example, lightning may strike someone at a golf course with a probability that may be fairly well known, but there is an element of randomness as to which golfer might get struck). These uncertainties must also be considered.

The QRA is comprehensive in that it estimates both public and worker risk, and also includes an evaluation of uncertainties. Uncertainties in the parameters and models used in the analysis are quantified to display the confidence in the results. In addition to the uncertainty analysis, selected sensitivity analyses are conducted. The sensitivity

nalyses determine how the risk results vary based on changes to key assumptions
n the risk model.

.1.6. Public Involvement

he risk management process also includes public involvement. Public involvement
ccurs through a number of avenues, some of which are mandated by federal and state
aw. The environmental permitting process includes provisions for notification of the
ublic regarding endeavours that could affect their communities. The public has specific
nechanisms for review and comment on permits and supporting analyses.

While the U.S. Army endorses and complies with these public involvement
ctivities, a more important effort is direct involvement of the public as an input to
ecision making. An extensive effort is focused on providing the public an opportunity
o share the information concerning the projects. Recent public involvement efforts are
ummarized in the Public Outreach and Information Office's annual report[22]. In
ddition to these public outreach efforts, specific activities to involve the public in risk
nanagement decision making have been initiated. The most comprehensive effort is
ublic involvement in the change management process throughout the facility lifecycle.
he process includes public participation in decisions concerning major facility changes
nat could impact risk, including procedural and equipment changes. This participation
rocess includes sharing with the public the risk inputs that form part of the basis for
ecision making.

.1.7. Uses of the QRA in Risk Management

he way that the QRA is used in risk management is a function of how site contractors
nplement risk management requirements. The development of a risk management
vorkstation was a goal coupled to the completion of the QRA. To meet that goal, SAIC
as developed the Quantus risk management program. Quantus is an easy-to-use,
ntegrated suite of risk assessment and management tools. Quantus was developed for
vo audiences. First, it meets the exacting needs of the risk engineers for accurate
evelopment and solution of complex probabilistic models. Second, it provides decision
nakers with access to results in usable and understandable formats. Decision makers
lso have the power to do "what-if" analyses to investigate changes. Because all models
re developed and stored in Quantus, the program and the QRA are integrated.

The QRA uses will evolve, but there are a number of demonstrated areas where they
ave proven their usefulness to decision makers.

The QRA has been used to examine the design of the facilities. For example, the
Tooele facility QRA resulted in a redesign of a portion of the disposal facility
structure to reduce possible earthquake damage. The amount of liquefied petroleum
gas stored near the facility was also reduced based on risk findings.

The QRA has been used to assess the scheduling of disposal operations. Along with
efficient plant operations, PMCD has a goal of eliminating the storage risk as
quickly as possible. Reducing equipment change-outs to accommodate different
types of munitions and reducing the need to clean the plant to switch between
different chemical agents are important considerations.

- The QRA has been used to make other operational changes. Residence times for metal parts in the furnace airlock at the Tooele facility were minimized based on a QRA finding of a potential for an explosive build up of agent vapour. Disposal of one type of munitions was delayed because of the potential for a munition-specific risk that required additional study before processing. The QRAs even had broader impact in that the U.S. Army-wide accident planning guidance for munitions handling (called maximum-credible events) was redefined using the QRA models of accident frequencies.

- The QRA was used to identify potential risk-reduction opportunities for storage of chemical weapons and agents. This included lowering the VX rocket pallet stacks to reduce earthquake damage potential at the Deseret Chemical Depot. All storage structures housing rockets have had additional electrical bonding done to offer increased protection against lightning based on QRA findings.

- The QRAs have also played a role in other management activities. The QRAs provide information in support of regulatory and legal activities. The emergency planning community uses the QRA accidents as a planning base to allow preparations for probabilistically significant accidents. The QRA has proven useful in accident investigations and in pre-operational surveys. Other related issues have been addressed. For example, on-base land re-use proposals at Pine Bluff, Arkansas, and Pueblo, Colorado, have been studied from a risk perspective.

In summary, the QRA has found many useful applications in responding to day-to-day management needs, both internally and in response to Pentagon and other inquires.

8.2. Quantitative Risk Assessment Methodology Overview

Like most modern industrial facilities and processes, chemical disposal facilities and demilitarisation activities have been designed with careful consideration of safety. A QRA may be used to further enhance safety through development of models that enable an integrated assessment of equipment and operations. The quantification of these models provides insights concerning the frequencies of potential accidents and the relative safety importance of different equipment and activities. Thus, a QRA is a good adjunct to the engineering design and operation practices that ensure plant safety. The quantitative results are used to understand risks to the public and facility workers, allowing comparison to other risks for further perspective on the safety of the overall process.

The methods used in this analysis were based on QRA approaches used on other facilities and technologies. The methods have been customized and extended for chemical disposal to reflect the specific nature of the activities and to ensure maximum benefit in terms of insights and feedback that could be used to understand risks and improve the processes. The QRA process is summarized in the following paragraphs and illustrated in Figure 8.1.

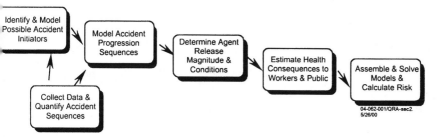

Figure 8.1 Steps in the Quantitative Risk Assessment process

.2.1. Identification of Initiators

Accidents can be systematically examined as a progression of events, called an accident sequence, which describes how a facility or operation moves from a normal, safe state to an accident condition in which the public or workers are exposed to potential health consequences. Given that risk is examined in terms of accident sequences, it is essential that the identification and modelling of these sequences be as complete as possible. The first step in the QRA, therefore, is an exhaustive consideration of the potential events that could initiate an accident sequence.

Each accident sequence can be described as beginning with an initiating event, or initiator, that starts an off-normal progression of events that could result in agent release. For analytical convenience, events are usually categorized as either *internal* or *external* events. Internal events occur within the process system, such as an operational error or equipment failure. External events occur outside the process system or have widespread effects. Thus, an operational error or a failure of a piece of equipment is an internal event, while earthquakes, fires, floods, or aircraft impacts are external events.

Identification of possible initiators is generally based on past analyses. For chemical disposal facilities, these analyses include the Tooele facility QRA[17], analyses associated with operations at that facility and elsewhere, and technical evaluations of operations and equipment. In addition, QRAs of other facilities have developed lists of initiating events, which are used to ensure completeness[23-26].

Figure 8.2 illustrates the initiator identification process. Internal initiators are identified through a systematic evaluation of the entire disposal process, from loading munitions at the storage yard to final disposal of the munitions and their agent. The evaluation is aided by the use of process operations diagrams (PODs), which delineate the steps of a process and the possible deviations from normal processing that might occur at each step. The thorough consideration of the process and past evaluations result in a comprehensive assessment of potential initiating events.

After identification of the initiators, fault tree models are developed to quantify the various combinations of failures that could lead to the initiator. A fault tree is a logic structure that determines the possible combinations of events that can lead to a specific outcome. In this case, the fault trees model the basic causes of various types of initiators. For example, a POD might show that a munition could be dropped during

132

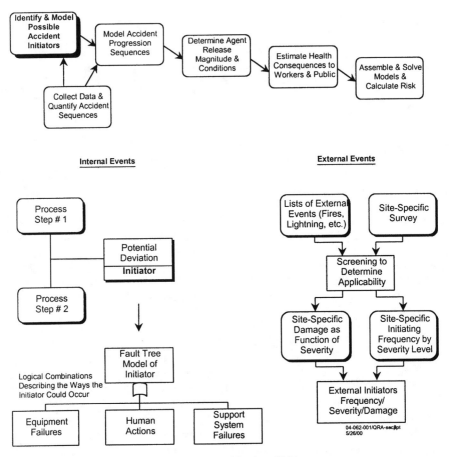

Figure 8.2 Identification of initiators

handling in the unpacking area of the facility. The fault tree then identifies the specific combinations of equipment and/or human failures that could result in a munition drop. Some events identified on a POD did not require detailed fault tree models because they could be described in a single event. Other events, however, required detailed system modelling along with support system models to fully identify all combinations of failures that could cause the event.

The search for external initiating events began with an exhaustive list of potential events and an initial evaluation to determine if each event is possible. As noted previously, other sources provide extensive lists of possible external events. The initial evaluation of an event is based on applicability to the site (e.g., a tidal wave is not possible in the interior of the U.S.), frequency relative to cut-off criterion of the 1×10^{-8} per year accident sequence frequency, and susceptibility of the site and facility to the postulated hazard.

For events not screened, it is necessary to determine the frequency and magnitude of the hazard. In general, historical records are used to estimate the frequency and magnitude of the events of concern. The method for this part of the analysis depends on the specific external event, but the basic steps are similar. For example, weather records can be used to generate the frequency of events (such as tornadoes) of different severities. Other events (such as earthquakes) require combinations of historical information and analytical techniques to estimate the hazard at the site. Estimation of some external events with no data relies primarily on analysis. If an external event can occur, it is also necessary to understand the level of damage that might be induced. For example, different structures and equipment will respond differently to the same earthquake. The response of structures and equipment to wind must also be analysed. This information, coupled with the frequency and magnitude information, allows the identification of specific initiating events that could cause the potential for agent release.

8.2.2. Modelling of Accident Progression

After the initiators are identified, it is necessary to describe the potential accident sequences that could result in a release of agent and subsequent public or worker risk. The initial concern is whether an initiator could progress to the point where agent is released from its intended confinement. (Some initiators may be so severe that the initial confinement is breached directly.) It is also important to consider the conditions associated with the initial release (e.g., agent leak or spill, munition explosion, or fire with agent involvement). Thus, the initiator analysis may identify the drop of a rocket pallet from a forklift, and the accident progression analysis will identify the possible outcomes (e.g., no agent release, agent leakage or spill, or rocket explosion). The outcomes are most often probabilistic assessments of physical phenomena, such as a rocket leak probability after a drop.

In some scenarios, the initial release may be compounded by further failures. Two types of events are generally considered in modelling accident progression: mitigative and propagative events. *Mitigative* events are those actions or systems that operate to reduce or prevent an eventual release, such as filters, blast gates, and human actions. *Propagative* events are those events that account for physical phenomena (e.g., explosive effects) that cause the accident to involve additional agent sources or to fail barriers. Additional agent sources are generally other munitions in the area.

The analysis of potential accident progression is accomplished through the use of an accident progression event tree (APET), shown schematically in Figure 8.3. The goals of APET modelling are to delineate the full range of sequences that could result in agent release and to characterize the sequence in sufficient detail to permit analysis of the amount and characteristics of agent release. The APET is a probabilistic model for postulated accidents that lead to agent release. The APET considers accident progression from initiation to agent release and includes potential propagation to other munitions. The APET also models the status of barriers to release (e.g., room confinement) and mitigation systems (e.g., the filter system). The APET provides a consistent framework for the accident progression analysis.

134

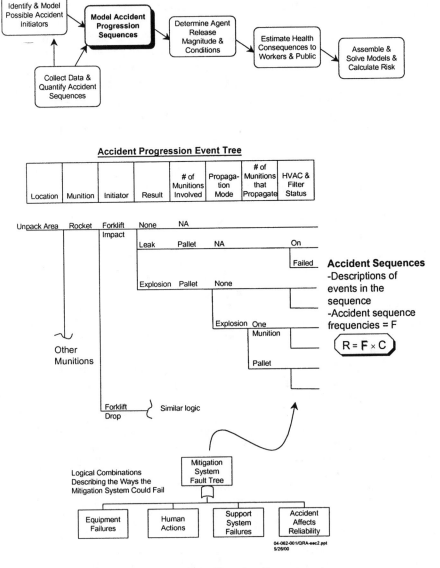

Figure 8.3 Modelling of accident progression

The APET consists of a series of questions and potential answers (or outcomes) that define how the accident might proceed. Frequencies are assigned to the initiating event in each sequence. Probabilities are assigned to all subsequent outcomes based on their relative likelihoods. The probabilities used in the APET are determined by several different approaches, including fault tree analysis, mechanistic analysis, past experience or experiment data, and engineering judgment. The APET logic specifically includes ny dependencies among events so that each accident sequence is appropriately quantified. For example, the potential occurrence of an explosion following an initiator would influence the availability of the air ducting and filter system as a potential mitigation of the accident. Each path through the tree (or accident sequence) has a frequency of occurrence equal to the product of the initiating event frequency and the probabilities assigned to each outcome.

3.2.3. Quantification of Accident Models

The goal of a QRA is to obtain a probabilistic estimate of risk by quantifying the events in the models described previously. This quantification requires assigning probabilities or frequencies to each event in the accident sequences. Data collection and model development are closely coordinated because the extent to which a model can be developed is governed, to some degree, by the availability of relevant data. Similarly, the accident progression phenomena in the APET need to be modeled at the level of detail matching available mechanistic calculations. Figure 8.4 is a schematic of this analysis activity.

The fault tree and event tree models require three types of quantitative input:
. Equipment Reliability. The equipment (and the components making up the equipment) are modeled in fault trees for initiators, mitigation systems, and support systems. Quantification of the models requires assigning failure frequencies or probabilities to each event in those models. For some components, past reliability data are sufficient, while for others industrial data must be included. Industrial reliability data are developed from a combination of generic data derived from process industries and nuclear facilities, U.S. Department of Defense, U.S. Department of Energy, and other sources. The equipment reliability database developed for the Tooele facility QRA from the information collected during operations at Johnston Atoll Chemical Agent Disposal System are also used.
. Human Reliability. Human performance affects the potential for accidents. While some data for equipment performance might include human failures, unique events associated with process operations require an assessment of human reliability. QRA techniques developed to assess human reliability can be used while considering specific operations, procedures, and facilities. The human reliability events are initially assigned conservative screening values to determine if the events, in combination with the other events in the accident sequence, are important to public or worker risk. Only the significant events are analysed in more detail.
Probabilities for Mechanistic Phenomena. The accident sequence models include many events whose quantification depends on mechanistic analyses. For example, the responses of furnaces to various perturbations are considered, as are explosive propagation phenomena involving structural damage. Some values are developed based on models drawing on basic chemical or physical principles. Other values may draw on existing experiment or operational experience. The probabilities for these events were assigned after mechanistic analyses had been performed, and considered both available probabilistic data about the phenomena and engineering judgment. Consistent with other data efforts and program goals, the probability

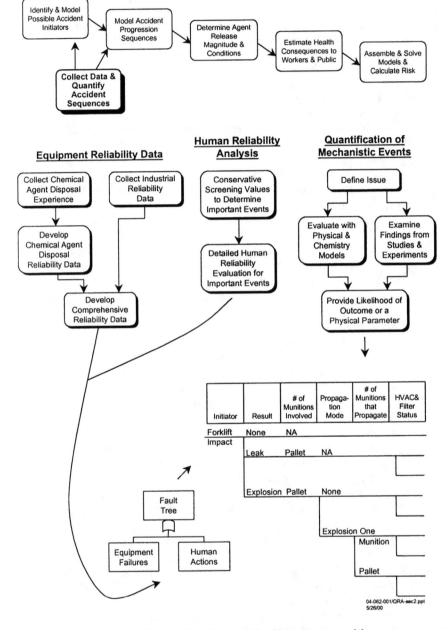

Figure 8.4 Quantification of accident sequence models

assignments frequently involve conservative assessments with refinements of the values that were found to be important to risk.

The external event tasks also require data such as frequencies of the natural phenomena that initiate the accident sequences. As described previously, the data are derived from historical information or from models reflecting historical and analytical data.

8.2.4. Characterizing Agent Releases

The goals of APET modelling are to define the types of sequences that could result in agent release and to characterize the sequences in sufficient detail to permit analysis of the agent release. The factors that significantly affect a release are therefore determined, and the APET logic is designed to explicitly reflect these factors for every release sequence. This information can then be used to develop a source term that characterizes a release for evaluation of consequences. The expression source term refers to the following information characterizing a release of agent: 1) the type(s) of agent released; 2) the quantity of each type; 3) the physical state of the released agents (vapor or liquid aerosol); 4) the rate, timing, and duration of the release; 5) the elevation of release; and 6) the time of day at which the release is possible. (Because some operations are limited to daylight hours and because weather patterns are different day to night, it is necessary to consider when the accident could have occurred in order to develop a reasonable estimate of health consequences.) Taken together, these characteristics define the *source term* for agent release.

Figure 8.5 illustrates the source term task. Based on the description of the accident, a source term is defined. A source term function uses the information defining an accident progression sequence to estimate a source term for the sequence. The source term function can be automated through development of a computer code function in Quantus. For purposes of development and for use as a stand-alone source term evaluator, the source term algorithm can be developed in Excel™ spreadsheets. The source term algorithm defines a source term for each sequence by assembling the information needed to estimate each of the source term parameters listed previously. The source term algorithm includes modelling necessary to specify the actual release expected from the accident sequences. For example, the model includes an evaporation model that determines the amount of release based on evaporation rates for the agents and the conditions of the accident. An explosive release model is included that is used to determine the release associated with various types of explosions. The source term function also considers the effect of mitigation systems such as carbon filters. The release for an accident sequence is the sum of releases from all of the phenomena and all of the agent sources involved in the accident.

The source terms developed for each accident progression sequence form the basis for the next steps of the analysis, including atmospheric dispersion modelling, which can be computer-resource intensive. Because many calculated source terms have nearly identical consequences, it can be more efficient to calculate one set of consequences that applies to a group of similar source terms. A function is available to allow grouping of

138

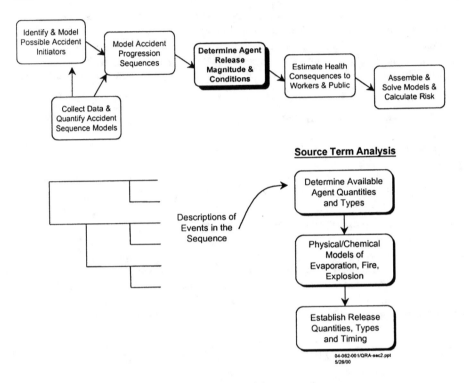

Figure 8.5 Characterizing agent releases

like source terms, if necessary. These source term groups are the input required to assess consequences.

8.2.5. Estimating Consequences

The final technical evaluation step in the QRA process is the assessment of potential public or worker health consequences. As indicated in Figure 8.6, it is necessary to estimate health consequences for each source term. The consequences of an accident are estimated by evaluating the dispersion of agent in the environment, determining the population exposure to agent (doses), and estimating the probable number of persons who would experience the health consequence of interest (in this case, fatality from agent exposure, or increased cancer risk from exposure to mustard agent). To obtain a probabilistic evaluation of potential consequences, the evaluation considers the variability in weather.

The U.S. Army has developed a dispersion model contained within the Army's D2PC computer program[27]. The model has been incorporated in a consequence analysis code that was originally developed for QRA in the nuclear industry; the result is a code specifically applicable to chemical agent risk assessment. This code, CHEMMACCS, includes the appropriate D2PC models for chemical agent in a structure

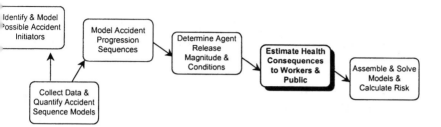

CHEMMACCS Dispersion and Consequence Model: Public and Workers

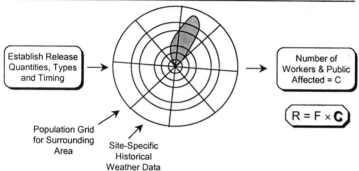

Worker Risk: Close-In Effects
(CHEMMACCS does not model impacts to workers in the immediate vicinity of the accident)

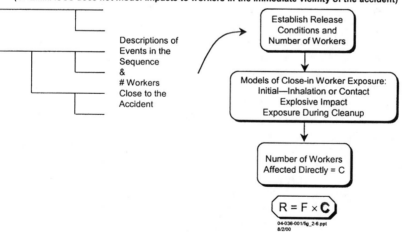

Figure 8.6 Estimating public and worker health consequences

that is suited for QRA. The CHEMMACCS code permits input of the local population distribution and an hourly set of site-specific weather data over 1 year.

Using CHEMMACCS, public and worker health consequences for the source term groups are calculated. The consequences include estimates of acute fatality and excess cancers.

As indicated in Figure 8.6, there is another consequence evaluation associated with close-in effects. To comprehensively cover worker risks, it is necessary to consider the effects of the accident on the workers close to the accident. An atmospheric dispersion model is not applicable in this circumstance, because workers could be affected directly through such mechanisms as splashing or explosion. Thus, another function is used, similar to the source term function, that estimates close-in effects. Included in the function are calculations of inhalation or skin contact, explosive effects, and possible exposures during cleanup of an accident. Consequences are calculated for these close-in risk effects and then added to the consequences calculated for other workers who might be exposed to agent as it is dispersed from the immediate area, as calculated in CHEMMACCS.

8.2.6. Assembling Risk Results

Figure 8.7 illustrates the overall risk assessment arranged as a process from initiator identification through risk assembly. The process of assembling the risk from thousands of individual accident sequences is complex, but is implemented in the Quantus risk management workstation.

The risk assembly process combines inputs and outputs from the fault tree analysis, the APET solution, the source term analysis, and the consequence analysis. The source term production proceeds as previously described. A source term is estimated for each accident progression sequence. Source terms that are similar enough to produce similar consequences may be combined into source term groups. The relationship between the individual accident progression sequence and the source term is tracked in a set of computer files used in the final risk assembly.

Consequences from each source term are estimated using the CHEMMACCS dispersion model and also using a separate algorithm for close-in risks. The results of these calculations are the numbers of various types of consequences (fatalities or cancers) that would be expected for each accident sequence. This is the second element of the risk equation.

The frequency and consequences associated with each sequence are combined to estimate risk. The risks of sequences are summed to arrive at the total risk.

This description of the risk assembly process is somewhat idealized. The consequence values described in the previous paragraphs are actually produced as curves of probability and consequence, and the frequencies of the accident progression have probability distributions in the uncertainty analysis.

Once the risk is assembled, the relationships of the model inputs are carefully evaluated for insights. Insights are derived from the quantitative assessment of the importance of various plant features, operations, or individual failures. The release

141

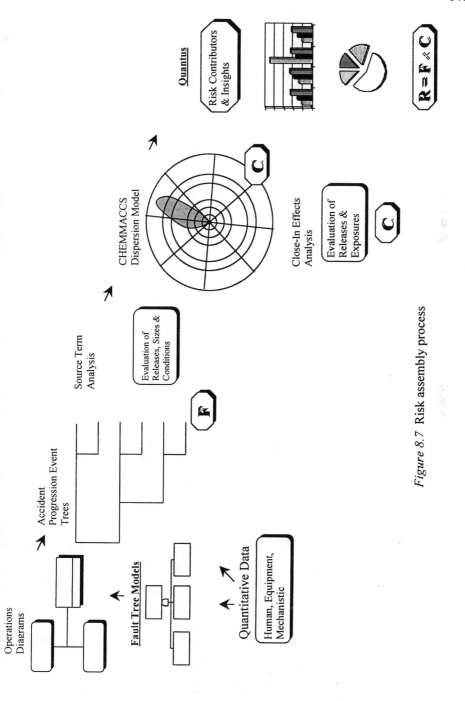

Figure 8.7 Risk assembly process

characterization process yields insights concerning mitigation features. The consequence assessment will also help to identify the accidents with the most significant potential for public or worker health consequences. The risks of different activities may also be compared. For example, the risk of the disposal processes can be compared to the risk of continued storage. Sensitivity analyses are used to investigate the most important aspects of the facility and its operations and highlight important uncertainties.

The QRA process also provides qualitative insights. These insights are sometimes derived from the quantitative results, such as ranking of the relative importance of events in the models. Given a consideration of relative risks, it is possible to derive conclusions about the system and its operations. The QRA process itself often yields engineering insights that are not based on quantitative assessments, but instead result from the assembly of an integrated model of the entire process and its operations. For example, POD development can generate insights concerning operational steps and uncertainties in the exact nature of the activities. The integrated assessment of support systems can suggest means to reduce common dependencies. The investigation of the systems and operations also often identifies procedures or support information that could be refined or improved.

8.3. Presentation of Risk Results

The QRA produces a great deal of information concerning risk. The presentation and communication of those risks is an entire field of endeavour.

8.3.1. Discussion of Numerical Estimates of Risks

The QRA provides the U.S. Army and systems contractors with a tool for evaluating the relative importance of equipment and operations, as measured by the risk to the public and workers. While decision making based on the relative importance of different contributors is a primary objective, the QRA also produces risk estimates that can be viewed on an absolute basis. This estimate includes many numerical values that may represent new ways of expressing risk to some audiences.

One of the ways of representing risk is the product of the frequency of an accident (an estimate of the ratio of the expected number of times that the accident would occur in a set of repeated trials of the process) and the severity of the consequences (e.g., how many fatalities could be expected). Risk can be expressed as either *societal* [the risk to a population as a whole (a society)], or *individual* (the risk to a single person within a population). Both measures are important to decision makers who need to minimize impact to the population while also ensuring that no individuals are unduly exposed.

Table 8.1 lists a few examples of risks, and illustrates how risks seen in everyday life translate to numerical estimates of risk, either societal or individual. The entries are just examples of risks—there are obviously many other risks that could be included. The discussion also highlights the need to carefully describe the desired risk measure. For example, the individual risk from hurricanes over the entire U.S. population provides little insight concerning risk to the individuals most exposed. Decisions regarding the acceptability of the risk from hurricanes should be made by considering

TABLE 8.1 Examples of risks

Cause	Average Societal Fatality Risk (per year)	Average Individual Fatality Risk (per year)	Discussion
Motor Vehicles[a]	49,000	2×10^{-4}	Individual risk assumes 200,000,000 population is exposed to motor vehicle risk.
Hurricanes[a]	41	2×10^{-7}	Individual risk for the U.S. population as a whole.
		2×10^{-6}	Individual risk for the approximate 10% of the population that are more exposed to hurricanes. One could also calculate individual risk to coastal residents in hurricane-prone areas, which would indicate considerably higher individual risks.
Lightning[a]	141	7×10^{-7}	Individual risk assumes 200,000,000 population is exposed to lightning. Individual risks in specific areas would vary substantially due to differences in exposure.
Canvey Island Industrial Accidents[b]	1.3	4×10^{-5}	Based on the revised assessment of the risks from industrial activities on Canvey Island in the United Kingdom. In this case, societal risk is associated with the 33,000 residents of the island, and the individual risk is the average individual risk for all residents.

Notes: a = Societal risk data from Cohen, 1991; b = Canvey results from Safety and Reliability Directorate, 1981.

the individual risk near specific locations. The last entry in Table 8.1 is an estimated risk value, based on a QRA of industrial activities on an island in the Thames River in the United Kingdom. This is provided as an illustration of how QRA results can be used to examine risks to a very specific population. In this case, the societal risk is limited to the island's residents.

Table 8.1 does not provide all the necessary insights on risk, however, because it does not directly indicate the magnitude of the consequences associated with individual accidents. Accidents that happen very infrequently but that yield large numbers of casualties may have the same numerical average risk value as those that happen relatively often with comparably fewer casualties. Average risk values combine all accidents and, in the process, important insights may be lost. For example, although hurricanes result in 41 deaths per year on the average (based on casualty data from the recent past), a decision regarding the acceptability of hurricane risk might also focus on the fact that about once every 100 years a hurricane could result in approximately 10,000 deaths. However, advances in storm tracking and emergency management make it unlikely that 10,000 deaths would occur in any current U.S. hurricane. Risk is typically illustrated on a complementary cumulative distribution function (CCDF) plot indicating frequency of exceedance (i.e., frequency of exceeding a given level of consequence) versus number of consequences. Figure 8.8 is such a plot, which shows how often a given level of consequence might be expected for some common risks.

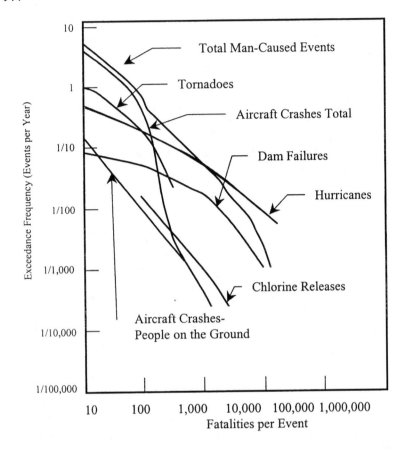

Figure 8.8 Representative societal complementary cumulative distribution function risk curves (derived from [29])

Interpretation of these curves must include consideration of the data used to produce them. As discussed in the previous paragraphs, the hurricane curve shows that the frequency of exceeding 10,000 deaths in the United States is approximately 1 in 100. The upper end of the curve, however, is formed from events that occurred in the early part of the 20th century, and it is likely that the current annual probability of exceeding 10,000 deaths is much lower than 1 in 100. Thus, while the average of the hurricane curve in Figure 8.8 would indicate an expected number of several hundred deaths per year, the recent statistical evidence supports a lower value. Note that a curve indicating historical frequency of exceeding various sizes of hurricanes would probably be reasonably accurate; it is prediction and evacuation capabilities that have changed over time.

8.3.2. Presentation of Risk Results for the Chemical Disposal Facilities

Risks to the public and workers are presented for the chemical disposal facilities in terms of acute fatalities for all agent types and exposure-induced cancers for mustard agent. Acute fatality is death from agent exposure associated with an accidental release. Acute refers to the fact that the death occurs soon (within days) after the exposure, as opposed to any potential for effects, such as cancer, that could cause death a long time after exposure. The risk of acute fatality is described as the frequency of fatalities over the duration of interest. In risk studies of industries such as nuclear power, risk is usually described as the frequency of fatalities per year. Per-year results are also provided for the chemical disposal facilities, but the other emphasis is "frequency over the facility or campaign lifetime." (A campaign is a period of processing devoted to one type of disposal activity. Because most sites have several types of munitions with different agents, there are several campaigns.) This presentation allows an integrated examination of risk on a campaign-by-campaign basis and also allows calculation of risk for the entire disposal effort.

The public is defined as any member of the surrounding community that could be affected by the accident. Although effects would only be expected close to the facility, the calculations are conducted out to 100 km (60 miles) to ensure completeness. In addition to public risk, risk is also calculated for workers. The worker risk is further defined into risks for two groups: 1) Disposal-Related Workers, defined as workers within (or just outside) the security fences surrounding the disposal facility and storage yard and 2) Other Site Workers, defined as workers who are not within the previous category. The risk measures used in QRA studies are summarized in Table 8.2.

As described previously, risk to the public is calculated and reported as either societal or individual. Societal risk is calculated over an entire affected population as a whole. Individual risk is societal risk divided by the number of persons in the population at risk. Individual risk represents the risk to an individual in the population at risk. The individual risks are calculated for groups residing within various distance intervals from the facility.

These two different measures are provided because they can both be useful for decision makers. For example, when making decisions regarding safety programs, it may be useful to know that approximately 1,000 vehicle accident deaths occur each year in some U.S. states (societal risk). It is also useful, however, to know how likely it is that an average individual would die in a car accident–1,000 deaths per 4.3 million people in that state or 0.00023 (2.3×10^{-4}) deaths per person per year. The per-person death rate is generally used to compare to other causes of death and to gain an understanding of how a specific risk is likely to affect a specific population.

TABLE 8.2 Summary of risk measures and population types

Measure	Description
Societal Acute Fatality Risk	Probability of death over a unit time[a] in the surrounding population from one-time exposure associated with a postulated accidental release of chemical agent. Any death would occur soon after the exposure.[b] Public risk includes people who could be affected up to 100 km from the point of release. Societal risks are also provided for specific subpopulations, such as societal risk to those people residing within various distance intervals from the site. When this measure is applied to workers, *societal* refers to the total population of workers.
Individual Acute Fatality Risk	Probability of death over a unit time[a] per individual in an affected population. It is calculated as the appropriate societal acute fatality risk divided by the population of interest. Individual risks are often calculated for subpopulations residing within various distance intervals from the site (e.g., individual risk for persons living between 5 and 10 km from the site). Individual risks for the facility and site workers are also calculated.
Societal Latent Cancer Risk	Probability over a unit time[a] of cancers occurring in the future in the surrounding population from one-time exposure associated with a postulated accidental release of mustard agent.[b] Public risk includes people who could be affected up to 100 km from the point of release. Societal risks are also provided for specific subpopulations, such as societal risk to those people residing within various distance intervals from the site. When this measure is applied to workers, *societal* refers to the total population of workers.
Individual Latent Cancer Risk	Probability over a unit time[a] per individual of cancers occurring in the future in the surrounding population from one-time exposure associated with a postulated accidental release of mustard agent.[b] It is calculated as the appropriate societal latent cancer risk divided by the population of interest.
Population	**Description**
Public	Census-based population residing up to 100 km from the facility that contains the chemical dipsosal and munitions storage area.
Disposal-related workers	People working within the chemical disposal facility and storage area security fences, plus those workers in offices just outside the fence. Also included are those workers responsible for retrieving the munitions from storage for delivery to the facility.
Other site workers	People working at the site who are not in the disposal worker category as described above.

Notes: a = For this report, the unit time is the total duration of processing, unless otherwise noted; b = There are no nonacute (latent) effects (such as cancer) for nerve agent that would cause death after an extended duration from the postulated exposure.

One value presented for societal risk is a statistical quantity called *expected
atalities*. The annual expected fatality risk for automobiles would be approximately
,000 in a state, because about 1,000 people die each year. Because accidents involving
hemical weapons are very unlikely, the QRA is used to estimate expected fatalities.

These examples illustrate the information that may be associated with a value
uoted as the risk. QRA studies typically present cumulative distribution functions,
ecause they provide more information concerning severity as a function of probability.

.4.   Application to Cold War Legacy Facilities

he chapter describes a full-scale QRA effort. As with most techniques, it is possible to
cale the assessment to satisfy budgetary or schedule constraints. Although Quantus
reatly facilitates the creation, maintenance, and use of a QRA model, most of the
nalyses can be done using spreadsheets, especially for simple systems.

QRAs provide a structure to conduct integrated accident analyses of complex
rocesses. Their strength lies in the ability to assess situations and the associated
ncertainty for which little data exist. Additional benefits include:

An ability to assess the relative risks of multiple sites to determine a priority in
cleanup or disposal activities

A basis for evaluating competing technologies and assessing their ability to meet
risk management requirements

An ability to evaluate details of a selected process and identify weaknesses

Use as an integral tool in a comprehensive risk management program.

.5.   References

Public Law (PL) 99-145. (1986) *Department of Defense Authorization Act*, Title 14, Part B,
Section 1412.

Organization for the Prohibition of Chemical Weapons. (1993) *Convention on the Prohibition of the
Development, Production, Stockpiling and Use of Chemical Weapons and on Their Destruction*.
Organization for the Prohibition of Chemical Weapons, The Hague, Netherlands.

Covello, V.T., and Mumpower, J. (1985) Risk analysis and risk management: an historical perspective,
Risk Analysis, 5(2), 103-120.

Keeney, R.L. (1995) Understanding life-threatening risks, *Risk Analysis*, 15(6), 627-637.

Morgan, M.G. (1993) Risk analysis and management, *Scientific American*, July, 32-41.

Nathwani, J., and Narveson, J. (1995) Three principles for managing risk in the public interest, *Risk
Analysis*, 15(6), 615-626.

Somers, E. (1995) Perspectives on risk management, *Risk Analysis*, 15(6), 677-684.

Van Mynen, R. (1990) Risk management concepts in the chemical industry: one large manufacturer's
approach, *Plant/Operations Progress*, 9(3), 191-193.

U.S. Department of the Army. (1990) *System Safety-Engineering and Management*, AR 385-16, U.S.
Department of the Army, Washington, D.C.

Occupational Safety and Health Administration (OSHA). (1993) Occupational Safety and Health
Standards, *Code of Federal Regulations*, 29 CFR 1910.

U.S. Environmental Protection Agency. (1996) Risk Management Programs for Chemical Accidental
Release Prevention, *Code of Federal Regulations*, 40 CFR 68.

12. Program Manager for Chemical Demilitarization. (1997) *Guide to Risk Management Policy and Activities*. U.S. Army, Washington, D.C.

13. Program Manager for Chemical Demilitarization. (1996) *Chemical Agent Disposal Facility Risk Management Program Requirements*. U.S. Army, Washington, D.C.

14. National Research Council Committee for Review and Evaluation of the U.S. Army Chemical Stockpile Disposal Program. (1993) Letter to the Assistant Secretary U.S. Army for Installations, Logistics, and Environment. U.S. Army, Washington, D.C.

15. National Research Council. (1994) *Recommendations for the Disposal of Chemical Agents and Munitions*. National Academy Press, Washington, D.C.

16. Program Manager for Chemical Demilitarization. (1987) *Chemical Stockpile Disposal Program Risk Analysis of the Disposal of Chemical Munitions at Regional or National Sites*. Report SAPEO-CDE-IS-87008, U.S. Army, Washington, D.C.

17. SAIC. (1996) *Final Tooele Chemical Agent Disposal Facility Quantitative Risk Assessment*. SAIC, Abingdon, Maryland.

18. National Research Council. (1996) *Review of Systemization of the Tooele Chemical Agent Disposal Facility*. National Academy Press, Washington, D.C.

19. National Research Council. (1997) *Risk Assessment and Management at the Deseret Chemical Depot and the Tooele Chemical Agent Disposal Facility*. National Academy Press, Washington, D.C.

20. National Research Council. (1999) *Tooele Chemical Agent Disposal Facility Update on National Research Council Recommendations*. National Academy Press, Washington, D.C.

21. SAIC. (2000) *Anniston Chemical Agent Disposal Facility Quantitative Risk Assessment*. DRAFT, SAIC, Abingdon, Maryland.

22. Program Manager for Chemical Demilitarization. (1999) *Public Involvement in the U.S. Army's Program to Destroy Chemical Weapons, Fiscal Year 1999*. U.S. Army, Public Outreach and Information Office, Washington, D.C.

23. U.S. Nuclear Regulatory Commission. (1983) *PRA Procedures Guide-A Guide to the Performance of Probabilistic Risk Assessments for Nuclear Power Plants*. Final Report, Volume 1, NUREG/CR-2300, Office of Nuclear Regulatory Research, Washington, D.C.

24. U.S. Nuclear Regulatory Commission. (1990) *Severe Accident Risks: An Assessment for Five U.S. Nuclear Power Plants*. NUREG-1150, Office of Nuclear Regulatory Research, Washington, D.C.

25. U.S. Nuclear Regulatory Commission. (1991) Aircraft hazards, in *U.S. Nuclear Regulatory Commission Standard Review Plan*. NUREG-0800 (Formerly NUREG-75/087), Revision 2, Section 3.5.1.6, Office of Nuclear Reactor Regulation, Washington, D.C.

26. U.S. Nuclear Regulatory Commission. (1992) *Methods for External Event Screening Quantification: Risk Methods Integration and Evaluation Program (RMIEP) Methods Development*. NUREG/CR-4839 U.S. Nuclear Regulatory Commission, Washington, D.C.

27. Whitacre, C.G., Griner, J.H., III, Myirski, M.M., and Sloop, D.W. (1987) *Personal Computer Program for Chemical Hazard Prediction (D2PC)*. CRDEC-TR-87021, Chemical Research Development and Engineering Center (CRDEC), Aberdeen Proving Ground, Maryland.

28. SAIC. (1996) *CAFTA for Windows Version 3.2*. SAIC, Palo Alto, California.

29. U.S. Nuclear Regulatory Commission. (1975) *Reactor Safety Study: Assessment of Risks in US Commercial Nuclear Power Plants*. Wash-1400, NUREG75/014, U.S. Nuclear Regulatory Commission Washington, D.C.

Environmental Radiation Dose Reconstruction for U.S. and Russian Weapons Production Facilities: Hanford and Mayak

Another way to look at Cold War legacies is to examine the major environmental releases that resulted from past operation of Cold War-related facilities for the manufacture of nuclear weapons. Examining these historical releases and the resultant radiation dose to individuals living near these facilities is called environmental dose reconstruction. Dose reconstructions have been performed or are underway at most large Cold War installations in the United States, such as the Hanford facility; several are also underway in other countries, such as at the Mayak facility in Russia. The efforts in the United States are mostly based on historical operating records and current conditions, which are used to estimate environmental releases, transport, and human exposure. The Russian efforts are largely based on environmental measurements and measurements of human subjects; environmental transport modelling, when conducted, is used to organize and validate the measurements.

Past operation of Cold War-related facilities for the manufacture of nuclear weapons has resulted in major releases of radionuclides into the environment. Reconstruction of the historical releases and the resultant radiation dose to individuals in the public living near these facilities is called environmental dose reconstruction. Dose reconstructions have been performed or are underway at most large Cold War installations in the United States; several are also underway in other countries. The types of activity performed, the operating histories, and the radionuclide releases vary widely across the different facilities. The U.S. Hanford Site and the Russian Mayak Production Association are used here to illustrate the nature of the assessed problems and the range of approaches developed to solve them.

Different approaches to dose reconstruction have been taken at the Hanford Site and at Mayak. The U.S. efforts are mostly based on historical operating records used to estimate environmental releases, transport, and human exposure. Historical environmental measurements have been used to validate the models. The Russian efforts are largely based on environmental measurements and measurements of human subjects. Environmental transport modelling, when conducted, is used to organize and validate the measurements.

Of the dose reconstruction projects that have been conducted in the United States, the Hanford effort was by far the most expensive. This Hanford effort was the first major dose reconstruction related to Cold War activities that was undertaken in the United States. As such, the project had to define new processes for both scientific analysis and public involvement in that scientific analysis. The approaches and computer tools developed in the Hanford effort have been used in subsequent dose reconstruction at other sites in the United States.

The following overview of these dose reconstruction projects includes a comparative discussion. These approaches to defining dose and its uncertainty should be carefully studied by those planning new reconstruction efforts. Also the references to specific computer models will help define the types of such tools that may be needed.

9.1.   U.S. and Russian Production Facilities

In 1943, the U.S. Army Corps of Engineers selected an area of nearly 1450 km^2, in semiarid southeastern Washington State, to produce plutonium and other nuclear materials supporting the United States' effort (known as the Manhattan Project) in World War II. This area, called the Hanford Site, was used for uranium fuel preparation, nuclear reactor operations, fuel reprocessing, plutonium recovery, and waste management operations (Figure 9.1). Nine nuclear reactors for the production of plutonium were eventually constructed. Reactor operations began in 1944; the last production reactor was placed in cold standby in 1987. Additional support facilities were constructed in the 1940s and 1950s; some of these facilities continue to operate. There are 149 single-shell tanks and 28 double-shell tanks for the storage of high-level radioactive wastes. Hanford Site operations developed and changed as U.S. defence needs and the understanding of nuclear energy changed. The Hanford Site operated from 1944 through 1988; releases were highest in the early years to the atmosphere and during the late 1950s through mid-1960s to surface water, primarily the Columbia River which flows through the Site.

In what was then the Soviet Union, construction of the Mayak facility began north of the city of Chelyabinsk in November 1945; the first reactor became operational in June 1948 (Figure 9.2). The complex covers an area of about 90 km^2 and currently employs about 17,100 people. There used to be six reactors at Mayak for the production of weapons-grade plutonium. Of these, five were graphite moderated while the sixth was originally heavy-water moderated. The graphite-moderated reactors have now been shut down; the heavy-water reactor was later modified to a light-water reactor and remains in operation today. A seventh reactor is also operational for civilian isotope production. Other facilities currently operating include a reprocessing facility, a vitrification plant for liquid wastes, and about 100 storage tanks for high-level waste.

The designs of the main Russian plutonium production reactors are similar to the U.S. reactors at the Hanford Site (Figure 9.3). Both U.S. and Russian reactors were

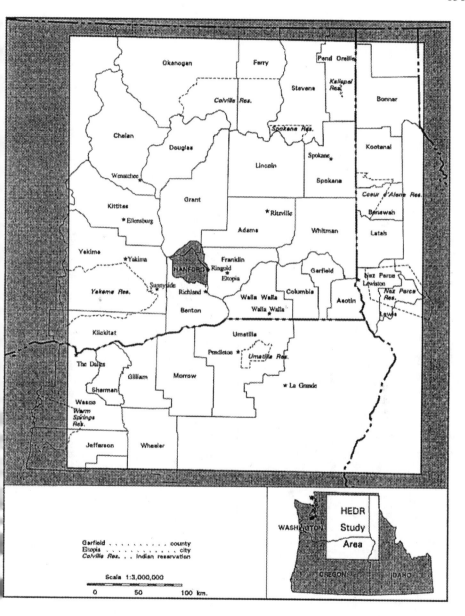

Figure 9.1 Location of the Hanford Site and the Hanford Environmental Dose Reconstruction (HEDR) study area in the Pacific Northwest of the United States

152

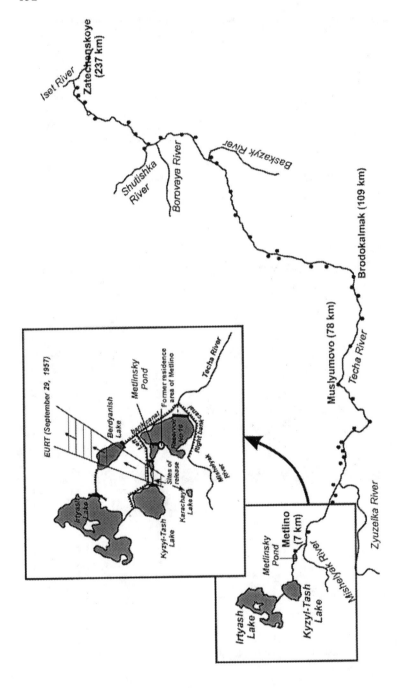

Figure 9.2 Location of the Mayak Production Association and map of the Techa River and major villages in the Russian southwestern Ural Mountains

Figure 9.3 B and C reactors in the early days of the Hanford Site in southeast Washington State

directly cooled with surface water that was returned to the water body from which it came. However, the initial Mayak reprocessing facility differs in many respects from its U.S. counterparts. The first U.S. plants used a bismuth phosphate process, which was replaced later with the "REDOX" and then the "PUREX" processes. Plant "B," the first Russian reprocessing plant, used an acetate precipitation process to separate the plutonium from the uranium, and then the plutonium was purified by precipitation out of fluoride solutions (this plant ran from 1949 through 1960). A variant of this process was used in the second plant (Plant "BB") from 1959 through 1987. A third plant, RT-1, which is still operational, began processing in 1972.

The procedures used for waste management also differed greatly between U.S. and Russian facilities. The Hanford Site had more long-term waste storage whereas Mayak had more direct releases to the environment. The plant processes and waste management differences not withstanding, both facilities released radioactive and chemical materials to their local environments.

Near the Mayak facility, water flows from a series of lakes at the foot of the Ural Mountains through the Techa River to the Iset River, a distance of about 240 km. In the late 1940s and early 1950s, the Techa River (Figure 9.2) was used as a discharge point for low-level and intermediate-level liquid radioactive wastes. However, accidental releases of highly radioactive wastes also occurred, primarily between March 1950 and November 1951. Total releases were about 76 million m³, with a total activity of about 1×10^{17} Bq (3 million Ci), of which about 98% was released during that time period[1]. Activity of the released waste was defined, mainly, by emissions from the fission radionuclides, which possess half-lives of from some days to dozens of years. The discharged material was inferred to have had an average age of fission products (from cessation of irradiation to time of release) of roughly 1 year.

The actions to minimize spread of contamination were primarily implemented in the region between Lake Kyzltash at the headwaters of the river and just below the confluence of the Mishelyak River. The reactor operating areas are situated around Lake Kyzltash. The reactor operations result in some release of primarily short-lived radionuclides to the lake and thus to the upper river. Releases of low-level and intermediate-level liquid wastes from the reprocessing plant occurred at the discharge point at the indicated location. Discharges of waste tank cooling water also occurred at this point, along with the accidental releases, which went with the cooling water. In 1951, the magnitude of the accidental releases to the Techa River was discovered. At this time, the bulk of the releases were shifted away from the Techa River into the Lake Karachai, a closed lake with no surface water outlets. From 1951 through about 1953, the villages of Metlino and Techa Brod, as well as several others further downstream, were evacuated.

The continuing flow of water from the upstream lake system through the ponds on the Techa River resulted in continuing transport of the contamination downstream. Therefore, the river flow system was extensively modified in 1956. This modification included construction of Reservoir 10 on the Techa River and a series of bypass canals to reroute most of the water flow away from the contaminated ponds (see Figure 9.2). The left bank canal routes most of the outflow from the upper lake system around the Mayak area. This canal is also connected to Lake Bernadesh to the north, through which water from the upper lake area was routed around Reservoir 10 to minimize the flow through the upper portion of the left bank canal. These actions change the nature of the upper Techa River, which complicates the analysis of the quantity and distribution of the release[2].

A second major accident occurred at Mayak in 1957: the so-called "Kyshtym explosion," which occurred in a waste storage tank. This accident released an additional 7×10^{16} Bq of radioactivity into the environment. A portion of this activity fell into the Techa River drainage area, and, in particular, some fell into the catchment of Lake Bernadesh, which had been incorporated into the hydraulic bypass canals around the upper Techa.

An additional dam and reservoir, Number 11, were added to the lower system in 1963. The left and right bank canals were extended at this time. An additional dam was added on the northernmost canal from Lake Bernadesh to prevent backflow into Lake Irtyash; routine drainage from Lake Bernadesh into the left bank canal was stopped. The intent of the entire upper Techa River reservoir system is to prevent flow of contaminated water out of the area. Under high water conditions and with special permission, releases from Lake Bernadesh were allowed to mix with associated releases to the Techa River. Some leakage of water occurs from the lakes into the canals and through the foot of Dam 11, also resulting in releases to the Techa. Residual contamination in the Azanov Marshes, caused by the initial releases, also continues to each into the river. Some radionuclides are also being transported through ground water from the Mayak site to the Mishelyak River and canal.

9.2. Approaches to Dose Reconstruction: The Hanford Environmental Dose Reconstruction Project

The Hanford Environmental Dose Reconstruction (HEDR) Project was initiated as a result of public interest in the historical releases of radioactive materials from the Hanford Site. Over 38,000 pages of environmental monitoring documentation from the early years of Hanford operations had been released to the public during 1985 and 1986. A special committee, the Hanford Historical Document Review Committee, was convened to review the documents and assess the significance of the data contained. This review was completed with a recommendation that potential health effects from these releases should be assessed to determine what other actions might be deemed appropriate. A second committee, the Hanford Health Effects Review Panel, was convened and completed its work by proposing about three dozen recommendations.

Two of those recommendations were to initiate a thyroid disease epidemiological study and to initiate a dose reconstruction study. The HEDR Project was initiated by the U.S. Department of Energy (DOE) in October of 1987. The Pacific Northwest National Laboratory, operated by Battelle Memorial Institute, was assigned the work. In early 1986, an 18-member, independent Technical Steering Panel (TSP) was formed to direct the work. The Hanford Thyroid Disease Study was initiated shortly thereafter by the Centers for Disease Control and Prevention (CDC) through a contract with the Fred Hutchinson Cancer Research Center in Seattle, Washington. The management of the HEDR Project was transferred to the CDC in 1992 under a memorandum of understanding between the DOE and the Department of Health and Human Services.

All aspects of the work–planning, budgeting, performance, and review–were conducted in a forum open to public participation and scrutiny. Communication of plans, progress, and results of work was a key and significant objective of the project. The scope of work included the search for and retrieval of historical operations and monitoring information, and demographic, agricultural, and lifestyle information necessary to 1) reconstruct source terms, 2) model environmental transport in the atmosphere and the Columbia River, 3) model transport and accumulation of radioactive materials in environmental media and food products, 4) determine food consumption and lifestyle patterns, and 5) estimate doses to real and representative individuals who may have lived near Hanford during its operation.

The key objective of the project was to estimate the radiation doses that real and representative individuals may have received from releases of radioactive materials from historical operations at the Hanford Site. Dose estimates include the uncertainties of information, such as the lack of information, regarding facilities operations, environmental monitoring, demography, food consumption and lifestyles, and the variability of natural phenomena. Other objectives of the project included[3]:

- Developing an integrated system of state-of-the-art computer codes with databases that can be used to calculate doses from several radionuclides released to the atmosphere and surface water to people of both sexes and different age groups, at different times, in different locations, eating various foods, and living various lifestyles.
- Supporting the Hanford Thyroid Disease Study through calculation of cumulative doses to the thyroid of about 1,200 individuals who were born in the mid-1940s in the three immediately downwind counties.
- Searching for, retrieving, evaluating, and declassifying (if necessary) Hanford-originated documents and information needed to reconstruct doses, and making this information available to the public.
- Performing high-quality, technically defensible, and credible science that would instil confidence and trust in the public and was acceptable to the technical community.
- Conducting the project in an open, public forum where individuals who have an interest in the project can acquire the information they want and need and participate in planning, conduct of work, and decision-making processes.

Meeting all project objectives required innovation, dedication, and persistence on the part of management, technical contributors, and the public. Science in a fish bowl takes on new meaning when the public and their representatives are invited to openly and fully participate in all activities. The TSP was made up of technical experts, representatives of state governments, a representative of Native American tribes, and a representative of the public. For the first 5 years of the project, the TSP was literally independent and exercised full responsibility for directing the study. The DOE commissioned the formation of the panel because of the pubic outcry of conflict of interest on their part. When the spectre of conflict of interest was not eliminated, management of the project was transferred to the Department of Human and Health Services. The TSP was maintained in the new contract, with Battelle as the Technical Director of the contract.

Battelle also opened its doors to the public in an unprecedented way for the project. Dr. William R. Wiley, then Laboratory Director, personally invited any person with interest in the project to visit the laboratory and discuss the ongoing work with the scientists and engineers performing the work. Workshops conducted to discuss technical approaches and to resolve key issues were open to the public and questions and comments were recorded and responded to. The public was invited to review draft reports, and individual comments were responded to and included in the final reports. Any working papers or other preliminary materials provided to the TSP, regardless of state of development, were available to the public. No information or communication between Battelle and the TSP was reserved from the public.

The initial phase of the project demonstrated the feasibility of dose estimation[4,5]. However, analysis of the initial dose estimates, based on available data and models, revealed several weaknesses in the approach used for modelling[6,7]. As a result, environmental dosimetry was substantially advanced. The final codes use a set of source term, transport, environmental accumulation, and dose models that are intimately linked, allowing transfer of information in such a way that spatial, temporal, and distributional

haracteristics of the data can be transferred across models. The following
ubsections describe these codes in more detail.

.2.1. HEDR Source Terms

ource terms for the HEDR Project came primarily from atmospheric releases and
eleases to the Columbia River. Each type had unique codes associated with it.

tmospheric Source Terms

coping studies indicated that the primary radionuclide of interest from the atmospheric
athway was iodine-131[8]. The project relied on original records generated during the
me period under study. These records were supplemented with other reports and
ummaries. Knowledge of the physical processes, monitoring techniques used, and
ompleteness of records allowed the uncertainty to be estimated for each value. The
roject generated estimates of the iodine-131 releases on an hourly basis for 1944
rough 1949; releases were estimated on a monthly basis for 1950 to1971.

The creation of iodine-131 in the reactors was calculated from reactor daily power
ecords and took into account the day-by-day changes in the amount of iodine-131
resent in the fuel. When irradiated fuel is discharged from the reactors, iodine-131
ecays with an 8-day half-life; the decay time, known as cooling, was inferred from
ecords showing when fuel was discharged from the reactor and when it may have
ntered the dissolving process.

Dissolving the fuel in the separations plants was a two-step process. First, the
luminium cladding was dissolved with a caustic solution of sodium hydroxide, then the
uel was dissolved with nitric acid. The iodine-131 was released during this step and
lso during processing steps after dissolving. Detailed plant records on the dissolution
f batches of fuel were correlated with reactor discharge records to determine the
mount of iodine-131 present during dissolving. The fraction of iodine that was released
irectly to the stack as well as during subsequent processing was taken into account.
he estimated amount of iodine-131 along with other radionuclides of interest released
 the atmosphere between 1944 and 1947 are summarized in Table 9.1[9,10]. The
stimated total release of iodine-131 from 1944 to 1971 is 2.8×10^{16} Bq (760,000 Ci).
ecause of the wealth of original documentation and redundant sources, there is a high
egree of confidence that the actual values fall within the computed ranges.

The source term release model[10] provides estimates of the hourly releases to the
EDR computational system. Uncertainties in the actual amounts released are
ldressed through use of multiple Monte Carlo simulations, each of which represents an
ternative release history that is consistent with existing knowledge. Together, these
ternative release histories represent the range of releases that could have occurred.
ne hundred separate realizations of the complete hourly release history were prepared
ith this source term code. Thus, the uncertainty in the amount of each hourly release is
presented by a distribution of possible released amounts.

TABLE 9.1 Mean estimated monthly iodine-131 releases from Hanford separations plants,
1944-1947 (Ci/month)

Month	1944	1945	1946	1947
January		1,221	11,753	6,158
February		2,126	7,399	3,835
March		2,082	7,952	5,617
April		28,746	11,680	4,853
May		74,482	13,820	3,989
June		46,466	4,609	1,652
July		47,036	5,558	2,297
August		72,090	8,642	1,249
September		88,682	7,670	1,206
October		92,066	4,819	472
November		37,752	5,525	261
December	2,139	62,340	7,398	261
Total	**2,139**	**555,089**	**96,284**	**31,848**

Columbia River Source Terms

The Columbia River passes through the Hanford Site and served as the source of cooling water for the original plutonium production reactors. The river water was drawn directly through the reactor core and returned to the river after a short retention time. The Columbia River is the major pathway for water-borne radionuclides. Radionuclide composition and activity level in the discharged cooling water varied considerably as a result of several factors[11], including the number of reactors and their power levels, seasonal changes in the parent elements in the raw river water (i.e., the elements activated as they passed through the reactor core), chemicals used in water treatment, corrosion rates of piping and fuel element cladding, occasional purging of radioactive film from reactor components, and the length of time effluent was retained in basins before discharge. Another factor was radionuclide releases from episodic fuel element failures. The wide variations in these factors, together with the hydrographic variables of the Columbia River and dam construction, produced a complex combination of river water and reactor effluent during the years of reactor operation. Scoping studies have indicated that the radionuclides of greatest interest to the HEDR Project are zinc-65, phosphorus-32, sodium-24, neptunium-239, and arsenic-76[12]. These radionuclides provide about 94% of radiation doses to people using the river. Chromium-51 emissions, although insignificant to dose, were reconstructed to serve as information for model validation purposes. Table 9.2[9,10] summarizes the releases of the radionuclides of interest.

The source term river release model[10] provides estimates of the monthly releases to the HEDR computational system. Uncertainties in the actual amounts released are addressed through use of multiple Monte Carlo simulations, each of which represents an alternative release history that is consistent with existing knowledge. Together, these alternative release histories represent the range of releases that could have occurred.

TABLE 9.2 Total releases of radionuclides to the Columbia River from Hanford reactors, 1944-1971

Radionuclide	Curies released	Half-life
Sodium-24	12,582,196	15 hours
Phosphorus-32	229,239	14.3 days
Zinc-65	409,993	245 days
Arsenic-76	2,519,734	26.3 hours
Neptunium-239	6,309,150	2.4 days

pproximately 100 separate realizations of the complete monthly release history was epared with this source term code. Thus, the uncertainty in the amount of each onthly release is represented by a distribution of possible released amounts.

2.2. Environmental Transport Modelling

he environmental transport models are linked directly to the source term model outputs. ach is designed to continue the stochastic simulation begun at the source term level.

tmospheric Transport

he model developed for the HEDR atmospheric transport calculations is called the egional Atmospheric Transport Code for Hanford Emission Tracking (RATCHET) 3,14]. The RATCHET computer code is a Lagrangian-trajectory, Gaussian-puff spersion model. Sequences of Gaussian puffs are used to represent plumes released om ground level and elevated sources. Time-integrated air concentrations and surface positions are calculated at nodes in the model domain by summing the contributions om puffs as they move past the nodes. Movement, diffusion, and deposition of aterial in the puffs are controlled by wind, stability, precipitation, and mixing-layer pth fields that vary in both time and space.

The current project model domain extends about 500 km from north to south and 0 km from east to west. Geographically, the area covered extends from central egon State to northern Washington State, and from the crest of the Cascade Mountains the eastern boarder of northern Idaho State. The area includes essentially all of the gion known as the Columbia Basin and is bounded on all sides by mountains or other ghlands.

Atmospheric transport, diffusion, and deposition calculations are based on observed eteorological data. Data are available for about 25 reporting stations in or near the odel domain. RATCHET prepares fields for the entire domain by interpolating the servations from the stations to a gridded coordinate system. The model is capable of ating four types of material—noble gasses, non-reactive gasses, particulates, and active gasses. Iodine is treated as a special type of material; it may be partitioned into active gas, nonreactive gas, and particulate components. RATCHET treats uncertainty three ways. Uncertainties in wind direction, wind speed, atmospheric stability class, onin-Obukov length, precipitation rate, and mixing-layer height are treated explicitly thin the code. Uncertainties in surface roughness length, source terms, and rtitioning among physical and chemical forms are treated explicitly in the model input.

The explicit treatment of uncertainty in the variables and parameters leads to implicit treatment of uncertainty in all model calculations using these variables and parameters.

Columbia River Transport

The model used for analysis of transport of radionuclides in the Columbia River is called WSU-CHARIMA. The CHARIMA code[15] is a commercial surface water hydrology and sediment transport model; WSU-CHARIMA is an adaptation created at Washington State University. It uses daily river discharge and water surface elevation data to predict dilution and travel time to downstream locations. The model is basically one-dimensional, but the HEDR Project has added empirical corrections for lateral dispersion at some locations near reactor outfalls.

The river source term release model prepared 100 realizations of the monthly time-history of the Hanford Site releases. The project used the CHARIMA model in a deterministic sense and prepared 100 realizations of the downstream concentrations of radionuclides in water without varying the parameters of this transport model[16].

Terrestrial Environmental Transport

The environmental accumulation model provides Dynamic Estimates of Concentrations and Accumulated Radionuclides in Terrestrial Environments (DESCARTES)[17]. The DESCARTES model tracks and estimates the accumulation and transfer of radionuclides from initial atmospheric deposition and interception through various soil, vegetation, and animal products compartments. This model contains a set of four coupled linear differential equations that give the model its dynamic nature, generating daily soil and vegetation concentrations. Other portions of the model use these daily concentration data and equilibrium-type equations to estimate time-dependent radionuclide concentrations in animal products. The model also performs ancillary estimates required by the core estimations. Environmental concentration data needed by the subsequent individual dose model are stored in large binary files.

The model function may be visualized as a series of sequential operations. The biomass submodel generates daily biomass values for each plant type modelled. These values are then used in the soil and vegetation submodel to determine the daily concentrations of radionuclides in soil and vegetation. Results are estimated for every grid node, providing the concentration in vegetables, grains, and fruits directly consumed by people and in plants (grass, alfalfa, silage, grain) used for animal feed. Animal feed concentrations are then used to determine concentrations in animal product (beef, venison, poultry, eggs, milk), also on a grid basis. Finally, the radionuclide concentrations in commercially distributed vegetables and milk are estimated.

The commercial food distribution systems were reconstructed from records and reports available from the U.S. Bureau of Census, the Washington State Dairy Herd Improvement Association, the Washington State Dairy Products Commission, and other governmental and dairy industry organizations[18]. These sources provide some information on the amount of milk produced and sold in each county, the locations of individual dairies and distributors, and dairy industry practices in the 1940s. Additional information was obtained through discussions with dairymen, farmers, ranchers, and gricultural extension agents. These key contacts provided information that was hen supplemented and organized by local experts into a detailed source/distribution etwork by project domain grid cell. A similar undertaking was needed for the istribution system for fresh leafy vegetables[19].

Like the preceding source term and transport models, the DESCARTES code reates 100 realizations of the environmental conditions at each node for each time step. 'alues of radionuclide concentrations are stored for later use by the individual dose 1odel.

quatic Pathway Modelling

*quatic organisms in the Columbia River were extensively monitored during the latter ears of Hanford Site operations. Many thousands of river water and fish samples were ollected. The HEDR Project has catalogued this information and used it to develop *cation-, seasonal-, and species-dependent bioconcentration factors[20]. The ioconcentration factors were developed for three types of resident freshwater fish; mnivors, first-order predators, and second-order predators. Factors were also eveloped for ducks and other birds that might have been contaminated via the 'olumbia River pathway and hunted by sportsmen in the area. The Columbia River ipports major stocks of anadromous salmon and steelhead. These fish return to the ver to spawn. However, the limited monitoring data indicate that they do not eat while :turning upstream, and so their radionuclide concentrations are representative of the ortions of the Pacific Ocean where they lived before returning to the Columbia River. nnual estimates of concentrations of radionuclides were assembled and used for stimating doses for all locations along the river for people who caught and ate salmon r steelhead. Because of the sparse data, an upper-bound concentration was assumed ased on the bioconcentration factors for fish near the top of the aquatic food chain.

Along the Columbia River downstream of the Hanford Site, only the three large *wns immediately adjacent to the Site (Richland, Kennewick, and Pasco) used olumbia River water for domestic drinking water[21]. Drinking water, and potential r removal of radionuclides by municipal water treatment plants, was also considered.

.2.3. **Individual Dose Modelling**

he primary purpose of the HEDR modelling effort was to prepare a complete system by hich individuals may receive estimates of their doses from past Hanford Site)erations. The project estimated doses for representative individuals who lived in the oject domain. Doses have been estimated for real individuals included in the Hanford hyroid Disease Study. Doses for other real individuals who request them are being ilculated with the same models through a separate project run by the states of 'ashington, Oregon, and Idaho called the Individual Dose Assessment project.

In the individual dose models, the human receptor is introduced into the estimations. he terrestrial dose model CIDER[17] calculates dose for four pathways: submersion in *ntaminated air, inhalation of contaminated air, irradiation from contaminated surfaces, id ingestion of contaminated farm products and vegetation[22]. The CIDER code :ats people differently as they age, including during prenatal and nursing periods. The

Columbia River Dosimetry[23] model calculates dose via water immersion, drinking, and consumption of resident fish, game birds, salmon, and ocean shellfish.

9.2.4. Uncertainty and Sensitivity Analyses

The HEDR Project included concepts of uncertainty and sensitivity analysis from its inception[24]. A definitive uncertainty and sensitivity analysis plan [25] was prepared and peer reviewed to guide these analyses.

Uncertainty analyses have been conducted for essentially all dose estimates. These analyses lead to the most appropriate interpretation of the estimated doses because they provide a measure of the precision of the estimates. A Monte Carlo technique was used to estimate all dose uncertainties because it can be applied consistently across all the HEDR models, because it is cost effective and accurate, and because it is the appropriate technique for such complex models. The sampling strategy used was Latin Hypercube stratified random sampling for those model parameters that were infrequently sampled. For those parameters that were frequently sampled, for instance on a daily basis, simple random sampling was used.

Sensitivity analyses were performed on all HEDR estimating models. Sensitivity analyses provided a method for 1) effectively interpreting the dose estimates, and 2) prioritising individual parameters according to the uncertainty they contribute to the estimated doses. The results of the sensitivity analyses allowed development of the most cost-effective strategy for evaluating the uncertainties in the value of model parameters. Sensitivity analyses were performed using measures from multiple regression (coefficient of partial determination and the standardized regression coefficient). Multiple regressions were performed on both the original parameter values and the rank transformed values. Multiple regression was cost-effective because the software was readily available and the approach was not labour intensive. The appropriateness of the multiple regression approach was measured with the coefficient of determination. For the HEDR codes, sensitivity analyses using multiple regression were successfully demonstrated for DESCARTES and CIDER. When the coefficient of determination was small, sensitivity analyses were performed by holding subsets of the parameters constant and measuring the reduction in the uncertainty for each subset. For the complex set of HEDR models, the sensitivity analyses were done hierarchically, starting with the dose results and working backward through the various pathway, transport, and source term models.

9.2.5. HEDR Model Validation

Validation can be said to consist of four steps: peer review of the models as they are being developed, verification of the computer implementations as the codes are developed, verification of the assumptions and parameters going into the codes, and comparisons of the results to actual measurements. The HEDR models were subjected to numerous reviews by the TSP and others (e.g., TSP/CDC review of the RATCHET code, extensive discussions with the TSP during the development of the surface water modelling effort). Independent testing of the various codes was completed and documented to ensure correct implementation of the models. The assumptions and parameters were published separately and have continued to undergo review. A pre-approved plan was developed and implemented for comparison of calculated estimates with historical measurements.

The HEDR models are used to estimate the potential for radiation dose to individuals living in a large spatial area, over long periods of time, by a number of potentially important exposure pathways. It would be highly desirable to validate the various models at points throughout the spatial domain, in areas of high deposition, light deposition, and sporadic or minimal deposition. It would be desirable to observe the variation in time of radionuclide concentrations in each of the pathways at these various locations. A high level of coverage of the various space/time/pathway combinations used in the primary dose calculations would lead to the most rigorously defensible validation. Data are not available to support such an ambitious validation program. Contemporaneous data do not address all the necessary pathways, over space or over time, needed to provide a complete validation. The data sets that were selected for validation were chosen to provide the best examples of coverage of the domain in time, in space, and for as many pathways as possible. The tests defined provided a reasonable set for the needs of the project, and sufficient coverage of the spatial, temporal, and pathway variables was achieved for the demonstration of the adequacy of the HEDR approach and implementation.

Evaluation of the results of the validation tests was a necessary component of the validation[26,27]. The general HEDR philosophy was to compare the estimated values of dose, or of the surrogate measurement closest to dose available (e.g., concentrations of radioiodine in sagebrush), with measurements. The purpose was to understand the differences between the estimated doses and the measurements. Thus, the statistical methods used were aimed at describing these differences so that the causes could be understood and recommendations for any improvements made.

9.3. Approaches to Dose Reconstruction: The Techa River Dosimetry System 2000

There were 40 villages on the Techa River downstream from the Mayak Production Association when the discharges occurred (see Figure 9.2). The population of the contaminated territories was chronically exposed to external and internal irradiation. Villagers were exposed through a variety of pathways; the more significant included drinking water from the Techa River and external gamma exposure from proximity to the Techa River bottom sediments and shoreline[28]. After the extent of the contamination of the Techa River became known, all villages on the upper part of the Techa River (<78 km from the site of release) were evacuated. Some villagers on the lower part of the Techa have remained in their homes up through the present time (Figure 9.4).

In 1968 the Techa River Registry was created with the goal of including residents of the Techa riverside villages who lived there during the periods of high exposure from 1949 through 1952. Twenty-five thousand persons who lived in these 40 villages during the period of the larger releases and for whom residence records were available were

Figure 9.4 Techa River valley

enrolled in a fixed cohort known as the original Techa River Cohort (TRC). In addition, 5,000 persons who migrated to the villages after the main period of exposure, but before 1960, have been added to form the Extended Techa River Cohort (ETRC). This extended cohort consists of 30,000 persons from whom subcohorts are being drawn for epidemiological studies and for whom it is desirable to calculate individual internal and external doses. Study of this population will likely provide a good opportunity to determine whether a dose-rate-reduction factor exists for the induction of cancer in human populations. The TRC is one of a few that represents an unselected population; the presence of two distinct ethnic groups (Russians and Tartars-Bashkirs) also provides the opportunity to examine the population variability of risk factors.

Historical evidence indicates that the main contributor to internal exposure among the radionuclides released into the Techa River was strontium-90, which is accumulated in bone tissues and retained for many years. In vivo beta-ray measurements, which have been performed since 1959, on teeth and a large number of strontium-90 measurements in whole body (based on the measurement of bremsstrahlung, a type of radiation emission) have been the basis of internal dose reconstruction[29,30]. The reconstruction of internal dose depends on both estimates of the intake and models for the metabolism of ingested radionuclides. Beta-ray measurements on teeth are utilized to deduce the annual levels of intake of strontium-90 in the different villages in different age cohorts. The ingestion of other radionuclides (strontium-89 and cesium-137, predominantly) occurred mostly with water in the first 3 years of the river contamination. The intake rates of these two radionuclides were therefore derived from estimates of the ingestion of strontium-90 scaled in terms of the radionuclide composition of the river water. These data were used to estimate age-dependent intake rates for all Techa River villages[31]. Calculation of absorbed doses in tissues as a result of radionuclide incorporation was based on age-dependent metabolic and dosimetric models and the corresponding ingestion rates. A large number of measurements of strontium-90 body content made with a whole-body counter (WBC) were used to validate the metabolic model for strontium retention in human bone[32]. Absorbed doses in red bone marrow and bone surfaces have been calculated for all age cohorts; these absorbed doses are substantially higher than those in other tissues because strontium-90 is a bone-seeking radionuclide.

The absorbed doses from external exposure were estimated on the basis of systematic measurements of gamma-exposure rate along the banks of the river and the typical lifestyle patterns of the inhabitants of the riverside villages. This approach has given the average annual absorbed doses from external sources for different age groups in each village.

Russian and U.S. scientists have been involved in collaborative research programs since 1995. As part of studies under the sponsorship of the Russia–U.S. Joint Coordinating Committee on Radiation Effects Research, the authors are currently engaged in a comprehensive program to develop improvements in the existing dosimetry system for the members of the TRC by providing more in-depth analysis of existing data, further search of existing records for useful data, model development and testing, evaluation of uncertainties, verification of procedures, and validation of current and planned results. This work is the result of a first year's pilot study[31] and extensive meetings and discussion among the participants in the dosimetric and epidemiological studies. The specific aim of this project is to enhance the reconstruction of external and internal radiation doses for individuals in the ETRC. The purpose of the enhanced dose reconstruction is to support companion epidemiological studies of radiogenic leukaemia and solid cancers.

The details of the methods that are being used in this enhanced dose-reconstruction effort are described below.

1.3.1. Techa River Dosimetry System

The Techa River Dosimetry System (TRDS) is a modular database processor. That is, depending on the input data for an individual, various elements of several databases are combined to provide the dosimetric variables requested by the user. The TRDS databases consist of three modules. The first module is an environmental module that contains, for each of the Techa Riverside settlements, age-dependent mean annual-intake levels of radionuclides and mean annual external doses in air near the shoreline, outdoors in the residence areas, and indoors.

The second module is a metabolic module that contains the results of age-dependent model calculations of doses in different organs per unit intake for all radionuclides ingested (dose-conversion factors). The third module is an individual-data module that

contains the following information for each of the ETRC members: identification
code, year of birth, year of entry to the epidemiological catchment area, year of
migration from the catchment area, vital status, year of vital status determination, and
residence history within the contaminated areas. This third module is prepared and
updated by epidemiologists working on companion studies.

The method being used for the TRDS basic dose calculations is relatively simple
and can be written as a single equation:

$$D_{o,Y} = \sum_{y=1}^{P} \sum_{L} M_{y,L} \left[\left(\sum_{r} I_{y,r,L} \cdot DF_{r,o,Y-y} \right) + A_o \left(D_{Riv,L,y} \cdot T_1 + D_{Out,L,y} \cdot T_2 + D_{In,L,y} \cdot T_3 \right) \right] \tag{9.1}$$

where

$D_{o,Y}$	=	Absorbed dose (Gy) in organ o accumulated to calendar year Y
Y	=	The calculational endpoint for a particular individual (can vary within the range 1950-2005)
y	=	Year of environmental exposure (external irradiation and intake of radionuclides)
P	=	The endpoint of external exposure and intake of radionuclides for a particular individual (can vary within the range 1950-1959)
L	=	River location (village) identifier
$M_{y,L}$	=	Months in year y spent in location L (relative to 12 months)
r	=	Identifier of ingested radionuclide (strontium-89, strontium-90, zirconium-95, niobium-95, ruthenium-105, ruthenium-106, cesium-137, cerium-141, or cerium-144)
$I_{y,r,L}$	=	Intake function (Bq/y) for year y, radionuclide r, and location L (function of age, related to y)
$DF_{r,o,Y-y}$	=	Dose-conversion factor (Gy/Bq) for organ o in year $Y-y$ from intake of radionuclide r in year y (function of age, related to y)
A_o	=	Conversion factor from absorbed dose in air to absorbed dose in organ o (function of age, related to y)
$D_{Riv,L,y}$	=	Dose rate in air near river shoreline at location L in year y (Gy/y)
$D_{Out,L,y}$	=	Dose rate in outdoor air within residence area at location L in year y (Gy/y)
$D_{In,L,y}$	=	Dose rate in indoor air at location L in year y (Gy/y)
T_1	=	Time spent on river bank (relative to whole year) (function of age, related to y)
T_2	=	Time spent outdoors (relative to whole year) (function of age, related to y)
T_3	=	Time spent indoors (relative to whole year) (function of age, related to y).

In this formulation, the term $M_{y,L}$ comes from individual-life-history information and
is a series of constants. The calculation's endpoint, Y, can vary according to the
analyst's wishes; for a particular individual it might be the year of death, the year of exit
from the cohort because of migration, the year of vital status determination, or the date
of "fixing" the cohort for analysis. Y could also be any or all of the above minus some
presumed latent period for cancer induction. All of the other parameter values are either
calculated or approximated and have associated uncertainty.

Parameters for Internal Dose Calculations

The basic data sets and models used to calculate internal doses for the Techa River residents were presented in a professional journal[33]. As described there, assessments of internal dose are based firmly on strontium-90 body burdens and tooth-beta counts that have been measured for about half of the ETRC members (including all age groups, all villages, and long periods after the onset of contamination).

Intake function. As indicated in Equation 9.1, a key parameter in determining internal dose is the annual average-intake function, $I_{y,r,L}$, of radionuclide r in year y at location L. The most important radionuclide from a dosimetric standpoint for the affected population is strontium-90; this radionuclide has received special attention for the determination of the intake function[30]. Data from beta-ray measurements of teeth surfaces for the residents of Muslyumovo Village are used as a reference, as this village has a significant population of 3,000 persons and has been investigated in most detail because it is the most contaminated of the unevacuated villages. Beta-ray measurements of permanent front tooth enamel have been very useful, as the formation of this enamel occurs within a short age interval in childhood, and the subsequent rate of metabolism is extremely slow. The principle of computation was to express the average values of the observations for different age cohorts in terms of a comparatively simple model that contained unknown dietary contents of strontium-90 for each year and unknown age-dependent uptake factors; the unknown parameters were then determined by a least squares fit of the model to the data. The ratios of strontium-90 intake in children to that of adults for different years were derived by analysing age-dependent contributions of different dietary components (water, milk, fish, etc.) to the total diet[30]. Table 9.3 exemplifies resulting values of annual strontium-90 intake for different age cohorts of Muslyumovo residents.

To reconstruct strontium-90 intake for other settlements, it was assumed that the ratio of average intake to the average intake at Muslyumovo is equal to the ratio of the mean age-standardized strontium-90 contents in the skeleton (by statistical analysis of

TABLE 9.3 Example values of annual strontium-90 intakes for different age cohorts of residents of the reference settlement Muslyumovo on the Techa River

Year	Intake of Strontium-90 for Cohort Members (by calendar year of birth), Bq/y					
	≤ 1940	1945	1946	1947	1948	1949
1950	2.08×10^6	1.37×10^6	1.19×10^6	9.78×10^5	7.50×10^5	4.99×10^5
1951	4.70×10^5	3.76×10^5	3.41×10^5	3.07×10^5	2.68×10^5	2.21×10^5
1952	4.62×10^5	4.15×10^5	3.93×10^5	3.66×10^5	3.39×10^5	3.10×10^5
1953	8.61×10^4	8.30×10^4	8.02×10^4	7.75×10^4	7.39×10^4	7.11×10^4
1954	1.92×10^4	1.91×10^4	1.89×10^4	1.85×10^4	1.82×10^4	1.77×10^4
1955	1.38×10^4	1.38×10^4	1.38×10^4	1.38×10^4	1.38×10^4	1.38×10^4
1956	6.90×10^3	6.90×10^3	6.90×10^3	6.90×10^3	6.90×10^3	6.90×10^3
1957	5.88×10^3	5.88×10^3	5.88×10^3	5.88×10^3	5.88×10^3	5.88×10^3
1958	4.93×10^3	4.93×10^3	4.93×10^3	4.93×10^3	4.93×10^3	4.93×10^3
1959	3.39×10^3	3.39×10^3	3.39×10^3	3.39×10^3	3.39×10^3	3.39×10^3

WBC data) for the relevant settlement relative to Muslyumovo. Representative values are shown in Table 9.4; as can be noted, the relative levels of intake depend on the distance from the site of release and the main sources of drinking water.

TABLE 9.4 Relative annual strontium-90 intake (relative to Muslyumovo) for several settlements

Settlement	Distance from the Site of Release, km	Main Sources of Drinking Water	Location Factor, f_L
Metlino	7	Techa River and wells	0.57
Asanovo	33	Techa River and wells	0.56
Nadyrov Most	48	Wells and Techa River	0.19
Ibragimovo	54	Techa River	1.39
Isaevo	60	Techa River and wells	0.56
Muslyumovo	78	Techa River	1.0
Kurmanovo	88	Techa River	0.65
Brodokalmak	109	Wells and Techa River	0.16
Russkaya Techa	138	Wells and Techa River	0.22
N. Petropavlovskoye	148	Techa River and wells	0.50
Lobanovo	163	Techa River and wells	0.35
Anchugovo	174	Techa River and wells	0.34
V. Techa	176	Techa River and wells	0.46
Pershinskoye	212	Wells and Techa River	0.16
Klyuchevskoye	223	Wells and Techa River	0.13
Zatechenskoye	237	Techa River and wells	0.27

Age-dependent mean-annual-intake levels for cesium-137 and short-lived radionuclides were calculated on the basis of the following assumptions. As most of the ingestion of radionuclides occurred with the consumption of river water in 1950-1952, intakes of cesium-137 and short-lived radionuclides were derived from estimates of age-dependent intakes of strontium-90 scaled in terms of radionuclide composition of the river water. The ratios of radionuclide concentrations to strontium-90 as functions of calendar year and distance downstream from the site of release were calculated using the Techa River Model[34].

Dose-conversion factors. Organ doses, $DF_{r,o,Y-y}$, for different periods of time following radionuclide intake were calculated using age-dependent biokinetic and dosimetric models. For strontium-90 (and strontium-89), the biokinetic model developed on the basis of measured strontium-90 contents of residents living on the Techa River was used, and dose coefficients to target tissues were derived on the basis of published data[35,36].

For radionuclides other than strontium, the age-dependent biokinetic models from ICRP Publication 67[37] were used. As this publication contains data for only six age groups, dose-conversion factors on a year-by-year basis for these radionuclides were calculated using the special software IDSS[38]. The latest version of the TRDS (TRDS-2000) contains dose-conversion factors for red bone marrow, bone surface, walls of the upper and lower parts of the large intestine, wall of the small intestine, stomach wall, ovaries, testes, and uterus.

Dose Rates Near the Shoreline

To evaluate external doses near the shoreline all available results of exposure-rate measurements on the shoreline were retrieved from the archives and databases[1,31]. Such measurements were carried out since 1952 during summer on the river bank near the water. To reconstruct external dose rates in air in 1949-1951 the model[38] describing radionuclide transport from the site of release along the river and the accumulation of radionuclides by bottom sediments was used. Dose rates in air were calculated on the basis of modelled radionuclide concentrations in bottom sediments and conversion coefficients obtained by Monte Carlo simulations of air for contaminated soil[39] with a dose-reduction factor for river shorelines. Figure 9.5 illustrates modelled and measured dose rates in air used for external dose calculations near the Techa River shoreline.

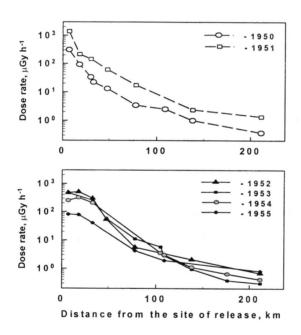

Figure 9.5 Absorbed dose rates in air along the shoreline of the Techa River. The top plot has modelled results, and the bottom plot has monitored data.

Indoor and Outdoor Dose Rates Within the Residence Areas

Typical locations of residence areas for the Techa Riverside settlements were on streets parallel to the shoreline. Schematic maps for the majority of villages were collected[40]. The decrease of dose rate with distance from the shore depended on the topology of the bank and was specific for each location. Dimensions and configuration of each residence area also influenced the distribution of doses. Gamma-exposure rates as a function of distance from the shoreline and within residence areas (streets, yards,

170

vegetable gardens, etc.) of several Techa Riverside villages were measured in 1952-1956[1]. Later, detailed surveys and maps of exposure-rate distributions were made for the upper Techa locations[41]. Figure 9.6 illustrates early and late measurements of dose rate in air for several locations within one example village. On the basis of these types of data, weighted ratios of shoreline dose rate to residence area dose rate were calculated for each settlement; weighting was determined by the number of houses located at different distances from the water's edge within a given settlement. Weighted ratios for the 40 settlements vary from 4 to 200[42].

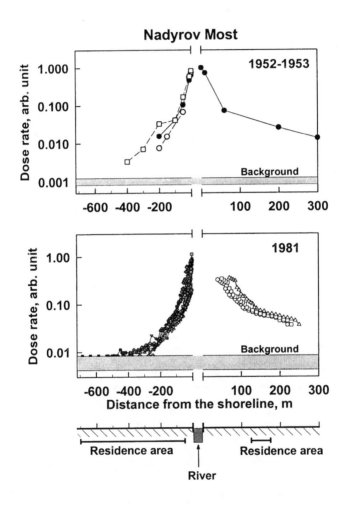

Figure 9.6 Measured absorbed dose rates in air as a function of distance from the shoreline. Data are for the example village of Nadyrov Most.

The ratio of dose rate indoors to dose rate outdoors was evaluated from measurements carried out in 1954 of indoor and outdoor dose rates for several dwellings in Metlino and Muslyumovo. The value for this average ratio is 0.45[42].

Model Behavioural Factors

As discussed above, the distributions of external exposure rates varied within residence areas and adjacent flood plains. To calculate doses for individuals, it is necessary to know the amounts of time spent by inhabitants at each location of differing dose rate. It is impossible to reconstruct accurately the behavioural patterns of the Techa Riverside residents 45 years after the fact; rather a simplified model was developed of typical life patterns for three types of specified locations within riverside settlements. The first location was the river shore, where people drew water, bathed, fished, rinsed linen, bathed horses, and bred waterfowl in summer time. The second location was the streets, yards, vegetable gardens, and other outdoor areas in the residence areas. The final location was inside the dwellings. Such a model was suggested based on questionnaires to evaluate periods spent on the Techa River shore[43]. Questions included what kinds of work and pastimes were conducted near the river, their duration, and their frequency. The study was conducted in the 1950s, and parameters were evaluated (conservatively) first for a critical group of subjects for radiation protection purposes[43]. Later, the author repeated the evaluation on the basis of the same data but for the purposes of average dose reconstruction; different estimates were obtained[2].

Typical life patterns for Techa riverside residents[2] include four age groups: young children, schoolchildren, agricultural workers, and pensioners. Agricultural workers and schoolchildren spent some time in fields, meadows, and forests outside of the contaminated areas. Young children and pensioners spent most of their time in residence areas and more time inside houses. Schoolchildren spent considerable time in school. Of course, such a model of life patterns is designed to represent the more conventional lifestyles; some kinds of workers (millers, teachers, physicians, etc.) would spend at least some of their working time within the residence area. However, for the current dose assessments, these generic life patterns have been used. It has been assumed that the variation of these estimates is about 20% to 35%[44].

Conversion Factors from Absorbed Dose in Air to Absorbed Dose in Organs

Age-dependent conversion factors for the organs of interest (as noted in the above section on internal dose calculations) were taken from the literature[39,45]. While such factors are a function of photon energy, there is a large plateau of values between about .08 and 1.3 MeV where the conversion factors can be considered to be essentially constant. This is the range that applies to most of the spectra of photons emitted by radionuclides absorbed by the Techa River sediments and flood-plain soils. Therefore, it has been assumed that dose-conversion factors are independent of energy and correspond to Monte Carlo simulations for a 500-keV monoenergetic source.

.3.2. Uncertainty Assessment

Examination of the sources and magnitudes of uncertainty in radiation doses to individuals in the ETRC is important. The analysis of uncertainty in the TRDS is incomplete, although the source of information for each term in the TRDS has been evaluated. The terms T_1 and T_2, while ideally coming from individual data, are currently assigned generic values, depending on the age of the individual in year y. The external dose rates $D_{Riv,L,y}$, $D_{Out,L,y}$, and $D_{In,L,y}$ are derived from measurements, or alternatively, from the radionuclide contents of sediment calculated from a model[34] multiplied by external dose-rate factors (such as those in Ekerman and Ryman[39]). The key term $I_{Y,r,L}$ is derived from information in the literature[30]; it has a very complex uncertainty structure. The variation of intake levels within the same village and age cohort depends mainly on the source of drinking water. Dose-conversion factors, $DF_{r,o,Y-y}$, are calculated using biokinetic models and their uncertainties are determined mainly by the variability of metabolic parameters. To estimate the uncertainty of the dose estimates calculated using the TRDS, a Monte Carlo version of the TRDS is under development.

9.4. Representative Doses to Members of the Public

Both dose reconstruction projects estimated doses to members of their respective publics.

9.4.1. Hanford

The largest doses resulting from Hanford operations occurred in the mid-1940s[46]. The most important radionuclide was iodine-131 released to the atmosphere. The most important exposure pathway was consumption of milk produced by cows on pasture downwind of Hanford. The iodine-131 releases were essentially routine and continuous during the first period of site operation. Infants and young children who drank milk from cows that ate fresh pasture are likely to have received the highest doses. Median doses for individuals in this group ranged from about 0.02 to 2.4 Gy to the thyroid. The uncertainty on the initial dose estimates is fairly large; for example, the 95th percentile reported for Ringold, the location for which the median dose was 2.4 Gy, was 8.7 Gy.

Recent work has given a better estimate of the overall pattern of iodine-131 deposition. An estimate of the extent of the deposition, scaled to thyroid dose to a reference infant drinking milk from a domestic cow on fresh pasture, is given in Figure 9.7[47]. This figure indicates that thyroid doses in excess of 0.085 Gy to infants with backyard cows could have extended to the Washington/Canada border.

Table 9.5 summarizes doses and their uncertainties to maximally exposed individuals at several locations throughout the study area. Cumulative radiation doses to maximally exposed individuals from releases to the Columbia River range from about 4.6 to 14.2 mSv for the period 1950 through 1970, which is the period of highest releases (Table 9.6). The major radionuclides contributing to doses from the river pathway are zinc-65, phosphorus-32, arsenic-76, and sodium-24. The range of doses largely depends on the amount of fresh, resident fish consumed. Drinking water contributes only a small dose, although nearly all residents of the local downstream communities received one.

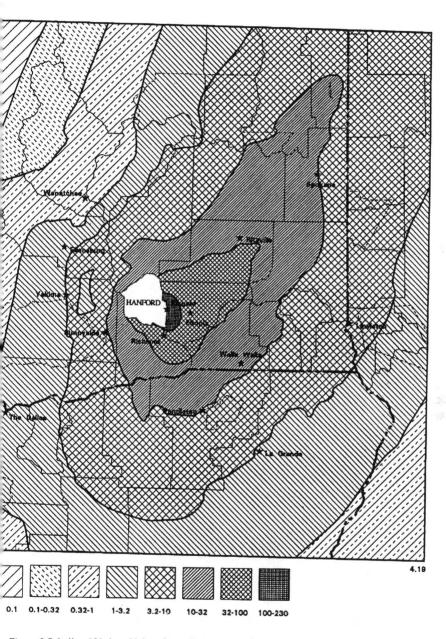

0.1   0.1-0.32   0.32-1   1-3.2   3.2-10   10-32   32-100   100-230

Figure 9.7 Iodine-131 thyroid dose from all exposure pathways--milk cows on fresh pasture

TABLE 9.5 Median and ranges of thyroid doses to infants in the study area drinking milk from backyard cows on fresh pasture (Gy)

Location	Median	Range
Ringold	2.40	0.54-8.70
Richland	0.93	0.24-3.50
Eltopia	0.73	0.19-3.00
Ritzville	0.28	0.074-1.20
Spokane	0.11	0.03-0.44
Walla Walla	0.13	0.04-0.44
Pendleton	0.09	0.02-0.30
Lewiston	0.04	0.01-0.15
Yakima	0.03	0.007-0.10
Ellensburg	0.02	0.005-0.07

TABLE 9.6 Cumulative doses to the maximally exposed individual from the river pathways, 1950-1970 (μSv EDE)

Location	Maximum	Typical
Ringold	14,200	510
Richland	13,900	290
Kennewick/Pasco	13,000	630
Snake/Wall Walla Rivers	8,800	440
Umatilla/Boardman	7,100	260
Arlington	6,800	240
John Day Dam/Biggs	6,700	230
Deschutes River	6,300	220
The Dalles/Celilo	6,200	200
Klickitat River	6,000	200
White Salmon/Cascade Locks	5,700	190
Bonneville Dam to River Mouth	4,600	150

9.4.2. Mayak

As described above, internal doses for members of the ETRC are calculated on the basis of age- and location-specific mean-annual-intake levels of radionuclides, age-dependent biokinetic models for radionuclides, and individual-residence histories for each subject. Figure 9.8 presents the distribution of internal dose in red bone marrow among the ETRC members. As seen, more than half of the people have internal red bone marrow doses between 0.1 and 0.5 Gy. Absorbed doses in cells on bone surfaces have the same distributions as do the bone marrow doses, but the values are about two times higher.

Example Case Histories
The process of dose calculation from initial data through TRDS results can be illustrated for two cases in the ETRC. A summary of pertinent information for these cases is

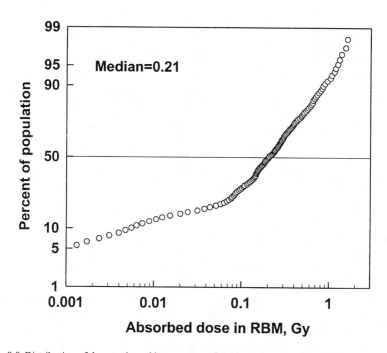

Figure 9.8 Distribution of dose to the red bone marrow from ingestion of radionuclides for members of the Extended Techa River Cohort

hown in Table 9.7[48]. As seen, both persons lived on the Techa River during their ntire lives. Calculations of lifetime dose to the red bone marrow are presented in 'ables 9.8 and 9.9.

Table 9.8 illustrates the calculation of external dose for Case 1. This person was xposed in Metlino during 5 years (1950-1954) plus 2 months before death in 1955. According to his age at exposure, his behavioural factors (T1, T2, and T3) correspond to hat of a pensioner. Because the range of attained age for this case belongs to a single ategory (>60 years) during 1950-1955, these behavioural factors don't change during he period of exposure. Annual and cumulative doses to the red bone marrow from xternal sources are shown in the last column.

Table 9.9 illustrates the calculation of internal dose from strontium-90 intake for 'ase 2. The individual calculational endpoint for this case is 1994. This person changed is place of residence in 1953; his levels of intake are calculated by multiplying f_L, $M_{y,L}$ nd reference annual intake for his age group. Dose-conversion factors corresponding to he age at intake and time interval between the year of intake and the calculational endpoint are shown in column 9; annual and total doses to the red bone marrow from strontium-90 intake are shown in the last column.

TABLE 9.7 Case history data for the two example cases; results are presented in Tables 9.8 and 9.9

Information	Case 1	Case 2
Identification code	611	65737
Date of birth	1881	4 June 1928
Sex	Male	Male
Residence history	Metlino: 1881-Feb. 1955	Ibragimovo: June 1928-June 1953 Muslyumovo: June 1953-Dec. 1994
Vital status	Died	Died
Date of vital status	7 February 1955	4 December 1994

Validation of TRDS-2000 Doses

As mentioned, the calculation of internal doses is based strongly upon the direct measurements of strontium-90 body burdens by a special WBC designed to measure bremsstrahlung from yttrium-90; such measurements have been made for half of the members of the ETRC. Strontium-90 body burdens have also been measured in samples collected at autopsy, and the two sets of data compare well[49]. Thus, a large body of data is available that can be used to verify the calculated body burdens (and, by extension, doses). Doses from the short-lived radionuclides have not been validated, and it seems unlikely that a direct method can be found for validation for the organs (gastrointestinal tract) of larger dose. The doses are being indirectly validated through measurements of concentrations of radionuclides in water and comparisons to strontium-90.

The validation of the new assessments of external dose is now an issue of major importance. The applicability of the use of "natural" dosimeters has been investigated within the framework of several international projects. Bricks from abandoned buildings located near the Techa shoreline were sampled, the quartz was extracted from the bricks, and doses were assessed by using the quartz as a thermoluminescent dosimeter[50]. This study has demonstrated the potential of the method in combination with Monte Carlo simulations of radiation transport at sampling sites for the validation of environmental doses in the upper and middle Techa region.

A pilot study[51] measured dose received by teeth as determined by electron paramagnetic resonance analysis. This study confirmed the applicability of that method

TABLE 9.8 Calculated external dose to red bone marrow (RBM) for Case 1 (see Table 9.7 for case history data)

Year of Exposure	Location	$M_{y,L}$	Age, years	$D_{Riv,L,y}$ Gy/y	$D_{Out,L,y}$ mGy/y	$D_{In,L,y}$ mGy/y	T_1	T_2	T_3	A_0	RBM Dose, Gy
1950	Metlino	1	69	2.72	27.2	12.2	0.017	0.284	0.699	0.73	0.045
1951	Metlino	1	70	12.0	120	54.0	0.017	0.284	0.699	0.73	0.203
1952	Metlino	1	71	4.12	41.2	18.5	0.017	0.284	0.699	0.73	0.069
1953	Metlino	1	72	4.07	40.7	18.3	0.017	0.284	0.699	0.73	0.068
1954	Metlino	1	73	2.19	21.9	9.86	0.017	0.284	0.699	0.73	0.036
1955	Metlino	0.17	74	0.70	7.0	3.15	0.017	0.284	0.699	0.73	0.001
Total dose											**0.42**

TABLE 9.9 Calculated internal dose to red bone marrow (RBM) for Case 2 (see Table 9.7 for case history data)

Year of Intake	Location	Location Factor, f_L	$M_{y,L}$	Reference ^{90}Sr intake, Bq/y	Individual ^{90}Sr Intake, Bq/y	Age at Intake, y	Time since Intake, y	$DF_{r,o,y,y3}$ Gy/Bq	RBM Dose, Gy
1950	Ibragimovo	1.385	1	2.08×10^6	2.88×10^6	22	44	2.44×10^{-7}	7.03×10^{-1}
1951	Ibragimovo	1.385	1	4.70×10^5	6.51×10^5	23	43	2.41×10^{-7}	1.57×10^{-1}
1952	Ibragimovo	1.385	1	4.62×10^5	6.40×10^5	24	42	2.33×10^{-7}	1.49×10^{-1}
1953	Ibragimovo	1.385	0.5	8.61×10^4	1.03×10^5	25	41	2.25×10^{-7}	2.32×10^{-2}
	Muslyumovo	1	0.5						
1954	Muslyumovo	1	1	1.92×10^4	1.92×10^4	26	40	2.18×10^{-7}	4.19×10^{-3}
1955	Muslyumovo	1	1	1.38×10^4	1.38×10^4	27	39	2.13×10^{-7}	2.93×10^{-3}
1956	Muslyumovo	1	1	6.90×10^3	6.90×10^3	28	38	2.10×10^{-7}	1.45×10^{-3}
1957	Muslyumovo	1	1	5.88×10^3	5.88×10^3	29	37	2.07×10^{-7}	1.22×10^{-3}
1958	Muslyumovo	1	1	4.93×10^3	4.93×10^3	30	36	2.04×10^{-7}	1.01×10^{-3}
1959	Muslyumovo	1	1	3.39×10^3	3.39×10^3	31	35	2.02×10^{-7}	6.80×10^{-4}
Total dose									**1.057**

for retrospective individual-dose evaluation. This method, based upon measurements of samples collected for dental health reasons, could also validate estimates of uncertainty in assessment of external dose.

9.5. Conclusions

The HEDR Project was based extensively on computer simulations. HEDR prepared a state-of-the-art set of computational tools for estimating historical doses to representative and real individuals. The tools incorporate significant advances in tracking spatial and temporal relationships of environmental dosimetry and in the application of uncertainty analysis. The models and computer codes were extensively peer reviewed, tested, verified, and validated for use in estimating the doses to representative individuals who lived in the Columbia Basin from 1944 through 1992. They also have been used to estimate doses that real individuals included in the Hanford Thyroid Disease Study may have received because of their locations, lifestyles, and food consumption patterns. The complete set of configured codes, parameter values, data files, and pertinent documentation have been turned over to the CDC, the local States, and other interested parties.

Specific conclusions of the HEDR Project are that:

- The largest doses were from iodine-131 released to the atmosphere between December 1944 and December 1947. The highest were in 1945.
- The most important radiation exposure pathway for iodine-131 was from drinking milk from cows on irrigated pasture close-in and downwind of the site.
- The median dose for a child at the maximum exposure location is about 2.40 Gy (with a range of 0.54 to 8.70 Gy). At the lowest exposure location, the estimated dose is 0.0007 Gy (with a range of 0.00012 to 0.0034 Gy).
- There is a 90% chance that a similar individual's dose would be within a range of one-fifth to five times the median values stated.
- Radiation doses from the release of radionuclides to the Columbia River were highest in the years 1956 to 1965. The peak was 1960.
- The most important means of exposure from the river pathway was the consumption of resident fish.
- A person who consumed 40 kg of fish per year (about three fish meals per week) at Richland would have received a dose of about 1,400 µSv EDE in 1960. Consuming the same amount of fish in the lower river below Bonneville Dam would have produced a dose of about 410 µSv EDE in 1960.
- A typical adult who ate no Columbia River resident fish would have received a dose of about 53 µSv in 1960 at Pasco, primarily from drinking water. The dose to a person not eating resident fish below Bonneville Dam would have received a dose of about 13 µSv in 1960.

For the Techa River study, the following important tasks were performed to develop the TRDS-2000:

- Development of a source term for releases to the Techa River and a river model that describes radionuclide concentrations in water and sediments. This combination allowed more realistic calculation of absorbed dose in air near the river shore during the 1950-1951 period of massive releases.

• Reconstruction of intake of short-lived radionuclides and reassessment of organ-specific doses for ETRC members from internal exposure from all radionuclides ingested.

• Study of the variations in gamma-exposure rate within residence areas of riverside settlements according to distance from the river. This study allowed calculation of more realistic weighted-average values of external dose within residence areas.

• Study of the effects of location and time spent in streets, gardens, and homes. Study of the first, third, and fourth factors has led to more realistic assessment of organ-specific doses for ETRC members from external gamma exposure.

• Development of a system to describe accurately the uncertainties (systematic bias and random errors) in all models and measurements and to propagate such uncertainties through to the final results with proper allowance for correlation structures within the data.

• Calculation of new individual organ doses for all ETRC members using the updated TRDS-2000 approach.

Studies of the possible effects of radiation on those exposed to the releases to the Techa River were started in Russia in the 1950s. The basis of the past dose-reconstruction efforts for the TRC has been summarized in several publications [30,31,32,48,50,51]. A preliminary report on the status of the follow-up of the TRC has also been published[53]. It is reported that, despite a number of limitations, there does appear to be a dose-related increase in risk of mortality from leukaemia and other cancers.

9.6. References

Vorobiova, M.I., Degteva, M.O., Burmistrov, D.S., Safronova, N.G., Kozheurov, V.P., Anspaugh, L.R., and Napier, B.A. (1999) Review of historical monitoring data on Techa River contamination. Health Phys. 76 605-618.

Saurov, M.M. (1992) Measurement of dose-rates of external exposure and survey on life-styles of inhabitants along the Techa River. Presented at the Russian-Japanese experts' meeting on epidemiological research of radiation effects in the Techa River Basin in Southern Urals. January 24-28, 1992, Tokyo, Japan.

Shipler, D.B., Napier, B.A., Farris, W.T., and Freshley, M.D. (1996) The Hanford Environmental Dose Reconstruction Project-an overview, Health Physics 71 532-544.

Pacific Northwest Laboratory. (1990) *Air Pathway Report: Phase I of the Hanford Environmental Dose Reconstruction Project.* PNL-7410 HEDR, Pacific Northwest Laboratory, Richland, Washington.

Pacific Northwest Laboratory. (1990) *Columbia River Pathway Report: Phase I of the Hanford Environmental Dose Reconstruction Project.* PNL-7412 HEDR, Pacific Northwest Laboratory, Richland, Washington.

Simpson, J.C. (1990) *Dose Estimate Variability Caused by Air Model Uncertainties.* PNL-7737 HEDR, Pacific Northwest Laboratory, Richland, Washington.

Simpson, J.C. (1991) *Effects of the Loss of Correlation Structure on Phase I Dose Estimates.* PNL-7638 HEDR, Pacific Northwest Laboratory, Richland, Washington.

Napier, B.A. (1991) Selection of Dominant Radionuclides for Phase I of the Hanford Environmental Dose Reconstruction Project. PNL-7231 HEDR, Pacific Northwest National Laboratory, Richland, Washington.

Heeb, C.M. (1992) *Iodine-131 Releases From the Hanford Site, 1944 through 1947.* PNWD-2033 HEDR Vol. 1-2., Battelle Pacific Northwest Laboratories, Richland, Washington.

10. Heeb, C.M., Gydesen, S.P., Simpson, J.C., and Bates, D.J. (1996) Reconstruction of radionuclide releases from the Hanford Site, *Health Physics* **71** 545-555.

11. Walters, W.H., Dirkes, R.L., and Napier, B.A. (1992) *Literature and Data Review for the Surface Water Pathway: Columbia River and Adjacent Coastal Areas.* PNWD-2034 HEDR, Battelle Pacific Northwest Laboratories, Richland, Washington.

12. Napier, B.A. (1993) *Determination of Key Radionuclides and Parameters Related to Dose From the Columbia River Pathway.* BN-SA-3768 HEDR, Battelle Pacific Northwest Laboratories, Richland, Washington.

13. Ramsdell, J.V., Jr, and Burk, K.W. (1992) Regional Atmospheric Transport Code For Hanford Emission Tracking (RATCHET). PNL-8003 HEDR, Pacific Northwest Laboratory, Richland, Washington.

14. Ramsdell, J.V., Jr., Simonen, C.A., Burk, K.W., and Stage, S.A. (1996) Atmospheric dispersion and deposition of ^{131}I released from the Hanford Site, *Health Physics* **71** 568-577.

15. Holly, F.M., Jr., Yang, J.C., Schwarz, P., Schaefer, J., Hsu, S.H., and Einhellig, R. (1990) *CHARIMA: Numerical Simulation of Unsteady Water and Sediment Movement in Multiply Connected Networks of Mobile-Bed Channels.* IIHR Report No. 343, Iowa Institute of Hydraulic Research, Iowa State University, Iowa City, Iowa.

16. Walters, W.H., Richmond, M.C., and Gilmore, B.G. (19960 Reconstruction of radioactive contamination in the Columbia River, *Health Physics* **71** 556-567.

17. Ikenberry, T.A., Burnett, R.A., Napier, B.A., Reitz, N.A., and Shipler, D.B. (1992) *Integrated Codes for Estimating Environmental Accumulation and Individual Dose for Past Hanford Atmospheric Releases.* PNL-7993 HEDR, Pacific Northwest Laboratory, Richland, Washington.

18. Beck, D.M., Darwin, R.F., Erickson, A.R., and Eckert, R.L. (1992) *Milk Cow Feed Intake and Milk Production and Distribution Estimates for Phase I.* PNL-7227 HEDR, Pacific Northwest Laboratory, Richland, Washington.

19. Marsh, T.L., Anderson, D.M., Farris, W.T., Ikenberry, T.A., Napier, B.A., and Wilfert, G.L. (1992) Commercial Production and Distribution of Fresh Fruits and Vegetables: A Scoping Study on the Importance of Produce Pathways to Dose. PNWD-2022 HEDR, Battelle Pacific Northwest Laboratories, Richland, Washington.

20. Denham, D.H., Thiede, M.E., Dirkes, R.L., Hanf, R.W., and Poston, T.M. (1993) *Phase I Summaries of Radionuclide Concentration Data for Vegetation, River, Water, Drinking Water, and Fish.* Battelle Pacific Northwest Laboratories, Richland, Washington.

21. Freshley, M.D., and Thorne, P.D. (1992) Ground-Water Contributions to Dose from Past Hanford Operations. PNWD-1974 HEDR, Battelle Pacific Northwest Laboratories, Richland, Washington.

22. Anderson, D.M., Marsh, T.L., and Deonigi, D.A. (1996) Developing historical food production and consumption data for 131I dose estimates: the Hanford experience, Health Physics 71 578-587.

23. Farris, W.T., Napier, B.A., Ikenberry, T.A., and Shipler, D.B. (1996) Radiation doses from Hanford Site releases, Health Physics 71 588-601.

24. Gilbert, R.O., Napier, B.A., Liebetrau, A.M., and Haerer, H.A. (1991) Statistical aspects of reconstructing the I-131 dose to the thyroid of individuals living near the Hanford Site in the mid-1940s. *Radiation Protection Dosimetry* **36** 195-198.

25. Simpson, J.C., and Ramsdell, J.V., Jr. (1993) *Uncertainty and Sensitivity Analysis Plan.* PNWD-2124 HEDR, Battelle Pacific Northwest Laboratories, Richland, Washington.

26. Napier, B.A., Simpson, J.C., Eslinger, P.W., Ramsdell, J.V., Jr., Thiede, M.E., and Walters, W.H. (1994) *Validation of HEDR Models.* PNWD-2221 HEDR, Battelle Pacific Northwest Laboratories, Richland, Washington.

27. Napier, B.A., Eslinger, P.W., Nichols, W.E., and Anderlini, L. (2000) Improvements in modeling sagebrush concentrations of radioactive iodine released from the Hanford Site," *Environmental Radioactivity,* in press.

28. Kossenko, M.M., Degteva, M.O., Vyushkova, O.V., Preston, D.L., Mabuchi, K., and Kozheurov, V.P. (1997) Issues in the comparison of risk estimates for the population in the Techa River Region and atomic bomb survivors. *Radiat. Res.* **148** 54-63.

29. Kozheurov, V.P. (1994) SICH9.1-A unique whole-body counting system for measuring Sr-90 via bremsstrahlung: The main results from a long-term investigation of the Techa River population. *Sci. Total Environ.* **14** 37-48.

30. Kozheurov, V.P., and Degteva, M.O. (1994) Dietary intake evaluation and dosimetric modelling for the Techa River residents based on in vivo measurements of strontium-90 in teeth and skeleton. *Sci. Total Environ.* **14** 63-72.

181

31. Degteva, M.O., Drozhko, E., Anspaugh, L., Napier, B., Bouville, A., and Miller, C. (1996) Dose Reconstruction for the Urals Population. Joint Coordinating Committee on Radiation Effects Research. Project 1.1—Final Report. Rept. UCRL-ID-123713, Lawrence Livermore National Laboratory, Livermore, California.
32. Degteva, M.O., and Kozheurov, V.P. (1994) Age-dependent model for strontium retention in human bone. *Radiat. Prot. Dosim.* **53** 229-233.
33. Degteva, M.O., Kozheurov, V.P., and Tolstykh, E.I. (1998) Retrospective dosimetry related to chronic environmental exposure. *Radiat. Prot. Dosim.* **79** 155-160.
34. Vorobiova, M.I., and Degteva, M.O. (1999) Simple model for the reconstruction of radionuclide concentrations and radiation exposures along the Techa River. *Health Phys.* **77** 142-149.
35. Le Grand, J. (1972) Contamination by osteotropic β-emitters – an evaluation of the doses in bone marrow and endosteum. In *Proceedings of 2nd International Conference on Strontium Metabolism*, 49-65, CONF-720818, Glasgow and Strontian, United Kingdom.
36. Spiers, F.W., Beddoe, A.H., and Whitwell, J.R. (1978) Mean skeletal dose factors for beta-particle emitters in human bone. Part 1: Volume-seeking radionuclides. *Brit. J. Radiol.* **51** 622-627.
37. International Commission on Radiological Protection. (1993) Age-Dependent Doses to Members of the Public from Intake of Radionuclides: Part 2: Ingestion Dose Coefficients. ICRP Publication 67, Pergamon Press, Oxford, England.
38. Berkovsky, V. (1992) Computer system for reconstruction and prediction of human internal doses. Doctoral Thesis (in Russian), Radiation Protection Institute, Kiev.
39. Eckerman, K.F, and Ryman, J.C. (1993) *External Exposure to Radionuclides in Air, Water, and Soil.* Federal Guidance Report No. 12, EPA 402-R-93-081, U.S. Environmental Protection Agency, Washington, D.C.
40. Shvedov, V.L., Goloschapov, P.V., Kossenko, M.M., Akleyev, A.V., Vorobiova, M.I., Degteva, M.O., Malkin, P.M., Safronova, N.G., Peremyslova, L.M., Yakovleva, V.P., Kozheurov, V.P., Nikolaenko, L.A., Rayt, M.K., Babina, T.D., and Kravtsova, E.M. (1990) *Radiation-Hygienic and Medico-Biological Consequences of Massive Radioactive Contamination of the River System.* Technical Report, Urals Research Center for Radiation Medicine, Chelyabinsk (in Russian).
41. Lebedev, V.M. (1982) *Issues in the Reconstruction of Individual Doses for the Population of the Upper Techa Riverside.* URCRM Technical Report, Urals Research Center for Radiation Medicine, Chelyabinsk (in Russian).
42. Vorobiova, M.I., Degteva, M.O., Kozyrev, A.V., Anspaugh, L.R., and Napier, B.A. (1999) *External Doses Evaluated on the Basis of the Techa River Dosimetry System Approach.* Technical Report, Urals Research Center for Radiation Medicine, Chelyabinsk.
43. Saurov, M.M. (1968) Radiation-Hygiene Assessment of Natural Movement of Population Exposed to Chronic Influence of Uranium Fission Products. Doctoral Thesis, Institute of Biophysics, Moscow (in Russian).
44. Kravtsova, E.M., Kolotygina, N.V., and Barkovsky, A.N. (1994) External irradiation of the residents of the Muslyumovo Village of Chelyabinsk Oblast. In V.N. Chukanov (ed.), *Radiation, Ecology, Health. Part II: Impact of Radiation on the Public Health*, 13-16. Ekaterinburg (in Russian).
45. Petoussi, N., Jacob, P., Zankl, M., and Saito, K. (1990) Organ doses for foetuses, babies, children and adults from environmental gamma rays. *Radiat. Prot. Dosim.* **37** 31-41.
46. Technical Steering Panel. (1990) *Initial Hanford Radiation Dose Estimates.* Washington State Department of Ecology, Olympia, Washington.
47. Napier, B.A., and Snyder, S.F. (1992) *Determination of the Feasibility of Reducing the Spatial Domain of the HEDR Dose Code.* BN-SA-3678 HEDR, Battelle Pacific Northwest Laboratories, Richland, Washington.
48. Degteva, M.O., Kozheurov, V.P., Burmistrov, D.S., Vorobyova, M.I., Valchuk, V.V., Bougrov, N.G., and Shishkina, H.A. (1996) An approach to dose reconstruction for the Urals population. *Health Phys.* **71** 71-76.
49. Tolstykh, E.I., Kozheurov, V.P., Burmistrov, D.S., Degteva, M.O., Vorobiova, M.I., Anspaugh, L.R., and Napier, B.A. (1998) Individual-Body-Burden Histories and Resulting Internal Organ Doses Evaluated on the Basis of the Techa River Dosimetry System Approach. Technical Report, Urals Research Center for Radiation Medicine, Chelyabinsk.
50. Bougrov, N.G., Göksu, H.Y., Haskell, E., Degteva, M.O., Meckbach, R., and Jacob, P. (1998) Issues in the reconstruction of environmental doses on the basis of thermoluminescence measurements in the Techa Riverside. *Health Phys.* **75** 74-583.

51. Romanyukha, A.A., Ignatiev, E.A., Degteva, M.O., Kozheurov, V.P., Wieser, A., and Jacob, P. (1996) Radiation doses from Ural Region. *Nature* **381** 199-200.
52. Degteva, M.O., Kozheurov, V.P., and Vorobiova, M.I. (1994) General approach to dose reconstruction in the population exposed as a result of the release of radioactive wastes into the Techa River. *Sci. Total Environ.* **14**:49-61.
53. Degteva, M.O., Kozheurov, V..P., Vorobiova, M.I., Burmistrov, D.S., Khokhryakov, V.V., Suslova, K.G., Anspaugh, L.R., Napier, B.A., and Bouville, A. (1997) Population exposure dose reconstruction for the Urals Region. In Proceedings of Symposium on Assessing Health and Environmental Risks from Long-Term Radiation Contamination in Chelyabinsk, Russia. Pages 21-23, American Association for the Advancement of Science, Washington, D.C.

10. Quantitative Risk Assessment Methods of Accounting for Probabilistic and Deterministic Data Applied to Complex Systems

Another aspect of understanding the risks of Cold War legacies is through the application of probabilistic safety assessments and probabilistic risk assessments, largely aimed at calculating and mitigating the risk and severity of accidents. This chapter provides a tutorial on accident risk assessment, which has been applied to a wide range of legacy weapons and weapon delivery systems. It discusses scenario development, merging of deterministic and probabilistic calculations, uncertainty, and facility risk management techniques.

Standard event tree quantification can be used to produce risk estimates and subsequent risk importance measures. These risk importance measures identify which component failures contribute most significantly to the overall risk for a system. Therefore, risks can be reduced or mitigated most effectively by controlling the failures of these components. However, this standard approach gives only a static, time-invariant picture of risk, which significantly limits the utility to a decision maker. Predictive analyses that identify risks over time are very difficult to build into this process. Moreover, in many cases, no information is available regarding the physical conditions that actually caused the failure to occur.

The methods presented in this chapter resolve these issues by taking the assessment down to the physical parameter level. The strict probabilistic treatment of event tree methodology is replaced with a hybridized method containing both deterministic and probabilistic components. One requirement for the application of these methods is availability of existing deterministic models that characterize the physical response to perturbations imposed by initiators. The availability of high-speed computational resources along with advances in the understanding of complex physical phenomena have allowed the approach presented in this chapter to become a realistic alternative to standard event tree methodologies.

10.1. Background

The objectives of a quantitative risk assessment (QRA) are to determine and quantify the risks associated with a given system. A system may be any facility or piece of equipment that presents a concern as a result of hazards inherent in its operation. The concern is often focused on potential system failures as well as on the effects of system failures on the environment and the general public. Examples of systems are:

- Nuclear power plants
- Coal-fired power plants
- Civilian aircraft
- Military aircraft
- Water treatment facilities
- Manufacturing facilities that use or produce hazardous materials
- Military equipment such as missiles, tanks, ammunition, etc.

The greatest benefit in using QRA is gained when the assessment is performed proactively, i.e., before an event occurs that leads to a system failure and a subsequent adverse consequence. It is important to note that QRA can also be applied in response to events that have already occurred. In either case, the resulting prioritised list of significant risk contributors allows analysts to propose measures to reduce or eliminate dominant risks. This, in turn, allows decision makers to focus limited funds on those areas that will most help to prevent or respond to a system failure. The results of a QRA can serve as input to a consequence analysis, which in turn can assess human health and environment effects.

10.2. Assessment Methodology

Risk assessment methodology for any complex system involves several steps and various levels of analysis to quantify the risk. A general flow diagram of the major risk assessment steps is shown in Figure 10.1, and a detailed flow diagram of selected steps is shown in Figure 10.2. These steps are discussed in the following subsections.

10.2.1. Data Collection

The objective of data collection is to compile the knowledge base necessary to conduct the study. Some of the data collected are used for various calculations, e.g., bounding cases for selected initiators and application of screening values. Some data are used to develop detailed phenomenological models. These data included drawings, finite element models, and existing test data for verification and validation. Other data are used to assess the frequencies of events and conditional likelihoods based on those events. Data specific to the system and its operations are used, if available. Applicable surrogate data are used when the specific data are limited.

The collection of system-specific data usually requires visits to the site(s) where the system operates. The primary tasks performed during the visits are to observe all aspects of system operations, obtain necessary and available information relative to the system,

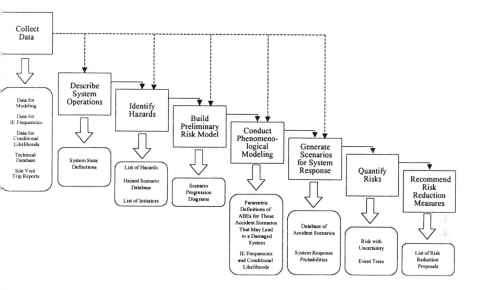

Figure 10.1 Flow diagram of the risk assessment methodology and outputs at each step

speak with key personnel about their activities in support of the system, and identify hazards that could possibly lead to adverse environments.

10.2.2. System Description

Before hazard identification can begin, it is necessary to understand the system operations so that the risk analysis covers all activities of interest. This understanding is achieved by reviewing applicable documents and drawings. The operations are grouped into System States. A System State is a logical grouping of activities and operations, such as transportation, storage, or maintenance. These System States are used to organize the risk analysis.

10.2.3. Hazard Identification and Hazard Scenarios

Hazard scenarios are defined based on the list of hazards identified during site visits. These scenarios describe how a hazard may lead to possibly adverse system consequences. Examples of hazards include mechanical events (vehicle collision, system drop), thermal events (fuel leak ignition), and electrical events (lightning). Hazard scenarios are developed for all identified hazards and combinations thereof. For example, a combined hazard scenario might be an aircraft crashing into the system, which presents a mechanical as well as a thermal event.

The hazard scenarios are organized into sets for each System State. Each hazard scenario includes information on the hazard, the system location and/or activity, a description of how an event is initiated and how it could lead to an adverse system

186

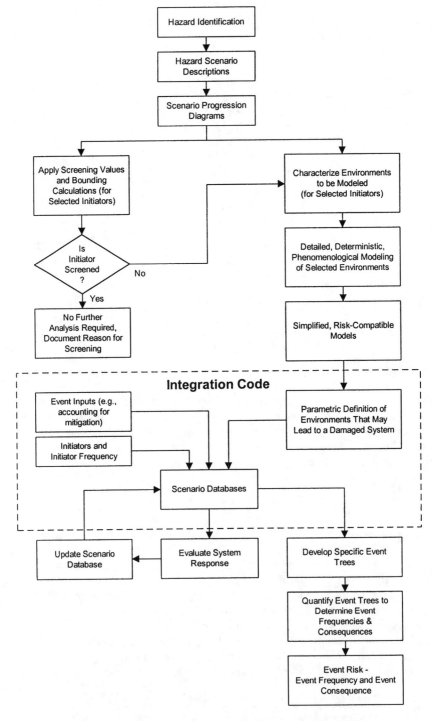

Figure 10.2 Detailed flow diagram of the risk assessment methodology

consequence, and a listing of any mitigation systems or procedures that could minimize the likelihood or severity.

The initial event that may subsequently cause a hazard to impact a system is called an initiating event, or initiator. Based on the identified hazards, observations made during site visits, knowledge of operations, and hazard scenarios, a full set of initiators is developed for each System State.

10.2.4. Preliminary Risk Model

Scenario progression diagrams (SPDs) are developed based on the identified hazards and the hazard scenarios. These SPDs are used to improve the understanding of the possible sequence of events following the occurrence of an accident for a system in a given System State. The SPD is a multi-branch tree, the purpose of which is to convey the accident progression in a brief, succinct manner. It is important to note that, unlike an event tree, the SPD is not used to quantify the frequency of the accident. SPDs are used to streamline the modelling process by identifying similar scenario environments, provide a structured approach to apply screening values, and identify what types of models are necessary for a given initiator. The hazard scenarios and SPDs together form the preliminary risk model, which guides the subsequent modelling effort.

10.2.5. Screening Values and Bounding Calculations

Once the hazards are identified, scenarios are developed, and SPDs are built, the level of analysis required to compute the risk is determined. This determination is a staged process that uses screening values, bounding calculations, and progressively more detailed models to understand the scenario environment.

Certain hazard environments to which the system may be exposed do not present a risk. This information is provided in the form of screening values. For example, there may exist a known velocity below which an impact to the system poses no concern, and by using this velocity, many scenarios may be eliminated from further analysis. Other screens may be based on the ability of a system to withstand a fire or a lightning strike.

In some instances, the worst-case scenario that can result from an initiator, regardless of the frequency of the event, is analysed for its impact on the system. If it can be shown that the worst, or bounding, case scenario does not cause an adverse consequence, then the initiator is fully or partially screened (depending on the calculation) from further analysis. The bounding calculation may focus on mechanical energy, thermal energy, electrical energy, or any other hazard source.

10.2.6. Phenomenological Modelling

For those initiators that cannot be screened out based on bounding calculations, it is necessary to conduct detailed modelling of the environments that may impact the system. Because of the complex nature of the environments, the use of sophisticated codes and finite element models is required to predict the environment to which the system is exposed. Throughout the modelling effort, engineering judgment is applied to modelling assumptions and results interpretation. To the extent possible, results should be benchmarked against test data.

For example, a fire exhibits complex behaviour that can only be predicted using two- or three-dimensional fire analysis codes which model parameters including fuel volume, pool size, fire duration, and heat flux. By understanding this behaviour, and knowing the proximity of the fire to the system, the impact on the system can be determined. For mechanical environments, detailed structural finite element models of the system and associated items such as buildings or aircraft may be needed. These models are used to assess various impact scenarios, including objects impacting the system directly or impacting nearby equipment, as well as the effects of a seismic event on the system.

These models are deterministic in nature rather than probabilistic. This means that each analysis focuses on a single environment, providing a point-value result. However, the time required to run a detailed phenomenological model is prohibitively long when quantifying the risk of numerous accident scenarios. Thus, based on the detailed modelling, simplified (i.e., risk-compatible) models are developed that can predict the response to an environment. These simplified models are used to generalize the results of the detailed calculations to all other scenarios in a form that is compatible with the risk model.

Two general approaches are taken for the development of these risk-compatible models. The first approach involves constructing a physical response model based directly upon the output of the detailed models and any existing test data. This regression approach yields a surface, which is then used to predict the system environment for those accidents that are not contained within the existing detailed model results and test results. The ability of the response surface to give reasonable results depends upon the choice and number of calculations performed with the detailed model and the extent of existing test data.

The second approach used in the development of risk-compatible models involves using the detailed models and test data to get an understanding of under what conditions certain portions of the physics of the problem dominate the result. This approach requires sufficient detailed modelling results to get a firm understanding of the sensitivities. Once the dominant physics are identified, they are written directly into the models. Therefore, those portions of the detailed models whose contribution only weakly impacts the results are dropped from the risk-compatible models.

10.2.7. Risk Analysis

The actual risk analysis process includes several substeps, including accident scenario description, accident matrix development, risk matrix development, accident scenario generation, risk quantification, and risk visualization. These methods are applied to each initiator that is not previously screened or that required detailed phenomenological modelling. They are the basis for performing the point estimate, or base case, analysis as well as the uncertainty analysis described later.

Accident Scenario Description
Accident scenarios are developed for those initiators that
- Remained after applying screening values
- Remained after performing bounding calculations

- Required detailed phenomenological modelling to understand the accident.

Accident scenarios consist of parameters that characterize the accident environment and the system environment. A large number of accident scenarios are required to adequately cover the parameter space of all possible accident and system environments that may result from a particular initiator.

The scenario development process results in a set of accident and system environments that cover the accident parameter space. The complete characterization of these accidents for a given initiator is referred to as an accident matrix. To quantify the risk, a corresponding risk matrix is defined. A description of the components of accident and risk matrices follows; Figure 10.3 illustrates how the components interface.

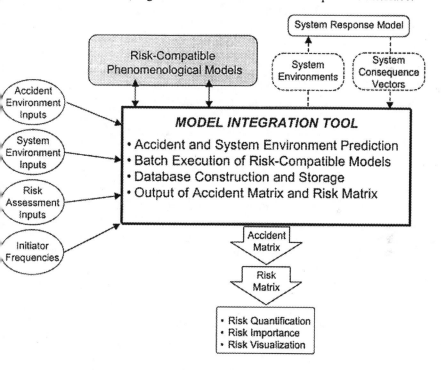

Figure 10.3 Integration tool structure for scenario development

Accident Matrix: Accident Environment Description

An accident environment is a set of parameters that completely defines the initial conditions of the accident. It contains all the necessary system-independent input data for modelling the accident progression. The accident environment parameters are developed using modelling input data and risk-assessment-specific input data. Figure 10.4 depicts the structure of the accident matrix with additional information provided below.

Modelling input data are used to represent the accident progression. These data are primarily developed through an examination of historical data, which are used to build distributions on the parameter variability as well as uncertainty. Also, modelling inputs can be provided through a combination of regression modelling and historical data. Regardless of the technique used, distributions on the variability and uncertainty for all applicable modelling inputs are developed.

Risk-assessment-specific data are used to organize the accident scenarios for risk quantification and presentation. Therefore, such data do not directly impact the accident progression, but they do have an impact on the display of the final risk results.

Accident Matrix: System Environment Description

System environment parameters are developed using system input data, phenomenological data, and consequence values. System input data indicate the state of the system just before an accident occurs.

For each of the accident scenarios used to characterize a complete accident initiator, more than one type of phenomenological data may apply (e.g., one collision and two fires). Additionally, multiple phenomenological data are often represented within a single system environment description (e.g., collision followed by fire).

The final parameters required to describe a system environment are the consequence values. A consequence value generally represents the conditional probability that an adverse system consequence will occur given the defined environment. Therefore, as defined in the accident matrix, the system consequence vector contains the probabilities for each outcome possibility, including the case of no impact to the system.

Risk Matrix

To perform risk quantification, a risk matrix must be developed for each accident initiator from the corresponding accident matrix. Two key differences exist between an accident matrix and a risk matrix: the presence in the risk matrix of the scenario frequency and scenario consequence. Figure 10.5 depicts the structure of a risk matrix.

Accident Scenario Generation

The complexity of the accident progression requires a tightly integrated framework to produce the many accident scenarios that compose accident and risk matrices. A risk integration tool is developed for each initiator to automate the construction of the matrices. An integral part of the computer tools is the use of Latin-Hypercube sampling to account for parametric variation, parameter dependency, and uncertainty replication.

Referring to Figure 10.3, inputs for the integration tools are accident environment data, system environment data, and risk assessment data, as previously described. Initiator frequencies, represented by probability distributions, are also inputs to the integration tools. In addition, the integration tools interface with the risk-compatible phenomenological models. With these data, the accident matrix for a given initiator is constructed by running its specific integration tool. The risk matrix is then generated by adding a system consequence vector and scenario frequency for each system environment. The scenario frequency is computed by dividing the initiator frequency by the number of accident scenarios developed for the initiator, given that each accident scenario is equally likely to occur.

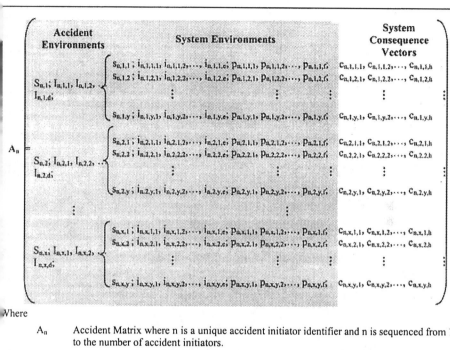

Where

A_n Accident Matrix where n is a unique accident initiator identifier and n is sequenced from 1 to the number of accident initiators.

$S_{n,x}$ Scenario indicator where n is a unique accident initiator identifier and x is the number of accident scenarios for a given accident initiator.

$I_{n,x,d}$ Accident environment modeling or risk assessment specific input where n is a unique accident initiator identifier, x is the number of accident scenarios for a given accident initiator and d is the number of required input parameters. The value of d varies across accident initiators.

$s_{n,x,y}$ System identifier where n is a unique accident initiator identifier, x is the number of accident scenarios for a given accident initiator and y is the number of systems involved in the specific accident scenario. The value of y varies across accident scenarios.

$i_{n,x,y,e}$ System environment input where n is a unique accident initiator identifier, x is the number of accident scenarios for a given accident initiator, y is the number of systems involved in the accident scenario and e is the number of input parameters. The value of e varies across accident initiators.

$p_{n,x,y,f}$ Phenomenon (mechanical, electrical, thermal, chemical, etc.) environment parameter where n is a unique accident initiator identifier, x is the number of accident scenarios for a given accident initiator, y is the number of systems involved in the accident scenario and f is the number of phenomenon environment parameters. The value of f varies across accident initiators.

$c_{n,x,y,h}$ Consequence value where n is a unique accident initiator identifier, x is the number of scenarios for a given accident initiator, y is the number of systems involved in the accident scenario and h is the number of consequence parameters. Therefore, a given system may potentially contribute to multiple outcomes.

Figure 10.4 Accident matrix structure

$$R_n = \begin{pmatrix} \textbf{Accident Environments} & \textbf{Scenario Consequence Vectors} & \textbf{Scenario Frequencies} \\ S_{n,1}; I_{n,1,1}, I_{n,1,2}, \cdots I_{n,1,d}; & C_{n,1,1}, C_{n,1,2}, \cdots, C_{n,1,h}; & F_{n,1} \\ S_{n,2}; I_{n,2,1}, I_{n,2,2}, \cdots I_{n,2,d}; & C_{n,2,1}, C_{n,2,2}, \cdots, C_{n,2,h}; & F_{n,2} \\ \vdots & \vdots & \\ S_{n,x}; I_{n,x,1}, I_{n,x,2}, \cdots I_{n,x,d}; & C_{n,x,1}, C_{n,x,2}, \cdots, C_{n,x,h}; & F_{n,x} \end{pmatrix}$$

Where

R_n — Risk Matrix n is a unique accident initiator identifier and n is sequenced from 1 to the number of accident initiators.

$S_{n,x}$ — Scenario indicator where n is a unique accident initiator identifier and x is the number of accident scenarios for a given accident initiator.

$I_{n,x,d}$ — Accident environment modeling or risk assessment specific input where n is a unique accident initiator identifier, x is the number of accident scenarios for a given accident initiator and d is the number of required input parameters. The value of d varies across accident initiators.

$C_{n,x,h}$ — Scenario consequence value where n is a unique accident initiator identifier, x is the number of scenarios for a given accident initiator and h is the number of consequence parameters. The equation for determining the values for independent systems is:

$$C_{.,.,.} = 1 - \prod_{i=1}^{r}(1 - c_{.,.,y,.})$$

where y is the number of systems involved in accident scenario n

$F_{n,x}$ — Scenario annual frequency where n is a unique accident initiator identifier, x is the number of accident scenarios for a given accident initiator.

Figure 10.5 Risk matrix structure

Risk Quantification

In addition to quantifying a numerical value for risk, which is the product of the scenario frequency and the scenario consequence, the contribution to risk from various sources is identified. This risk importance computation is essential to providing the decision maker with recommendations on how the quantified risk estimates can be reduced. This quantification is accomplished through the development of a ranked list of risk importance estimates. The parameters within the risk matrices are used directly and in combination to determine their importance to risk. As previously discussed, these parameters are grouped into accident environment parameters and system environment parameters. Contributors to risk from each grouping are identified, and their risk importance is computed.

Accident environment parameters that contribute to risk are identified using a risk quantification computer tool. The process to identify these risk contributors is outlined in Figure 10.6. This diagram shows, for the accident environment parameter of fire mitigation time, the calculation for computing the importance to risk. This calculation

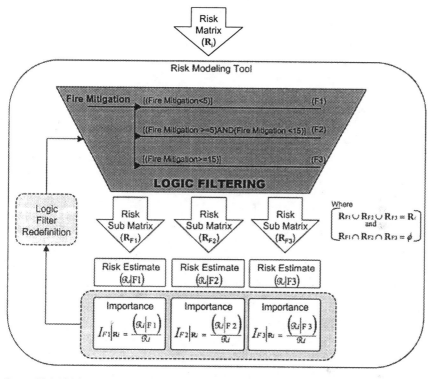

Figure 10.6 Risk importance calculation for accident environment contributors, fire mitigation example

egins by accessing the risk matrix for the accident initiator. The logic that breaks the
arameter space into mutually exclusive and collectively exhaustive bins is then defined.
hese bins act as filters in the process of importance evaluation. The filtering process
en divides the risk matrix into smaller sub-matrices, and risk estimates are produced
or these sub-matrices. Risk importance for each logic filter is then computed by
ividing through by the total initiator risk. The resulting set of importance values is
resented back to the user, at which time the user can determine if the logic filters (i.e.,
ins) identify a prominent parameter range that is key to influencing the risk for that
arameter. If the bin definitions do not produce insights into the key risk influences, the
ser can redefine the logic filters and compute new importance measures. Through this
erative process, risk trends within the parameters can be identified.

System environment parameters that contribute to risk are also identified using the
sk quantification computer tool. This process for identification of system environment
sk contributors as outlined in Figure 10.7 is slightly different from than the process for
ccident environment risk contributors shown in Figure 10.6. Because the risk matrix
oes not contain system-specific data, the accident matrix rather than the risk matrix is
sed as input to the risk quantification computer tool for system environment

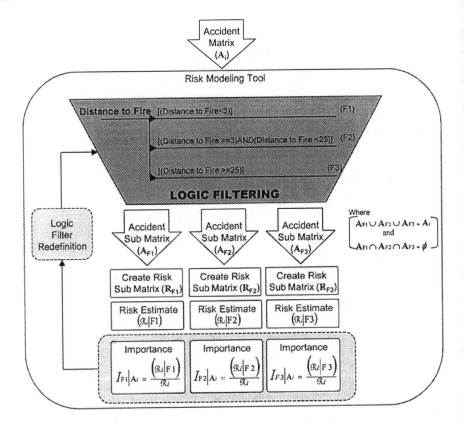

Figure 10.7 Risk importance calculation for system environment contributors, distance to fire example

parameters. As such, risk sub-matrices are developed from the corresponding accident sub-matrices with the addition of the scenario frequencies. This process begins by accessing the accident matrix for the accident initiator. The user defines the logic that breaks the parameter space into mutually exclusive and collectively exhaustive bins. The filtering process then divides the accident matrix into smaller sub-matrices. Risk sub-matrices are then developed from each corresponding accident sub-matrix, and risk estimates are produced for these sub-matrices. The remaining steps are the same as those presented for generating the importance measures for accident environment parameters.

The risk quantification computer tool allows not only for risk computations and importance evaluation for each initiator but also across multiple accident initiators. For any parameter across a single initiator or across multiple initiators, the importance computation is the same.

Risk Visualization
Up to this point, the discussion has focused on the mechanics of analysing and quantifying the risks. However, an equally important consideration is how these risk

estimates are presented to the decision maker. This process is referred to as risk visualization.

The goal of risk visualization is not to present an exhaustive accounting of all risk estimates that could possibly be derived from the data. Indeed, this approach would produce voluminous output with limited utility. Risk visualization presents only those risk results that can be used directly in the decision-making process. Generally, this includes providing both the current level of operational risk as well as indicators regarding which controllable factors impact this risk. Both graphical and tabular representations of risk data are valuable to a decision maker.

A commonly used tool is an event tree, which is defined as an inductive logic model that graphically shows the progression of an accident. The detailed structure of the risk assessment data for the type of analysis described thus far does not lend itself to standard event tree quantification. However, there are benefits in using the graphical aspect of an event tree in the visualization and communication of risk. In combination with a tabular presentation of the importance measures, an understanding of the risk is presented to the decision maker.

10.2.8. Uncertainty Analysis

The uncertainty analysis is based on a practical approach that considers uncertainties in all areas of the risk assessment at a level that is consistent with their overall importance. It uses information generated along the way to increase the level of detail in some areas and decrease it in others.

The process is outlined in Figure 10.8. It begins with a base case calculation of risk. It continues with two bounding calculations of risk (upper bound and lower bound), which are obtained by propagating bounding assumptions for a number of uncertainty issues through the analysis. The combination of base case and bounding risk analyses is used to identify regions of the parameter space where uncertainties have little effect on the results. For those regions, the base case risk results are considered to be satisfactory representations of the risk. In other regions, a more thorough analysis is performed to account explicitly for the effects of uncertainties on the risk. The end result is a 90% confidence interval (sometimes referred to as a "degree-of-belief" interval) for each important measure of risk. A brief description of each box in Figure 10.8 is presented below.

Calculate base case risk estimate. It is assumed that most of the base case risk analysis has been completed before beginning the uncertainty analysis. It is not assumed that the base case estimate is equivalent to a best estimate, although every attempt is made to use best estimate assumptions.

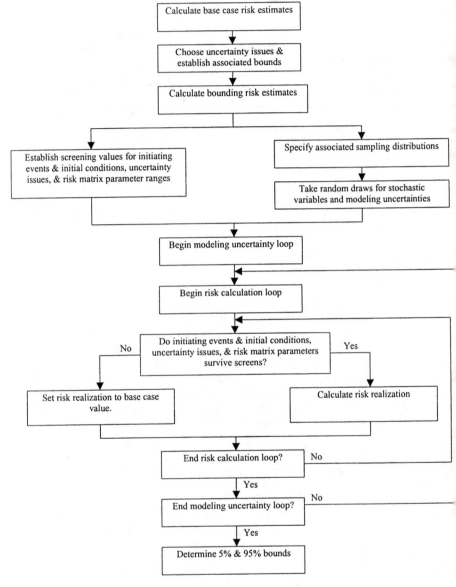

Figure 10.8 Uncertainty analysis process

- **Choose uncertainty issues and establish associated bounds**. A set of uncertainty issues is selected based on expert opinion of the parameters and assumptions in the analysis of the base case estimate that contains significant uncertainties. The uncertainty issues are defined at broad levels, such as the uncertainty in which of several competing models describes a particular set of phenomena or a particular set of statistical observations. A number of issues are selected from each phase of the analysis: initiator frequency modelling, event input modelling, environment modelling, etc. Using expert judgment, a range of plausible realizations is defined for each issue by specifying two bounding representations of the issue.

- **Calculate bounding risk estimates.** An upper bound point estimate of the risk is generated by propagating the upper bounds of the uncertainty issues through the risk calculation. Similarly, a lower bound estimate of the risk is generated by propagating the lower bounds of the uncertainty issues. The steps involved in determining the upper and lower bound risk estimates are the same as those for the base case estimate.
- **Establish screening values for initiating events and initial conditions, uncertainty issues, and scenario parameter ranges.** By examining the results obtained from the base case risk estimate and the two bounding estimates, it is possible to identify the events and conditions in the risk matrix that do not have any significant effect on the risk or on the uncertainty in risk. Hence, it is possible to identify combinations of initiating events and System States, specific uncertainty issues, and ranges of scenario parameters that should not require detailed evaluation.
- **Specify sampling distributions.** For each of the uncertainty issues, a sampling distribution is specified. The distribution is built around the base case and bounding representations of the uncertainty issue in question. The form to be selected for each distribution depends upon the strength of belief that the experts have regarding where, within the range of plausibility, the truth is likely to lie.
- **Take random draws for stochastic variables and modelling uncertainties.** Two separate matrices of randomly selected probability levels are developed. The first matrix contains levels for the parameters with stochastic variability. The second contains levels for the issues with modelling uncertainty.

Begin modelling uncertainty loop, and begin risk calculation loop. The outer loop of the risk analysis corresponds to the variation of uncertainty levels. The inner loop represents the variation of scenarios.

Determine whether initiating events and initial conditions, uncertainty issues, and scenario parameters survive screens. In a particular pass-through, the initiating event, initial conditions, uncertainty levels, and scenario parameters have specific values that are determined by the corresponding probability levels in the matrices of draws for stochastic variables and modelling uncertainties. The determination evaluates whether these values place the scenario within the more-important or less-important category for purposes of further evaluation.

Set risk realization to base case value. If the scenario at hand is in the less-important category, there is no further development of the scenario. The risk associated with the scenario is set equal to the value obtained in the base case risk calculation, as though there were no uncertainties for this scenario.

Calculate risk realization. If the scenario is in the more-important category, then the scenario is developed further by performing calculations to determine the phenomenological environments and the system response. The risk associated with the scenario is derived from these calculations.

End risk calculation loop, end modelling uncertainty loop, and determine 5% and 95% bounds. The risk calculation loop completes the evaluation of a set of risk measures for a particular level of uncertainty. The modelling uncertainty loop calculates a separate set of risk results for each uncertainty level. The results from the outer loop are used to obtain the 90% confidence interval for each measure of risk. The 90% confidence interval implies that the experts who participated in the study have a 90% degree of belief that the risk lies between the lower and upper bounds, a 5% degree of belief that it lies below the 5% bound, and a 5% degree of belief that it lies above the 95% bound.

When the uncertainty analysis has been completed, the final step (not shown on the chart) is to use the results to determine whether there are ways to reduce the risk that would be effective in light of the uncertainties.

10.2.9. Major Risk Contributors and Recommendations for Risk Reduction and Risk Mitigation

Once the risks are determined, the accident scenarios are ranked based on their contribution to the risk. This ordering provides a process by which the major contributors to the risk are identified both in terms of their initiating event as well as the sequence of events that make up the accident scenario.

With this information, the major contributors to the risk are identified as well as potential steps that may be taken to reduce or mitigate the risk. The possible options for risk reduction and/or mitigation are as follows:

- Eliminate the accident initiator (i.e., prevent the accident from occurring)
- Reduce the frequency of the accident initiator (i.e., make the accident less likely to happen)
- Reduce the likelihood of the negative events involved in the accident progression (i.e., make the major consequences less likely to occur)
- Reduce the severity of the environments given an accident occurs (i.e., eliminate an accident from resulting in unacceptable consequences or mitigate the potential consequences of an accident).

Processes are recommended whereby the major contributors to the risk may be reduced or mitigated by considering cost-effective, practical procedural or system changes. Suggested measures to reduce the potential of an event or minimize the effect of an event on a system are considered in three categories: operation-specific, environment-specific, and initiator-specific.

10.3. Conclusion

QRA methods help to ensure a complete analysis. The process discussed in this chapter focuses first on identifying all possible hazards, then describing potential hazard scenarios, followed by developing a preliminary risk model. The process continues in stages by applying screening values and performing bounding calculations to eliminate some initiating events from further analysis. Finally, for the remaining initiating events, it is necessary to conduct phenomenological modelling, the results of which then serve as input to a probabilistic model. The methods also allow the analyst to account for data voids and modelling assumptions by performing an uncertainty analysis.

As previously discussed, this analytical tool can be applied proactively or in response to events that have occurred. In either case, the resulting prioritised list of significant risk contributors can be used to propose measures for reducing or eliminating dominant risks. Thus, decision makers can focus limited funds on those areas that will most help to prevent or respond to a system failure. The results of a QRA can serve as input to a consequence analysis, which in turn can assess human health and environmental effects.

PART III. ANALYSES AND PROGRAMS APPLICABLE TO LEGACIES

We dance round in a ring and suppose
And the Secret sits in the middle and knows.
Robert Frost

"A skilful commander?" replied Pierre. "Why, one who foresees all
contingencies . . and foresees the adversary's intentions."
"But that's impossible," said Prince Andrew as if it were a matter settled
long ago.
Pierre looked at him in surprise. "And yet they say that war is like a game
of chess?"
"Yes," replied Prince Andrew. "But with this little difference, that in chess
you may think over each move as long as you please and are not limited for
time and with this difference too, that a knight is always stronger than a
pawn. . . . Success never depends, and will never depend, on position, or
equipment, or even on numbers, and least of all on position . . . [but] on the
feeling that is in me and in . . . each soldier. . . those . . . gentlemen won't
win the battle tomorrow but will only make all the mess they can, because
they have nothing in their. . . heads but theories not worth an empty
eggshell and haven't in their hearts the one thing needed tomorrow [which
is the feeling in each soldier]."
Leo Tolstoy, *War and Peace*

This literature from the East and West warns of the danger of being overly
confident. Details of the problem to be solved, not just the risk to be analysed, must
guide model and methodology selection. And the problem always includes issues of risk
perception and societal values, factors often ignored in technical analyses.

11. Environmental Risk Assessment of Installations and Sites Inherited from the Cold War Period in Bulgaria

Managing legacy waste risks often starts with effective use of resources. This chapter explains how one Eastern European country reorganized its resources to address issues of risk associated with new and existing facilities and contaminated sites. For example, they have moved chemical hazard expertise from military organizations to civilian authority. The chapter describes how they are analysing various types of risk and how they plan risk management efforts. The chapter further looks at how more democratic institutions are empowering the public to challenge government decisions and how information can be provided to the public in a way that allows them to join in its evaluation. A new national system of emergency response is also described.

Europe and the world have witnessed significant change over recent years. The world has entered a new era of international relations. The fall of the Berlin wall and the collapse of the Warsaw Pact started a period of warming of the relations between countries of the North Atlantic Treaty Organization (NATO) and central and eastern Europe. Within the framework of the United Nations (UN), a number of conventions were signed in the hopes of reducing environmental pollution:

.   Convention prohibiting the development and use of chemical weapons
.   Convention protecting waters and environment in a transboundary context
.   Convention confirming the transboundary impact of industrial accidents
.   Basel Convention for controlling transboundary movement of hazardous waste
.   European Agreement delineating transport of hazardous loads
.   Convention limiting long-distance transboundary air pollution
.   Conventions describing peaceful uses of nuclear energy.

Through agreements such as these, humankind is trying to reduce deterioration of the environment and secure a clean and safe planet for future generations. The issue of assessing environmental risks of installations and sites inherited from the Cold War period falls within the scope of these objectives.

This chapter focuses on the types of facilities left in Bulgaria as a result of Cold War activities, the environmental issues associated with the facilities, and the approaches that have been used to assess risks for these facilities. The chapter describes the differences between the group approach and the differentiated approach, as well as details some of the information necessary for conducting a risk assessment under either approach.

11.1. Types of Installations and Sites

Numerous sites have been left in Bulgaria as a result of Cold War activities. These sites are both industrial and military in nature. Some are still active. The sites can be divided into several main groups to better differentiate between the environmental problems and the respective approach to them:

1. **Privatised military complex sites** serve as sources of environmental pollution. Difficulties have arisen in liquidating old pollution from the reduction of their output or movement from military to civilian products.
2. **Military sites still existing and closed down** include those with environmental problems related to the existence of the armed forces or the operation of certain installations.
3. **Industrial sites outside the military complex** have become a source of pollution with lasting consequences because they served as a source of supply to the armed forces.
4. **Active military sites** can also have pollutants. These sites include those that house active units of the armed forces, sites for processing and storage of military stock, naval bases, polygons, airfields, missile sites, and others.

It is important to note that, as a result of certain circumstances, no troops of other Warsaw Pact countries have been deployed in Bulgaria. Because military activities were on a smaller scale, Bulgaria's position is more favourable than the position of other countries in central and eastern Europe.

11.2. Environmental Issues and Organisations to Deal With Them

If environmental problems in Bulgaria are viewed from their date of origin, it becomes evident that they come mainly from the Cold War period. The Republic of Bulgaria started industrialization after 1945, and this is precisely the time when the environmental problems started appearing. Unfortunately, the leadership of the country during that time allowed Bulgarian industry to develop imperfect energy practices and waste-intensive technology. Over a period of 45 years, with no environmental legislation in line with international standards, serious environmental problems have accumulated. Research in this area shows that 3% to 4% of the enterprises account for 96% of the hazardous waste in Bulgaria. The environmental programmes that were designed then resolved only some of the issues, then only in part, and the funds allocated for introduction of environmentally sound technology were extremely scarce. In addition, the coordination between the local structures and the industrial units was very poor in terms of environmental and health risk reduction.

Today in Bulgaria, environmental issues with the potential for catastrophic consequences or resulting from accidents or disasters are the province of Civil Protection. This organisation carries out a system of humanitarian activities of a social, economic, and techno-scientific character aimed at protecting the population. Activities include disallowing dangerous enterprises, mitigating harmful consequences of other enterprises, conducting relief and humanitarian operations, providing the necessary conditions for survival following a devastating emergency, and assisting when emergencies arise.

Over recent years, Civil Protection has proactively participated in a significant number of risk reduction initiatives for various industries in Bulgaria in the aspects of technology, environment, and health. These initiatives were jointly organized with the Ministry of Environment, Ministry of Health, Ministry of Agriculture and Forestry, Ministry of Labour and Social Policies, Ministry of Industry, and other organisations.

Many programmes were financed and carried out for surveying pollution and, on that basis, a number of environmental issues were resolved. For example, Bulgaria had a significant problem in the disposal of pesticides that are banned or beyond their safe use date. Bulgarian currently holds more than 300 tons of such substances. Currently staff from Civil Protection serve on local projects to collect and safely store pesticides until conditions are created for their disposal. Bulgaria also launched a joint project with the Netherlands to dispose of the first 50 tons of pesticides of the DDT and Lindan type. More broadly speaking, the actions related to reduction of risk to the population and the environment fall under the scope of two national documents of paramount importance:

.   National Programme for Prevention and Mitigation of consequences of natural disasters and industrial accidents
.   National Programme Environment and Health.

These problems were all treated in the Safe and Healthy Working Conditions Act.

1.3.   Approaches to Risk Assessment

Many of the environmental issues required the use of risk assessment to mitigate or eliminate. After Bulgaria's 1987 accession to the Partnership for Peace Programme, and subsequently after the country's application for NATO membership, the Bulgarian Armed Forces adopted a new approach toward environmental problems:

.   Establish environmental bodies within the Ministry of Defence, Civil Protection, and the structures of the Bulgarian Armed Forces
.   Introduce ecological training into military higher education curricula
.   Reorient some experts (chemical) to environmental issues
.   Provide conditions for access to military sites for representatives of the Committee on the Use of Atomic Energy for Peaceful Purposes and of the Ministry of Environment and Waters, as well as abandon maintaining an anonymity of their environmental problems

National System of Emergency Response in the Republic of Bulgaria

Civil Protection, a specialized structure for emergency response, has been established in the Republic of Bulgaria. It carries out a system of humanitarian activities of a social, economic, and techno-scientific character aimed at protecting the population, by refusing to permit dangerous enterprises and mitigating harmful consequences of other enterprises, conducting relief and humanitarian operations, providing the necessary conditions of survival, and assisting when emergencies arise.

The following list defines the specific tasks of Civil Protection:

1. The Council of Ministers, led directly by the Defence Minister and permanently by the Head of Civil Protection, carries out the general guidance of the Civil Protection Service.
2. National authorities, local governing authorities, and local administrations carry out activities on population and property protection in emergencies.
3. The Standing Committee for protection of the population in disasters, accidents, and catastrophes, with the Council of Ministers, carries out and coordinates the relief and emergency activities between the ministries, departments, and regional governing authorities as well as the preventive work for not permitting activities that would lead to negative consequences in emergencies.
4. The activity of the Standing Committee is assisted by a headquarters, the members of which are representatives of Civil Protection, ministries, departments, and nongovernmental organizations.
5. Standing departmental committees for response in emergencies have also been established at the ministries and departments. The respective chiefs organize and are responsible for the implementation of protection activities.
6. Standing committees for response in emergencies are established in regions and municipalities. Chairpersons of the committees are governors of regions and mayors.
7. Trading companies, enterprises, and firms also establish departmental standing committees for response in emergencies.

Management authorities and groups of Civil Protection, the forces and facilities of the ministries and departments, and volunteer groups carry out direct tasks to protect the population in emergencies.

As regards organisation, a National Crisis Management Centre has been established with Civil Protection. This centre is connected with the ministries, departments, territorial administrations, and groups of Civil Protection. Through it, coordination and interaction is effected with the local authorities and the management bodies and forces when an emergency arises. The centre incorporates an informational and analytical centre that collects, processes, analyses, and classifies the complete information on occurrence of an emergency situation and informs the national management authorities.

The system of Civil Protection maintains groups of regularly appointed professional search parties, available around the clock, with areas of operation proportionally positioned on the territory of the country. The Republic of Bulgaria has also developed a national plan to protect the population during disasters, accidents, and catastrophes; and to establish an organisation for timely prediction, management, and implementation of relief and emergency operations in extreme situations. The plan also defines the ensuing obligations and tasks for the preparedness and participation of the management authorities and the resources, the rules of provision, and its implementation. Adequate plans for protection of the population have been worked out in the ministries, departments, regions, municipalities, and projects of the national economy.

5. Allow participation of environmentalists from the Ministry of Defence interdepartmental commissions established for resolving certain environmental problems

Relief and emergency operations carried out by management authorities and forces of Civil Protection, ministries, departments, and local administrative structures are provided for by funds allotted for that activity pursuant to the State budget law. The activity to protect the population is largely financed in a centralized way by the state budget and a small part in a decentralized way by the municipalities. A standing committee with the Council of Ministers allocates and exerts control over the expenditure of the funds.

The national authorities responsible for protection of the population monitor the observance of the regulatory acts in the following basic directions: planning, maintaining, and implementing the plans for civil protection; organizing the activity of the management authorities and resources; organizing the protection activities and carrying out prevention activity; keeping the population informed; and providing logistical support of the management authorities and forces and interaction between the authorities, forces, and institutions.

Coordination between single institutions and Civil Protection is based on bilateral and multilateral contractual agreements for joint work. Many agreements are made at the regional level, where managers cooperatively solve emergency problems.

The practice of carrying out annual joint exercises and training enables the education and training of the management authorities of institutions and local authorities and forces that participate in the elimination or mitigation of accidents and emergencies.

The regulatory documents of participation and response in overcoming and eliminating the aftermaths of emergencies provide for the attraction of mass volunteer groups from nongovernmental organizations. These groups are used to protect the population and the environment, and for assistance in relief operations. Civil Protection in its activities cooperates with nongovernmental organizations such as the Bulgarian Red Cross, students' rescue parties, the ecological organization "Blue Flag," the children's Scout movement, and others. Civil Protection bears responsibility for elaborating and maintaining the National Plan and the database therein. All specialists who work on various protection problems have access to the plan in the part that concerns them. The database is computerized and stored at Civil Protection. It is updated at regular intervals when there are changes. The data are introduced into the plan with the participation of specialists from all concerned ministries and departments. The resources for response in emergencies are also computerized in certain ministries and departments.

Develop and implement programmes for reducing risks to the population and environment posed by liquidation of Armed Forces-related old pollution.

For this last item, Bulgaria then adopted the following approaches for environmental risk assessment for installations and sites from the Cold War period:

Group approach for certain types of sites and problems

Differentiated approach toward certain sites, production lines, installations, etc.

11.3.1. The Group Approach

The group approach makes it possible to assess the problem as a whole. This holistic approach also allows risk assessors to prioritise matters within the main problem, to determine the risk components and to decide on their order of priority, and to develop a list for resolving the problems. This approach is used when the problems are nationwide in scope and relate to entire sectors and branches of the national economy, for example problems in the following areas:

- Pesticides
- Ferrous and non-ferrous metallurgy
- Related enterprises of the chemical industry
- Waste water treatment
- Coal-based power generation
- Closure of sites with ionising irradiation sources
- Military polygons (within the Ministry of Defence)
- Military areas and bases of dissolved Armed Forces units.

11.3.2. The Differentiated Approach

The differentiated approach can be an element and a result of the group approach but in most cases it is completely independent, in view of the peculiarity of each site and its environmental problems. Examples of situations in which this approach was used include the Kozlodui Nuclear Power Plant, the National Radioactive Waste Depot, and the arsenic pollution in the area of Pirdop from the copper plant there.

11.4. Commonalities of the Two Approaches

Both the group and the differentiated approach require complex assessment, including adherence to the procedures for Environmental Impact Assessment, as per Ordinance No. 4 of the Ministry of Environment and Waters. This ordinance broadly considers the risk to the environment and the population, in all its aspects. This approach was accepted in the European Union member states and is regulated by a law in Bulgaria. Other commonalities include the fact that the two approaches are part of a decision-making process, and both approaches utilize a set of topics for evaluation and criteria by which they are evaluated.

11.4.1. Risk Assessment as Part of a Decision-Making Process

Of particular importance to the Bulgarian approach to risk assessment is the fact that Environmental Impact Assessments are part of a decision-making process. For new facilities, the Environmental Impact Assessments are widely discussed in the community, the benefits and the possible damage from a certain type of activity are carefully weighed, and, finally, the fate of a certain industrial site is decided upon. The process is a bit more specific for the Environmental Impact Assessment of an already existing site. Then the involvement of the environmental bodies and the community goes nore toward reducing the technological, environmental, and health risk. Demands are ften raised for introduction of state-of-the-art waste-free technology, for guarantees for afe operations by means of rehabilitation and modernisation programmes, and for losure of particularly hazardous and high-risk lines of production. In addition, when oncluding privatisation contracts, one of the tools for reducing risk to the environment nd people is to introduce clauses for compulsory investment on the part of the new wners to liquidate old pollution (if any) and to implement environmental programmes the future.

1.4.2. Topics for Evaluation

nvironmental Impact Assessments focus on identifying areas in which a particular nterprise can reduce risk to people and the environment. In this context, people can iclude workers as well as the public outside the boundaries of the facility. Risk can ccur through the system operating as usual or abnormally.

Assessment of the abilities of an enterprise to reduce risk can be carried out on a umber of topics, for example, the following:

. Capability to harness risks
. Preventive policies
. Movement
. Protection of the machinery
. Noise and vibrations
. Temperature and ambient air treatment
. Illumination
. Fire, explosion, and electricity-related risks
  Hazardous substances – health risk
). Collective and personal protection means
1. Heavy loads transportation
2. Maintenance
3. First aid
4. Interaction of the employees.

1.4.3. Criteria for Evaluation

ich item to be evaluated carries with it a set of criteria, which makes it possible for the iterprise to be given an evaluation score on each risk individually. For example, criteria r the first item in the list in Section 11.4.2 , capability to harness risks, could be:

Quality of information available to management with respect to risks related to the operations of the enterprise

General attitude toward detected risks (i.e., have efficient preventive measures been put in place)

Quality of the documentation regarding protective measures against various risks

Attitude toward existing or potential risk in the enterprise

Level of organisation of operations and selected production procedures

Safety of raw material used and effectiveness of training of personnel regarding that material

- Priority of collective or personal protection means and possibility to reduce risk before their introduction
- Availability of safety techniques briefing
- Possibilities to monitor all measures on the above matters
- Availability of consultations for the personnel or their representatives on all safety- and health-related issues.

The second item on the list, prevention policies, can be assessed according to the following criteria:

- Quality of division of responsibilities in the enterprise (competence, interaction)
- Level of compliance to the procedures and safety regulations in the enterprise
- Effectiveness of introduced changes following an occupational accident
- Effectiveness of disseminating initial information and providing update on accidents or calamities at the work place
- Effectiveness of safety control in the enterprise
- Effectiveness of operations procedure in the enterprise.

Similarly, based on a set of criteria, the other assessment items can be checked for the enterprise's capability to reduce the likelihood of serious consequences.

11.5. Conclusion

This chapter demonstrates that the Republic of Bulgaria has established a reliable system of risk assessment for all types of facilities remaining from Cold War activities. The system is flexible, in that topics and criteria can be tailored to meet the needs of a wide variety of situations, in both group and differentiated approaches. In addition, the system meets the international standards set for the European Union. This system is seen as a major step in fostering environmental decision making in Bulgaria and ensuring a prosperous economic future for the country.

12. Radiation Factors Risk Assessment Within the Chornobyl Nuclear Power Plant Exclusion Zone

One of the most challenging risk management cases recently is the management of risks associated with the 1986 accident at the Chornobyl Nuclear Power Plant. This chapter describes the application of risk modelling techniques to provide decision makers with information on managing radiation risks within the Exclusion Zone. The chapter also provides insight into the development of the nuclear power industry in the former Soviet Union, an industry closely related to the development of nuclear weapons production in terms of technology and lasting legacies.

At approximately 50 years old, the history of world nuclear energy development is very short when compared with other sources of energy. Nuclear energy, however, went through the stigma of association with the nuclear bomb as well as depression periods. Now, with the coming of the third millennium, this form of energy still has unsolved problems. Nations continue to seek the answer to the unbelievably complicated question of whether nuclear energy should be allowed to exist at all. This chapter provides a brief history of world nuclear power development and how it was affected by the Chornobyl nuclear catastrophe, information about the contamination spread by the catastrophe and the pathways along which it spread, and specific pathway modelling results of strontium-90. The discussion and calculations show how radioecological factors must be considered in effectively predicting and monitoring risks from Cold War legacies.

12.1. Background

The former Soviet Union was the first to use nuclear energy for peaceful purposes. Starting on June 27, 1954, when the first nuclear power plant (NPP) was opened at Obninsk, nuclear energy continued to expand. Reactor power increased to 5 MW, a variety of types was developed, and capacity increased.

A significant stage in nuclear energy development is connected with the Cold War. During this period, the former Soviet Union operated a series of nuclear reactors not only for peaceful purposes but for the production of weapons-grade plutonium as well.

Thus, in the former Soviet Union and other countries, the questions of nuclear power operation and development were the province of military authority and classified as secret until the end of the 1980s.

The catastrophe at the Chornobyl NPP in 1986 effectively divided development of nuclear power into two great stages: before and after the catastrophe. Before the accident, nuclear energy made a considerable contribution to electric and thermal energy production. Figure 12.1 shows the potential of nuclear facilities in 1986. Around the world, 26 countries contained 26 nuclear units for a total of 259 GW capacity; in addition, 157 units with a total of 142 GW of capacity were under construction.

The most intensive nuclear energy development took place from 1970 to 1980; during this time NPP capacity grew about 25% per year. However, after a series of accidents at NPP around the world, most visibly at Chornobyl, the rate of growth in NPP capacity considerably decreased to an average of 6% between 1980 and 1990.

After the Chornobyl accident, many countries reviewed the safety of their nuclear energy development. As a result, a number of NPP were closed, and less new plants were built. Thus, NPP capacity grew just less than 1% between 1990 and 1999.

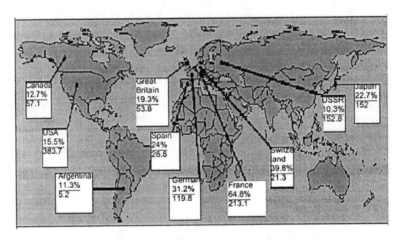

Figure 12.1 Countries with wide use of nuclear power, 1986
(% indicates portion of electricity production, number under the line indicates total terrawatt-hours)

According to data from the International Atomic Energy Agency (IAEA), world nuclear energy produces more than 17% of the total amount of energy consumed. Around the world, 440 nuclear energy units with a total capacity of 351.5 million kW are operated. In addition, 43 units (for a total capacity of 35.3 million kW) are under construction, and 51 units with a total capacity of more than 39 million kW are planned to be built. Table 12.1 presents information on the current status of nuclear energy use in different countries (according to IAEA data).

TABLE 12.1 World use of nuclear energy, 1997

Country	# Operating Nuclear Reactors	# Reactors Under Construction	1997 Electricity Production, terrawatt/year	Part in National Electricity Production, %
Argentina	2	1	7.45	11.40
Armenia	1	-	1.43	25.67
Belgium	7	-	45.10	60.05
Brazil	1	1	3.16	1.09
Bulgaria	6	-	16.44	45.38
Canada	16	-	77.86	14.16
China[a]	3	4	11.35	0.70
Czech Republic	4	2	12.49	19.34
Finland	4	-	20.00	30.40
France	59	1	376.00	78.17
Germany	20	-	161.40	31.76
Great Britain	35	-	89.30	27.45
Hungary	4	-	13.97	39.88
India	10	4	8.72	2.32
Iran	-	2	-	-
Japan	54	1	318.10	35.22
Kazakhstan	1	-	0.731	0.583
South Korea	12	6	930	4.08
Lithuania	2	-	10.85	81.47
Mexico	2	-	10.46	6.48
Netherlands	1	-	2.30	2.77
Pakistan	1	1	0.37	0.65
Rumania	1	1	5.40	9.67
Russia	29	4	99.68	13.63
Slovakia	4	4	10.80	43.99
Slovenia	1	-	4.79	39.91
South Africa	2	-	12.63	6.51
Spain	9	-	53.10	29.34
Sweden	12	-	67.00	46.24
Switzerland	5	-	23.97	40.57
Ukraine	16	4	74.61	46.84
USA	107	-	629.42	20.14
Total	437	36	2276.49	

Note: a = Taiwan is included in total amount: 6 nuclear plants, providing 26.35% of energy production.

12.2. The Chornobyl Catastrophe

On May 6, 1986, an accident occurred at the Chornobyl NPP, in what was then part of the Soviet Union (present day Ukraine). According to the last official data from Ukrainian scientists[1], the sum activity of radioactive material released during the accident consisted of 1.2×10^{19} Bq, including about 7×10^{18} Bq of inert gas. Releases included more than 3% of the fuel concentrated inside the reactor at the time of accident, up to 100% of the inert gas, and 20% to 60% of volatile radionuclides. This retrospective assessment of the activity at the moment of the accident was performed by recalculating the state of the plant before stopping the release process from the emergency reactor. The amount calculated exceeds the activity assessment officially presented by authorities of the former Soviet Union to the IAEA in 1986[1].

Meteorology conditions at the time of the accident led to a wide spread of radionuclides across the North Hemisphere. The most contaminated areas were in Belarus, Russia, and the Ukraine (46,500 km^2, 57,000 km^2, and 41,800 km^2, respectively), with concentrations of greater than 1 Ci/km^2 (37 kBq/m^2) of cesium-137. Areas of contamination were also found in Sweden, Finland, Germany, Australia, Switzerland, Rumania, Georgia, and other countries[2]. Figure 12.2 depicts data showing European radioactive contamination after the Chornobyl accident.

The most dangerous area with regards to the environment and human health is considered to be the Chornobyl Exclusion Zone (ChEZ) within the administrative part of the Ukraine. The ChEZ is shown in Figures 12.3 and 12.4. In 1986, all economic activity was stopped there, and the population was evacuated. The Ukrainian part of the ChEZ is made up of 2,044 km^2[1]. Currently, the most dangerous radionuclides that determine risk within the ChEZ are cesium-137, strontium-90, and alpha transuranium elements plutonium-238, 239, 240, and americium-241. Their levels within the ChEZ consist of 20×10^{15} Bq.

The following densities of radioactive contamination for cesium, strontium, and plutonium were accepted as criteria for the population evacuation from the contaminated area ($\approx$1,800 km^2)[3]:

- cesium-137 more than 555 kBq/m^2 (15 Ci/km^2)
- strontium-90 more than 111 kBq/m^2 (3 Ci/km^2)
- plutonium more than 3.7 kBq/m^2 (0.1 Ci/km^2).

Specialists from the Ministry of Emergencies summarized data on radionuclides distribution in different ChEZ objects[4], as shown in Table 12.2. Most radionuclides are concentrated in the "Shelter" object, which contains 180 tons of nuclear fuel with a total activity of 740 PBq (20 MCi). About 8.2 PBq (0.22 MCi) were released out of the destroyed unit.

Within the ChEZ, about 800 radioactive waste disposal sites and temporary radioactive waste storage sites contain about a total activity of 8.1 PBq (0.21 MCi).

Table 12.3 shows approximate amounts of radioactive waste that must be moved to special storage.

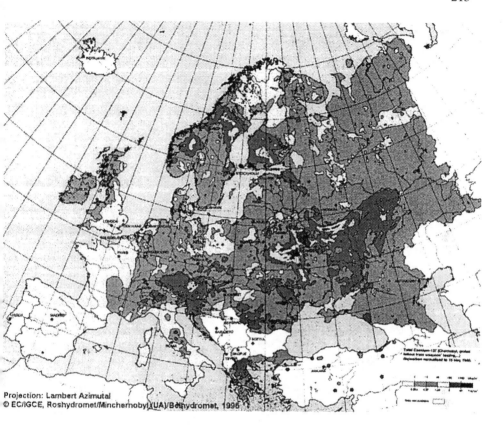

Figure 12.2 Radioactive contamination of Europe after the Chornobyl catastrophe

12.3. Transport of Contaminants Through the Environment

Radionuclides contained in the Chornobyl release penetrated into the environment and are now represented in all nature sub-systems. Figure 12.5 schematically depicts their migration in the environment and intake into the human body.

The pathways shown in Figure 12.5 are considered to be the risk factors and should be taken into account when developing risk assessment scenarios. These risk factors can be studied using radioecological monitoring data, which are often used to model radionuclides migration and to predict risk to the population and environment.

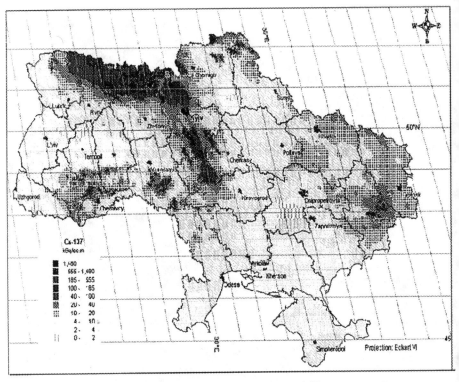

Figure 12.3 Contamination of Ukraine with [137]Cs

A multipurpose system for calculating these factors is the Multimedia Environmental Pollutant Assessment System (MEPAS), developed by the Pacific Northwest National Laboratory, operated by Battelle for the U.S. Department of Energy. MEPAS models radioactive and chemical contamination processes in different natural environments and, using different scenarios of radionuclide uptake, predicts risk to the health of the population. The calculations described below were, for the most part, performed using MEPAS[5,6].

The question on which radionuclide transfer pathways may result in contamination moving out of the ChEZ is of high importance. Figure 12.6 and Table 12.4 show potential pathways.

The data in Figure 12.6 and Table 12.4 show that the main potential source of radiation risk within the ChEZ is through surface waters. Radionuclides migrate from the ChEZ as a result of their wash-out from water catchments. Through the Prypyat and Uzh river basins, radionuclides migrate with surface and underground drainage. Indeed, water-borne contamination is the most dangerous radiation factor for the whole River Dnieper basin system, to which the River Prypyat estuary drains. Water from these sources is used as potable water for 35 million people in the Ukraine.

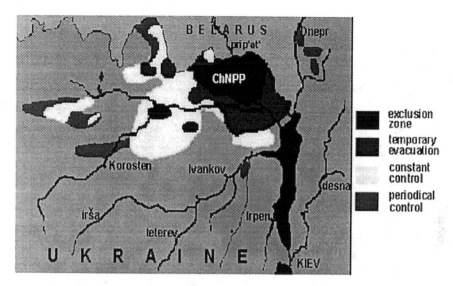

Figure 12.4 Zoning of the territory of Ukraine according to radioactive contamination rate

TABLE 12.2 Radionuclide distribution in Chornobyl Exclusion Zone objects

Object	Activity, Bq 10^{15} (PBq)			
	Σ	Cesium-137	Strontium-90	Trans-U elements
Exclusion Zone land	8.13	5.5	2.5	0.13
Cooling pond	0.27	0.16	0.1	0.005
Radioactive waste disposal sites	6.35	3.4	2.8	0.15
Radioactive waste temporary storage locations	1.84	1	0.7	0.04
Total	16.6	10.2	6.1	0.33
"Shelter" Object	740	480	260	10

Note: "Shelter" is the name of the fourth destroyed unit of the Chornobyl NPP, so called after activities on the sarcophagus building[7].

TABLE 12.3 Radioactive waste volumes in Chornobyl Exclusion Zone objects

Source	Radioactive waste volumes, km³	
	Minimum	Maximum
"Shelter"	120	400
"Shelter" territory	11	280
Radioactive waste disposal sites	21	37
Radioactive waste temporary localization sites	3	15
NPP operated and out of operation	3	
Treatment of used nuclear fuel of VVEP reactors	?	
Total	160	730

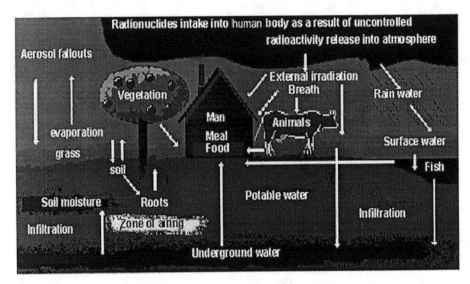

Figure 12.5 Radionuclide transport pathways

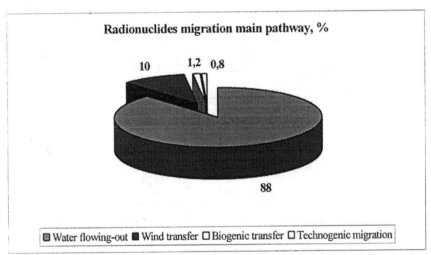

Figure 12.6 Potential radionuclide (mainly cesium and strontium) transfer pathways out of the Chornobyl Exclusion Zone (percent contribution, "," indicates decimal point)

The intensity of radionuclide migration with this water pathway depends on hydrometeorological conditions and primarily on the water level in the river drainage (e.g., during flood conditions, radionuclide outflow considerably increases).

The second most dangerous potential pathway for radionuclide transfer out of the ChEZ is wind transfer. This factor has a tendency to decrease with time because

TABLE 12.4 Radionuclide transport from different sources within the
Chornobyl Exclusion Zone[4]

Source	Activity, MBq/year		
	Σ	^{137}Cs	^{90}Sr
Chornobyl Exclusion Zone (flowing out of zone)			
River Prypyat (margins of fluctuation), 1990-1997	4.4-17.6	1.2-4.6	2.7-14.4
River Prypyat 1997	4.4	1.7	2.7
Underground filtration from cooling pool	0.37 (return 0.02-0.04)	7.4	2.96
Wind transfer	0.7	0.2	0.5
Biogenic transfer	0.07	0.055	0.015
Technogenic migration	0.029	0.021	0.08
"Shelter" (radionuclides released into environment)			
Planning release	0.011		
Non-organized release through cracks	0. 0006	0. 0005	
Chornobyl Nuclear Power Plant (radionuclides released into environment)			
Gas aerosol release	0. 048	0. 047	0. 001
Release into cooling pool with waste water	0. 035	0. 026	0. 009

Note: all flows are constant except wind transfer, which is occasional.

of radionuclides fixed in environmental objects at the expense of biological and geochemical processes in soil and biota.

Biogenic transfer is the third most important radionuclide transfer pathway. Figure 2.7 shows how different biological objects contribute to radionuclide transfer out of the ChEZ.

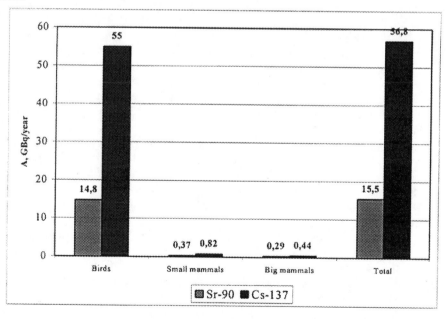

Figure 12.7 Radionuclide transfer out of the Chornobyl Exclusion Zone via biological objects

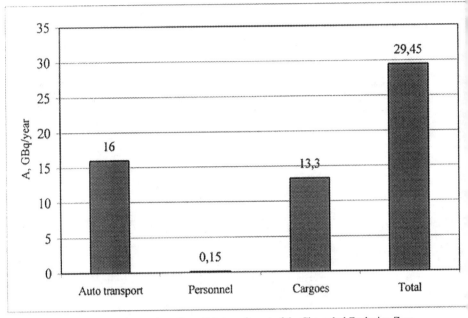

Figure 12.8 Technogenic radionuclide transfer out of the Chornobyl Exclusion Zone

Technogenic factors also allowed radionuclide transfer out of the ChEZ. In order of contribution, these factors include auto transport, cargo transport, and personnel transport (Figure 12.8).

A consideration of the contribution of various factors to radionuclide transfer out of the ChEZ, must emphasized that total potential risk to the health of the population and the environment from radionuclides concentrated in different objects within the ChEZ is higher than the possible risk posed by a serious accident at the "Shelter." Analysis of these data on potential pathway migration (shown in Figure 12.9) yields the following conclusions:

- When the "Shelter" is properly operating, the contribution to total dose made from the total radionuclide transfer out of the ChEZ via natural and technogenic pathways exceeds by 1 to 2 times the contribution to total dose from the "Shelter." The pathway of least importance proved to be imperfect emergency construction for high-activity radioactive waste storage, which requires further improvement to ensure safety.
- In case of an accident at the "Shelter," radionuclide transfer into the environment could lead to dose to the population of Ukraine at a level comparable to the sum of doses from all others sources of ChEZ radioactive contamination. At that level, these doses would be less than those already received from radionuclides flowing out during the 10 years following the Chornobyl catastrophe.

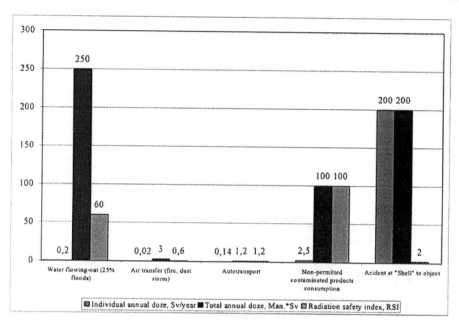

Figure 12.9 Approximate data on dose assessment for the population of Ukraine
for different scenarios of radionuclide transfer out of the Chornobyl Exclusion Zone

12.4. Modelling Strontium-90 Transport Through Exclusion Zone Water Systems

The results predicted above can be validated through use of computer modelling. An example of such modelling can be seen below to predict strontium-90 transport into surface reservoirs through the aeration zone and with underground waters. The results shown below were calculated using a Russian version of MEPAS (3.11RV, 1996-1998). This version was developed as a result of an agreement between the Pacific Northwest National Laboratory in the United States, Computer Technologies Association of the Kurchatov Istitute in Russia, and the State Scientific Center of Environmental Radiogeochemistry in the Ukraine. This version was used in the Ukraine to model and predict long-term radionuclide migration in different sub-systems of the ChEZ environment[8]. Whelan et al.[6] describe methods of modelling and prediction in MEPAS in detail.

The modelling considered two types of migration[9]:

Radionuclide migration from the ChEZ-contaminated catchments through the aeration zone and underground waters (area model of contamination)

Radionuclide migration from radioactive waste disposal sites and temporary storage sites (point-source contamination)

Strontium-90 migration in the aeration zone and underground water is described by the following equation:

$$\frac{\partial C}{\partial t} + \frac{u}{R_{f1}} \frac{\partial C}{\partial x} = \frac{D_x}{R_{f1}} \frac{\partial^2 C}{\partial x^2} + \frac{D_y}{R_{f1}} \frac{\partial^2 C}{\partial y^2} + \frac{D_z}{R_{f1}} \frac{\partial^2 C}{\partial z^2} - \lambda C \qquad (12.1)$$

where

$$R_{f1} = 1 + \frac{\beta_d}{n_e} Kd \qquad (12.2)$$

and

$$D = \alpha u + D_{mol} \qquad (12.3)$$

and where

R_{f1} = factor of delay (non-dimensional)
β = volume density (g/sm^3)
K_d = distribution coefficient (mL/g)
α = dispersity into x-, y-, or z- direction (sm)
D_{mol} = molecular diffusion.

Factor of delay was used as a measure of contaminant mobility in a porous environment. It represents the ratio of average velocity of porous water to average velocity of contaminated material transformation and could be represented several ways (for example, with agraph). In reference literature on underground water, the following equations for delay factor assessment are recommended:

$$R_{f2} = n/n_e + \frac{\beta_d}{n_e} Kd \qquad (12.4)$$

$$R_{f3} = 1 + \frac{\beta_d}{n} Kd \qquad (12.5)$$

$$R_{f4} = 1 + \frac{\beta_d}{\theta} Kd$$

(12.6)

where θ = moisture in the aeration zone (non-dimensional) .

Equations 12.2 and 12.6 are used in MEPAS, the former for saturated zones and the latter for partially saturated zone. After some combinations:

$$u^* = \frac{u}{R_f} \qquad (12.7)$$

and

$$D^* = \frac{D}{R_f} \qquad (12.8)$$

Equation 12.1 could be modified into simplified type:

$$\frac{\partial C}{\partial t} + u^* \frac{\partial C}{\partial x} = D_x^* \frac{\partial^2 C}{\partial x^2} + D_y^* \frac{\partial^2 C}{\partial x^2} + D_z^* \frac{\partial^2 C}{\partial z^2} - \lambda C \qquad (12.9)$$

To solve Equation 12.9 taking into account certain boundary and initial conditions, the set of semi-analytical equations was used in MEPAS. These equations allow predictive assessments of contaminant migration in saturated and non-saturated environments.

An exponential source model was used to describe radionuclide intake into the aeration zone:

$$Q(t) = Q_0 * \exp(- a * t) \tag{12.10}$$

where

Q_0	=	$K_1 * A_0 * S$, $\quad a = \lambda + K_1$, $\quad K_1 = \ln2/T_{\Pi B}$
$Q(t)$	=	integral for the source velocity of radionuclide intake into the aeration zone (Ci/yr)
A_0	=	density of surface contamination (Ci/m^2)
S	=	site area (m^2)
A_0*S	=	radionuclide reserve in source (Ci)
λ	=	constant of radioactive decay (year^{-1})
$T_{\Pi B}$	=	period of semi-decay (years).

Figure 12.10 shows contaminated catchments isolines of underground water levels and directions of strontium-90 migration with underground water currents into water systems of the ChEZ (Azbuchin Lake, Prypyat Creek, Semikhodovskiy Creek). From these locations, contaminated water flows into the River Dnieper, making it doubtless that the population of Ukraine consumed this water.

Tables 12.5 and 12.6, as well as Figure 12.10, show parameters for strontium-90 migration calculations as well as data on its reserve in these catchments, radioactive water disposal sites, and temporary waste storage sites.

Figures 12.11 to 12.17 show the results of predictive modelling. Here one can see the peculiar features of how strontium might be transported from groundwater to surface water of the ChEZ.

12.5. Conclusions

The catastrophe at the Chornobyl NPP showed the world that the "peaceful atom" can in reality be less than peaceful. However, through the use of predictive modelling, risk can be assessed to various ecological factors. This type of modelling allowed planning of a budget for radioecological monitoring within the ChEZ.

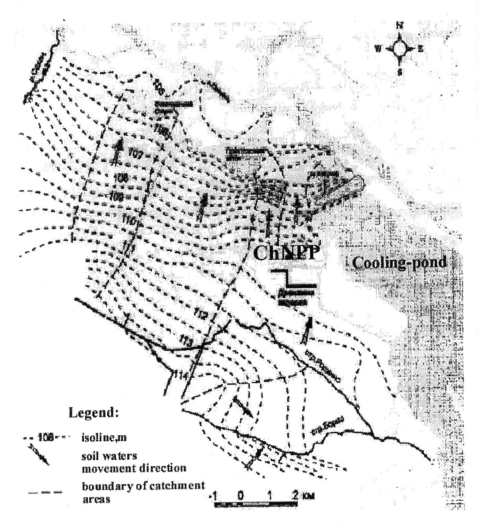

Figure 12.10 River Prypyat right bank catchments and radionuclide transfer pathways

TABLE 12.5 Hydrogeological and radiological parameters used for modelling with the Multimedia Environmental Pollutant Assessment System

Source	Loading Site	Catchment Dimensions, km x km	Distance from Catchment Center to Unloading Site, m	Average Flow Gradient for Ground-water	Aeration Zone Thick-ness, m	^{90}Sr Activity Reserve, Ci
Prypyat Creek catchment Site 1	Prypyat Creek	2 x 1	500	0.03	5	1000
Prypyat Creek catchment Site 2	Prypyat Creek	2 x 0.5	1,250	0.03	3	600
Prypyat Creek catchment Site 3	Prypyat Creek	2 x 2	2,500	0.03	1	1400
Azbuchin Lake catchment Site 1	Azbuchin Lake	1 x 1	500	0.043	1	300
Azbuchin Lake catchment Site 2	Azbuchin Lake	1 x 0.2	1,100	0.043	3	200
Azbuchin Lake catchment Site 3	Azbuchin Lake	1 x 0.6	1,500	0.043	5	500
ChNPP industrial site	Azbuchin Lake	0.5 x 0.5	1,500	0.043	5	1600
Semikhodovskiy Creek catchment Site 1	Semikho-dovskiy Creek	2 x 1.5	750	0.003	5	150
Semikhodovskiy Creek catchment Site 2	Semikho-dovskiy Creek	2 x 0.5	1,750	0.003	3	50
Semikhodovskiy Creek catchment Site 3	Semikho-dovskiy Creek	2 x 2	3,000	0.003	1	300
Radioactive waste temporary storage site "Red Forest"	Prypyat Creek	1 x 1	2,500	0.003	0	2000
Radioactive waste temporary storage site "Yanov"	Prypyat Creek	2 x 2	2,000	0.003	2	400
Radioactive waste temporary storage site "Prombasa"	Prypyat Creek	1.5 x 1.2	1,500	0.003	2	6000

TABLE 12.6 Filtration-migration parameters used in models

Parameter	Units	Value
Aeration Zone		
Infiltration feeding	mm/year	200
Volumetric moisture content in soils	doesn't matter	0.1
K_d ^{90}Sr	mL/g	1
Soil density	kg/dm^3	1.65
Aquifer		
Thickness	m	20
Filtration coefficient	м/сут	10
Efficient porosity	doesn't matter	0.2
K_d ^{90}Sr	ml/g	0.5
Soil density	kg/dm^3	1.65
Porosity	doesn't matter	0.3

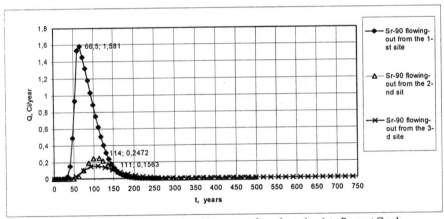

Figure 12.11 Maximum strontrium-90 transport from three sites into Prypyat Creek
(1.71 Ci/yr in 66.5 yr according to predictive modelling)

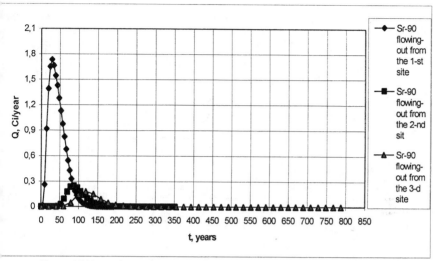

Figure 12.12 Maximum strontium-90 transport from three sites to Azbuchin Lake
(1.73 Ci/yr in 28.7 yr according to predictive modelling)

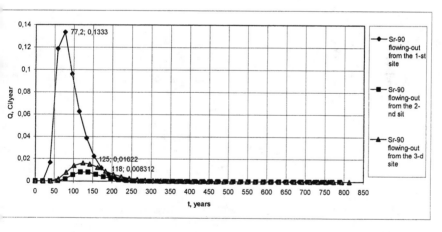

Figure 12.13 Maximum strontium-90 transport from three sites into Semikhodovskiy Creek
(0.14 Ci/yr in 77.2 yr according to predictive modelling)

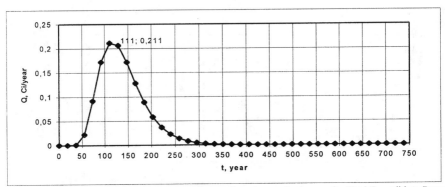

Figure 12.14 Maximum strontium-90 transport from radioactive waste storage site "Red Forest" into Prypya
Creek (0.2 Ci/yr in 110.7 yr according to predictive modeling)

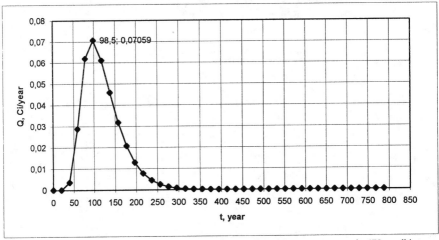

Figure 12.15 Maximum strontium-90 transport from radioactive waste storage site "Yanov" into
Prypyat Creek (0.207 Ci/yr in 98.5 yr according to predictive modelling)

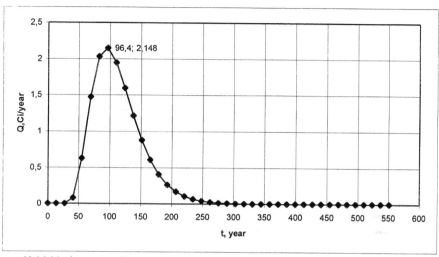

Figure 12.16 Maximum strontium-90 transport from radioactive waste storage site "Prombasa" into Prypyat Creek (2.15 Ci/yr in 96.4 yr according to predictive modelling)

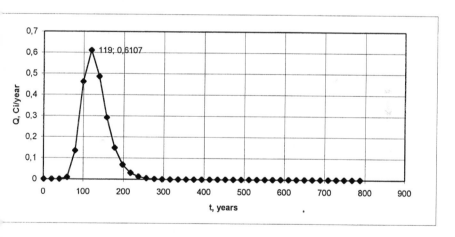

Figure 12.17 Maximum strontium-90 transport from the industrial zone of the Chornobyl Nuclear Power Plant into Azbuchin Lake (0.6 Ci/yr in 119.2 yr according to predictive modelling)

2.6. References

---- (2000) Conception of the Program on Chornobyl Catastrophe Consequences Minimization for the Period of 2000-2005. Tacis, Brussels.

Bar'yakhtar-Kyiv, V.G., editor. (1997) *The Chornobyl Catastrophe*. Nauk.Dumka, Kiev.

----(1996) *Ten Years After the Accident at the Chornobyl NPP*. National Report of Ukraine, Kiev.

Kholosha, V.I., Proskura, M.I., Ivanov, Y.O., and others. (1999) Radiation and ecological importance of natural and technogenic objects of the Chornobyl Exclusion Zone. *Bulletin of Ecological State of the Chornobyl Exclusion Zone and Zone of Obligatory Evacuation, Chornobylinterinform* **13** 3-8.

5. Droppo, J.G. Jr., Strenge, D.L., Buck, J.W., Hoopes, B.L., Brockhaus, R.D., Walter, M.B., and Whelan, G. (1989) Multimedia Environmental Pollutant Assessment System (MEPAS) Application Guidance, Volume 2-Guiddelines for Evaluating MEPAS Input Parameters. PNL-7216, Pacific Northwest National Laboratory, Richland, Washington.

6. Whelan, G., Buck, J.W., Strenge, D.L., Droppo J. G., Hoopes, B.L., and Aiken, R.J. (1992) Overview of the the Multimedia Environmental Pollutant Assessment System (MEPAS). *Hazardous Waste & Hazardous Materials.***9** 191-208.

7. Borovoy, A., Gorbachev, B., Kluchnikov, A., and others. (1988) *The Shelter's Current Safety Analysis and Situation Development Forecasts (Updated Version).* Tacis, European Commission, Brussels.

8. Lysychenko, G.V. (2000) Investigations on MEPAS System Possibilities for the Purposes of Ecological Prognosis Tasks. *Institute of Modelling Problems in Energetics* **9** 155-164.

9. Lysychenko, G.V. (2000) Modelling and Prognosis Assessments of Radionuclides Migration with Underground Waters of Near Zone of Chornobyl NPP. *Institute of Modelling Problems in Energetics* **5** 143-147.

13. Psychological Aspects of Risk Assessment and Management

Another major challenge to decision makers is communicating risk information to the lay public. Key to this communication is the set of perceptions held by the public regarding the particular risk. This chapter describes two studies concerning psychological aspects of risk perception, which provide insight into developing an information strategy that could be more effective in relaying risk information. The first study compares two groups of people—experts and lay people. This study can help risk managers understand the best ways to communicate with affected populations. The second study looked at the attitudes of three groups of residents in areas contaminated by the Chornobyl accident. Psychological attitudes are shown to create a motivation for relationships and actions. This study can help decision makers understand what motivates people.

The psychological problems related to radioactive contamination markedly intensify adverse radiation impacts on human health. To counter these problems, organizations and authorities often attempt to provide information to workers and the potentially affected population. Such was the case concerning the Chernobyl accident. Information prepared by experts on state measures to mitigate accident consequences was initially intended to create an attitude of cooperation within the population to overcome the accident impacts and normalise the social and psychological climate. However, when the population sees such information as inadequate, the results may be unexpected or even opposite to those intended. Research in the regions adjacent to Chernobyl, and later in the city of Slavutich, showed that psychological support is needed not only for power plant personnel but, to a greater extent, for lay people without professional knowledge of radiation but at risk from its effects. This chapter describes two studies concerning psychological aspects of risk perception and provides insight into developing an information strategy that could be more effective in relaying risk information.

13.1. Differences in Risk Perception Between Experts and Lay People

The first study involved a comparison of two different groups of people—experts, represented by personnel of the Chernobyl Nuclear Power Plant (CNPP), and lay people, represented by the residents of a radiation-contaminated area in the Kaluga District of the Russian Federation, located a considerable distance from the CNPP (further referred to as KA). A theoretical basis for the study was a risk perception concept by Rowe[1], which involved subjective assessment of an event possibility and hazardous consequences and evaluation of the scale of those consequences. This basis was supplemented by defining the constituents of the perceived risk according to the classification of Covello[2].

Objectively, risk is a non-linear multidimensional function of interrelated factors attributed to various sources. Practically, under specific circumstances, any social condition or process can be perceived by a poorly informed person as a risk factor. It is therefore important to evaluate a variety of factors to determine risk perception.

Data were gathered from the following two groups:

1. Personnel of the CNPP (720 persons), men and women age 23 to 55. They are informed on radiation impacts to the organism and experienced in handling radioactive materials. They have also experienced operating in an accident setting and high dosage loads. The group was studied from May to October 1986.
2. Population of three subregions in KA totaling 234, 148, and 120 residents, respectively (men and women age 17 to 64). With few exceptions, the population had no professional knowledge of radiation impacts to the organism. However, they live under low dosage loads for a considerable time. The group was studied from November 1990 to April 1991.

Both groups were studied using targeted interviews and questionnaires, standard techniques for evaluating personal qualities and psychological status of an individual. Table 13.1 shows some results of the study. Results are expressed in terms of a deviation indicator that represents the tendencies of the risk perceptions to be higher or lower than expected. The subtotal deviation indicator for each risk factor characteristic category is given in bold. Selected deviation indicators for subcategories of special interest are also included. The reader is referred to the source references for a more rigorous definition of the risk indicators and measures that are used in this chapter.

For both groups, risk perception is the more acute with less specific data on general features of the hazard. The results agree with the findings of Lee[3] on psychological mechanisms of exposure to risk: a person adapts to situations related to risk and repeated regularly; a cognitive rearrangement of information is oriented to the dissonance effect. Psychological protective mechanisms are activated that allow a person to balance the nervous system and psyche. The subconscious plays a leading role in ameliorating the effect of protective mechanisms.

In addition, the results show that the risk perception of professionals and other persons interested in facility operation has a sophisticated structure and is differently expressed before and after an accident. On the other hand, the psychological aspects of

TABLE 13.1 Comparison of risk factor characteristics between experts (CNNP workers) and lay people (KA residents) regarding the risks from the accident at Chernobyl (after V.T. Covello)

Risk Factor Characteristic	Deviation Indicator	
	CNPP	KA Residents
1. Awareness of the risk—Subtotal	0	1.5>
CNPP personnel were worried about lack of precise information. KA residents were worried about possible distortion or muffling of information on hazards.	1.2	>
2. Understanding—Subtotal	0	2.0>
For CNPP personnel, ignorance of mechanisms of radiation impacts intensifies a need for scientifically based information. For the population, difficulty in understanding the situation (hazardous or not?) results in either in diligent action in search of explanation or apathy, and, in acute cases, depression.	0	>
3. Indefiniteness—Subtotal	0	1.7>
When CNPP personnel personally assessed the situation as indefinite, their feelings of dissatisfaction were increased and they were more likely to become afraid. The population acutely reacts to an indefinite situation, particularly when assisting the accident victims is delayed.	0	>
4. Voluntary actions—Subtotal	1	1.5
CNPP personnel preferred perceived and voluntary risk. However, the rescue workers from other locations had an acute feeling of humiliation and fear when working in extremely dirty areas in the CNPP. The population was worried because it was impossible for them to independently resolve problems related to higher radiation impacts in their residential areas.	0	<
5. Personal involvement—Subtotal	0.5<	1.5>
CNPP personnel expressed guilt for threatening the health of their families and children. The population was more worried about clean food and aware of their desolate fate.	<	+>
6. Possibility of control—Subtotal	0	1.5>
Rescue workers were indignant at the lack of adequate dosimeter equipment and the impossibility of calculating the risks related to rescue activities. Residents mistrusted official information on the dosimeter situation and diseases in residential areas.	0	>
7. Disastrous potential—Subtotal	2.0	0
Acute radiation impacts to the personnel at Unit 4 of the CNPP during the night of the accident aggravated the perception that the situation was extremely risky. Before the accident, the population feared possible consequences of such an event, yet large-scale impacts of the actual event ameliorated fears and added a self-reassurance shade to risk perception.	2.0	0
8. Accident potential—Subtotal	2.5>	1.5>
Memories of the experience at Unit 4 aggravated risk perception among personnel; unpredictable nature of future events contributed to an anticipated recurrence of the situation. The population was depressed by fears that the radiation situation would remain negative for several years and that the accident might be repeated.	>	>
9. Immediate hazardous effect—Subtotal	2.0	0
Because of high dose loads, CNPP personnel and rescue workers were convinced of the actual danger of radiation impacts. Awareness of irreversible negative impacts is increasing. For residents, radiation impacts are not explicitly evident. However health problems were often associated with radiation impacts.	2.0>	0

TABLE 13.1 (continued)

Risk Factor Characteristic	Deviation Indicator	
	CNPP	**KA Residents**
10. Reversibility—Subtotal	1.1	0
Most CNPP personnel in late 1986 were concerned about personal problems (for example, should they continue working at the plant or retire). Meanwhile reversibility, in the form of possible irreversible physiological problems for young people, was crucial in corporate decision making. For residents, the problems of hazardous impacts of radiation are sophisticated and do not play a noticeable role in personal decision making.	1.1	0
11. Fear—Subtotal	0.5<	2.0
Professionals were relieved from fear when they acquired experience in handling radiation. On the other hand, habituating and ignoring a hazard were reported. Some residents (primarily physicians and teachers) left their dwellings at the first opportunity upon being informed of radioactive fallout. They were guided by fear and mistrust of official information.	0.5<	++>
12-13. Impact on children and future generations—Subtotal	1.5>	2.0>
The CNPP personnel felt extreme guilt for the environment (plants and animals) in the accident zone. For residents, the decision to leave the contaminated area was largely influenced by a fear for children, including unborn babies.	+>	++>
14. Specific nature of victims—Subtotal	2.0>	2.0>
CNPP personnel expressed by acute distress at news of the death of their friends. Perception by the population was rather abstract: they associated fatalities among the CNPP personnel with radiation contamination, which aggravated negative attitudes.	+>2.0	++>
15. Justice—Subtotal	1.5	1.5
CNPP personnel reacted emotionally to a biased mass media coverage of their activity and fault for the accident. The population was more concerned by conflicts between themselves and authorities on unjust (in the population view) distribution of bonuses among the accident victims.	1.5	1.5
16. Profits—Subtotal	1.3>	2.5>
Problems of status and future prospects among the CNPP personnel were considered priorities in May 1986. A month later, when the accident situation continued, the priorities markedly shifted. In the months to follow the profits from continuing work were evaluated by the professionals as a choice between health and welfare. The problems of residents were mostly confined to obtaining government compensation.	1.3>	+++>
17. Trust in organizations—Subtotal	1.5>	1.5>
CNPP personnel had no trust in the organizations responsible for technological safety at the NPP. The population felt organizations were muffling information on the radiation level in residential areas and extent of hazard to people's life and health.	1.5>	1.5>
18. Interest of mass media—Subtotal	1.5	0
Any information supporting personnel fault in the CNPP accident was considered personal assault and aroused a feeling of bitterness and despair.	1.5>	0

Note: The deviation indicator shows the tendency for the risk perception to be less than (<) or greater than (>) the indicated risk levels; a value of 0 indicates agreement The non-zero numbers, when given, indicate the factor by which the risk perception is greater or less than indicated risks.

the health of the population in the Kaluga region are largely related to social problems typical of the entire country's population. In other words, it is often difficult to differentiate influence from the radiation factors from influence of background social concerns.

It is also important to note that the constituents that make up perceived risk for professionals and lay people are significantly different. Subsequently, risk controls their behavior in a different manner, modifying their attitudes and motivations. The most important factors for creating distress (in different content aspects) in both groups were the following:

- Ambiguity
- Voluntary nature
- Personal involvement
- Possibility-impossibility to control
- Impacts on children and future generations
- Justice.

The perception of risk is related to the extent the situation is seen as ambiguous. This perception is particularly true when the threat has unknown boundaries.

Intensifying ambiguity simultaneously aggravates the perception of a situation as a dangerous one. As these perceptions expand in an avalanche-like pattern, the process could result in psychological tension and panic-like behavior patterns. Adequate and timely awareness could stop this process.

13.2. External Factors and Motivation in Risk Perception Among the Population

The second study looked at the motivations and attitudes of three groups of residents in areas contaminated by the Chernobyl accident. Motifs are directly related to psychological attitudes of people and create a motivation for their relationships and actions. Attitudes imply a person's readiness to act in a particular way. An attitude also expresses a person's position relative to a risk source. For instance, a need to earn money may be pressing, particularly in the general population, but when there are no conditions to realize this need, its utilitarian motivation cannot be met. Likewise, when there are conditions to earn money, but a person does not need money, the utilitarian motivation is absent.

The study identified five principle motivations: cognitive (M1), evading problems (M2), acquisition (M3), prestige (M4), and utilitarian (M5). Residents were evaluated through interviews with specialists and special methodology designed by the Prognosis Research Center (Obninsk) for this particular study. This methodology on the level of satisfying motivation was based on an integration of indicators in replies to questions concerning a specific motivation. When results were greater than 60%, the motivation was considered active and a determinant of a person's actions. Results less than 40% indicated that satisfaction of the motivation was hampered or the motivation was absent.

The study was carried out in three districts of the Russian Federation with contamination levels from 1 to 5 Ci/km^2. The total population in the Bryansk District (BD) was 236,400 people, with 140,900 in the Kursk District (KD), and 820,200 in the Kaluga District (KA). Population sampling in contaminated areas was based on multi-stage random selection. Sampling in clean areas was based on quota distribution related to gender and age of the recipients.

Table 13.2 summarizes the results of studying the motivational structure of this population. A range of the appropriate indicator expresses the average degree of realization of each motivation. The indicator is an "average degree of interest realization" expressed in percentage. Several conclusions can be drawn from the results:

- The most realized motivations are evading problems and demand for interesting activity. In other words, contrary to common opinion, the population feels adequately protected.
- Age played an important part in perception. For each indicator, data increasingly deviated up to the age range of 20 to 24 years. There was a further stable decreasing trend after that point.
- Unexpectedly high was the extent of realization of the cognitive motivation. This rating testifies that the population is confident in their awareness of risks. The most trustworthy sources of reliable information were usually gossip, retelling the stories heard from acquaintances, etc. While developing information for the population in the studied areas, one should take into account a relatively low level of interest and confidence in government sources of information.
- Extremely important, and the least expressed motivation, is the utilitarian one, i.e., satisfying a demand for personal welfare. This information suggests to decision makers that, without due account for attitudes related to difficulties in realizing a utilitarian motivation, any initiatives to inform the public may only yield negative results.
- Another infrequently realized motivation is that of prestige. The population in the studied areas is not satisfied with its social status. Hence, motivation of prestige may be the mechanism to compensate an unsatisfied utility motivation.

Of course, a motivational structure is heterogeneous for population groups of different sexes, ages, and residences; therefore, these motivation features should be taken into account when developing social protection measures and presenting information. Overall, the study found that a balanced system of motivations for personal activity that is relatively fully realized allows a person the flexibility to adapt to environmental transformations (both natural and social), including risk-related critical transformations.

TABLE 13.2 Range of motivation realization in three districts of the Russian Federation contaminated by the Chernobyl accident

Motivation Category	BD	KD	KA	Clean Areas	All Study Regions	Deviation Indicator
Cognitive motivation (M1)	3	2	2	2	2	0
Evading problems motivation (M2)	1	1	1	1	1	>2.5
Acquisition motivation (M3)	2	3	3	3	3	>1.5
Prestige motivation (M4)	4	4	4	4	4	<2
Utilitarian motivation (M5)	5	5	5	5	5	< 3
Average degree of interest realization	38.55	54.24	53.81	44.20	49.12	<1.7
Standard deviation of motivation realization	14.27	13.21	11.02	12.41	14.89	

Note: The indicator for the first five data columns is an "average degree of interest realization" expressed in %. The deviation indicator in the sixth column shows increased (>) or decreased (<) tendencies relative to a isk indicator; 0 indicates agreement with the indicator. The numbers are the factors for the indicated tendency which is a slightly different definition than in Table 13.1).

13.3. Developing an Information Strategy Based on Psychological Aspects of Risk Perception

Providing information is an important aspect of effective decision making on socio-economic rehabilitation and psychological support of the population. Information on events in the surrounding world create an information medium shaping an individual's outlook, opinions, system of values, and readiness to act. *Thoroughly planned and adequately presented information is a tool in a system of measures to reduce psychological and emotional stress.* Three parts of that tool include using dialogue, providing for independent review of population characteristics, and developing an effective information strategy.

13.3.1. Using a Process of Dialogue

An information strategy should be based on an active dialog between authorities and the population. This dialog consists of two units:
.   Direct flow of information from authorities in the form of instructions, programs, and publications
.   Feedback from the population to authorities.
The latter comprises the following information:
   Socio-psychological climate, demands, and necessities of target groups
   Thoughts on current decision making, possible consequences, and implementation of specific programs
   Expected effects of planned instructions, programs, and publications.
Thoughts on decision making are particularly important (Figure 13.1). Analysis of the current status of the problem is the core and starting point in decision making as well as in developing information.

238

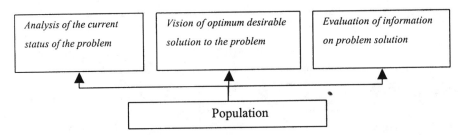

Figure 13.1 Simplified diagram of population feedback

Perception of the optimum desirable solution to the problem predetermines the program of actions to attain the desirable target as well as the format to present information to the population. This format will be more precise if a thorough and comprehensive account is made of the psychological factors of information perception (i.e., people's attitudes and interests, socio-demographic features, and psychology of perception).

Transition from the current to desirable state requires a series of practical actions to shape adequate information in terms of preciseness, timeliness, and ease of understanding. When the population perceives the information as adequate, tension is relieved. Information adequacy can be achieved by:

- Precisely presenting information (Figure 13.2)
- Taking into account general psychological regularities of information perception by an individual and differences in information perception by various population strata
- Addressing the nature of information to a person's individual features (e.g., social status, intellectual level, occupation)
- Taking into account social and psychological climate and making information content correspond to population expectations and needs
- Presenting information in a timely manner.
- Using expert sources on actuality and importance of the problem when preparing information.

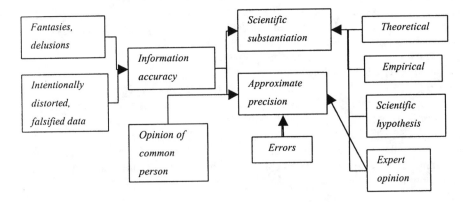

Figure 13.2 Criteria for evaluating information precision

Criteria for information precision are related to the information's correspondence to the objective reality (e.g., absence of errors, fantasies and delusions, and particularly distortion of risk factor data). The data supported by scientific research are considered the most precise. The more adequate the scientific substantiation of information (the more profound the theoretical study of the problem, the greater the number of empirical data that confirm the theory), the more precise the data are suspected to be. When information delivery is critical, criteria for information precision could be tested and confirmed by experts. Great danger lies in erroneous judgments based on everyday experience or even on specialists' practical work.

It is important to note that people make decisions based on how adequately their expectations have been met (Figure 13.3). A personal decision unexpected by decision makers rarely gives rise to positive emotions among the population because people do not believe decision makers consider their concerns in the process.

It is also important to note that timeliness of information plays a key role in determining its adequacy. Figure 13.4 shows two situations–stable and unstable or critical. The same information may cause different reactions in a particular situation. A stable situation allows time for contemplating, and a delayed reaction is possible. A critical situation characterized by a time deficit aggravates such psychological effects as attitude to the perceived information, neurotic reaction, negativism, and nihilism. Information is timely if it arrives the moment when it is most effective. If information arrives too early, people are unprepared for it. If it arrives too late, people are tired of waiting.

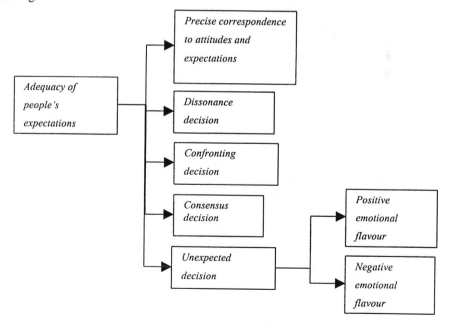

Figure 13.3 Criteria for evaluating the adequacy to meet people's expectations

Hence, a dialogue technique that includes direct information flows and feedback is necessary to communicate risk. Mutual understanding can be achieved through dialogue when a source is aware, through constant and timely communication, of the actual status, demands, and problems of target groups and others who contribute to psychological and emotional climate.

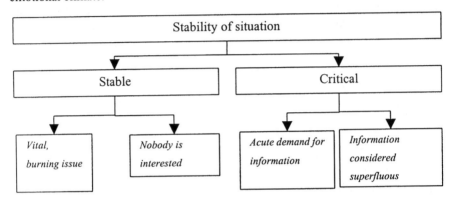

Figure 13.4 Criteria for evaluating timely submission of information

13.3.2. Gaining a Psychological Perspective

Socio-psychological expertise is an indispensable component of practically all activities related to development and implementation of an information strategy. To ensure an appropriate program of information provision, officials can implement an independent psychological review of messages. This review requires competence in economics, knowledge of actual specific features of the risk, and status in specific regions. It is necessary to obtain data on standard of living, people's attitude toward the problem, and public opinion on the problem. Based on results of this review, officials can then correct information, if necessary, in content, method of delivery, and, possibly most importantly, time of provision.

Such a review starts with the classification of the population to whom the information is addressed. Expert assessment is aimed at evaluating the information against age, gender, education, professional skills, and social status of the addressed individuals (Figure 13.5).

Personal factors influencing evaluation	Criteria for Information Clarity	Clarity
Age Gender Education Occupation Social Status	Psychological aspects of perception Socio-psychological climate Confusion, false scientific nature, incoherence Interesting presentation	Easy to understand Addition to previous information Links to previous information

Figure 13.5 Factors used to evaluate clarity of information

The Figures 13.4 and 13.5 show logical links between criteria-like features of timely and actual information. Individual decisions to adopt or reject government decisions concerning actions are related to a personal comprehension of the received information.

13.3.3. Developing an Information Strategy

Information from authorities addressed directly to the population is transmitted through mass media or regional authorities, staff at industrial facilities, and/or experts at scientific institutions fulfilling state contracts for preparing regulatory and information documents, statutes, and analytical materials on expected effects of planned programs. These methods can be grouped into an information strategy, based on principles of organizational psychology.

Within this strategy, information for the population should:

- Be easy to understand (expressed in plain language)
- Complement a comprehensive idea of a stress source (otherwise a person would develop his or her own, usually incorporating one's personal and generally inadequate meaning of the information)
- Incorporate the existing information in the context of a new information block so that the new will be easy to understand (simple absorbs complex)
- Carry additional information expanding the old information
- Meet population expectations

Be presented within a period of time not exceeding the threshold of interest and the threshold beyond which psychological and emotional tension could reach its peak following a delay.

Hence, an information strategy should be:

Oriented at actual demands of the population

Supported by timely and reliable information on population readiness to perceive social decisions

Attract broad population strata to discussing and solving social problems on the regional level through mass media and creating feedback from the population to regional authorities.

An effective information strategy should also take into account social features and psychological aspects of people's perception of information.

Throughout the implementation of the strategy, psychological expertise as well as risk-specific expertise should be used to evaluate the timeliness and effectiveness of draft information messages. Based on the evaluation results, experts should decide on the message expedience and recommend amendments to content or format. If the population is unprepared for innovations, negative results will ensue. In this case, it is essential to create adequate public opinion.

13.4. Conclusions

Research into psychological aspects of risk assessment and management[4-11] has resulted in the following conclusions:

- At a complete absence of knowledge on sources of hazards, people have no perception of risk. The most general indication of a hazard induces various patterns of consciousness and perception of risk among people with varying educational levels and awareness in risk factors. These patterns are manifested by various behaviors: from random action hoping for a lucky chance, to target-oriented search, to specific actions to reduce ambiguity, evaluate the hazard extent, and work out a program for behavior with minimum risk.

- In a situation when only an approximate evaluation of the known hazard is possible, risk was identified as a method of action with unpredictable consequences. In such cases, a distress reaction is inevitable (e.g., fear, negative emotions). When it is possible to precisely calculate consequences of activities and phenomena, people do not perceive a situation as risky. Emotions are smoothed, and people are aware of the advantage of their position and the importance of an individual. In such cases, a person is capable of acting reasonably and productively.

An informed person has a sense of self respect and confidence, and his or her actions are target-oriented and productive. One can say that an informed person is psychologically prepared for a potential hazard and modifies his or her behavior according to a plan in agreement with a person's information status. It is therefore important to clearly inform people in a timely manner of specific features of living in areas near Cold War legacy facilities and hazards so that people can live a life of full value while enduring stress. One positive trend in scientific research of the impacts of the Chernobyl accident is an increased attention to psychological and social aspects of these consequences. When making important social decisions it is essential to carry out a preliminary operational psychological study of population motivation dynamics. International cooperation is needed to search for adequate methods of working with populations, administrations, and mass media to relieve the serious psychological consequences of the Cold War.

13.5. References

1. Rowe, W.D. (1977) An Anatomy of Risk. Wiley, New York.
2. Covello, V. T. (1985) Social and behavioural research on risk: uses in risk management decision-making. In Environmental Impact Assessment, Technology Assessment and Risk Analysis. NATO ASI Series G, Vol. 4. Springer Verlag, Berlin.
3. Lee, T. R. (1983) The perception of risk. In Risk Assessment. A Study Group Report. The Royal Society, London.
4. Abramova, V.N. (1988) A psychologist outlook on the Chernobyl accident. Nauka i Zhizn Journal 11 78-81 (in Russian).
5. Abramova, V.N. (1992) Features of perceiving radiation risks by population in Kaluga region. In Chernobyl Legacy: Proceedings of Kaluga Scientific and Practical Conference, 17 April 1992, Kaluga, Obninsk, 106-114 (in Russian).

5. Abramova, V.N., and Volkov, E.V. (1999) A longitude verification of the organisational factor's influence on nuclear power plant's reliability. In Identification and Assessment of Organisational Factors Related to the Safety of NPPs. Nuclear Safety NEA/CSNI/R(98)17, Feb 1999, Expanded Task Force on Human Factors Principal Working Group No. 1, 2 21-24.

7. Abramova, V.N., and Matveenko, E.G. (1993) Socio-psychological aspects of public relations to prevent radiofobias and reduce psychological and emotional tensions. In Problems of Ameliorating the Impacts of the Chernobyl Accident, Proceedings of an International Workshop, Bryansk, Znanie Society, 1 64-70 (in Russian).

8. Abramova, V.N., Melnitskaya, T.B., and Kushneruk V.P. (1997) Analysis of reasons for citizen addressing social and psychological advisory centers. Regional Scientific and Practical Conference on Implementing Family and Youth Policy in Modern Conditions. Kaluga, 6-12 (in Russian).

9. Abramova, V.N., Melnitskaya, T.B., and Pavlova E.A. (1994) Motivation aspects of socio-psychological status of population in radiation contaminated areas in Russia. Scientific report by the RF Ministry of Emergency Situations. Prognosis Research Center, Obninsk, (in Russian).

0. --- (1990) Summary Report of the Working Group on Psychological Effects of Nuclear Accidents. Soviet Union, Kiev.

1. --- (1992) Chernobyl Trace: Medical and Psychological Aspects of Radiation Impacts. Collection of articles, Vol.1-2, Votum-psi Publishers, Moscow.

14. Utilizing a Multimedia Approach for Risk Analysis of Environmental Systems

The environmental legacies of the Cold War have raised concerns over impacts to air, water, and soil. In the past, most countries, including the United States, have considered each of these media as separate, relatively unconnected, issues. Nature, however, does not recognize this artificial compartmentalization. In the 1980s and 1990s, multimedia modeling developed as a means to obtain a complete picture of risk across these media. This chapter describes the need for and the current status of computer-based frameworks for conducting integrated multimedia modelling.

The Cold War environmental legacies include risk-based concerns related to air, water, soil, and biomass contamination. Past practice in most countries including the United States has been to consider each of these media as separate relatively unconnected issues. Nature, however, does not recognise this artificial compartilization. In the 1980s and 1990s, integrated models were developed to allow multimedia analysis of the total risk accounted for processes within and between these media.

To adequately address risk issues posed by technological legacies, scientists and decision makers need input from environmental systems modelling that can address both increasing technical scope and complexity. This modelling requires the integration of existing tools and the development of new databases and models, based on a comprehensive and holistic view of risk assessments. To meet these needs, scientists have developed multiple-media-based modelling systems using advanced computer hardware and software to view and assess risks from a comprehensive environmental systems perspective, crossing the boundaries of several scientific disciplines.

The need to perform an integrated multimedia approach to fully understand the potential ramifications of complex systems of environmental contamination/releases is gaining widespread recognition. The results of some of the early integrated risk applications utilizing multimedia models are described in Chapter 6.

This chapter describes the history and need for integrating risk modelling capabilities across disciplines. An example of a holistic framework for assessing risks is the Framework for Risk Analysis in Multimedia Environmental Systems (FRAMES). FRAMES was developed by the Pacific Northwest National Laboratory with support from the U.S. Department of Energy (DOE), U.S. Environmental Protection Agency (EPA), and U.S. Nuclear Regulatory Commission (NRC). Versions of this FRAMES software have been used for regulatory and risk analysis applications.

14.1. Initial Development of Risk Modelling Capabilities

For over 40 years, medium-specific models were developed in an effort to understand and predict environmental phenomena, including contaminant release, fate, and transport. In the past, government agencies like DOE and EPA simulated contaminant release and subsequent fate and transport, exposure, and risk for a single chemical within a single environmental medium. Most recently, these kinds of models were combined for either sequential or concurrent assessments [1-5].

In 1994, the NRC [6] recommended a three-tiered approach to risk modelling, ranging from screening-level assessments to detailed characterization to estimation of costs of remediation, alternatives, and risk reduction. EPA is also moving toward a multiple-tiered approach, as illustrated by the use of Risk-Based Corrective Action for managing contaminated release sites. This approach compares alternatives related to resource allocation, urgency of response, target clean-up levels, and remedial measures based on reasonable potential risks to human health and environmental resources.

The tiered approach represents a step-wise protocol for establishing validity of the potential risks posed, quantifying the risks by collecting additional data, and applying more science-based tools to assess the risks. As the NRC noted in 1998 [7], however, in many instances the tiered levels in the assessment are disconnected and therefore results are not as useful. Therefore, DOE and EPA are moving toward a more comprehensive phased approach by integrating more simplified analyses (e.g., using analytical models), which efficiently use fewer resources and help to focus assessments, with the more resource-intensive complex analyses (e.g., using numerical models and extensive databases).

14.2. Need for More Complex Systems

Increasing complexity in risk assessment has led U.S. government agencies to develop and implement computer-based tools that view the environment from multiple dimensions, accounting for various waste forms, environmental media, and relationships between the waste sites and the surrounding sensitive receptors. These tools are integrated methodologies based on principles of physics and utilizing latest computer advances to view the environment from a more holistic, systematic point [1,2]. Table 14.1 illustrates the dimensionality involved in simulating environmental systems and the evolving increase in complexity associated with risk assessment.

A number of motivating factors led to the design of more comprehensive risk-based frameworks, which can account for increasingly complex modelling systems. First,

TABLE 14.1 Dimensionality involved in simulating environmental systems[8]

Dimension	Attributes
Spatial	Local, regional, global
Temporal	Short-term/acute, seasonal, long-term/chronic
Chemical	Organics (pesticides, dioxins, furans, HCH, PAHs, PCBs, etc.) inorganics (organo-metals, lead, cadmium, mercury, tin, etc.)
Environmental media	Air, water (precipitation, ground water, surface water), soil, sediment, biota (food chain)
Environmental settings	Agricultural, industrial, residential
Chemical/biological fate characteristics	Speciation, reactivity, degradability, volatility, phase equilibrium constants, complexation, bioaccumulation, biomagnification
Environmental transport and transfer	Advection, dispersion, deposition, washout, degradation, partitioning, erosion, runoff, volatilization, resuspension, sedimentation
Receptors	Human (children, occupation sensitive, general population), wildlife (fish, birds, reptiles, mammals)
Exposure routes	Inhalation (gases, particulates), ingestion (plant, meat, milk, aquatic food, water, soil), dermal contact, external dose (radionuclides)
Risk endpoints	Human (cancer, non-cancer), ecological (individual, species, communities, habitats)

there is a need to assess risks in an increasingly complex and "realistic" manner, involving multiple disciplines. Second, there is a need to be consistent across levels of assessments (i.e., screening to detailed). These two needs spawned the concept of a modelling platform, which allows for both screening and complex models to be developed and applied within a single system. In such a system, the logical link between first-step screening analyses and more complex assessments is clear. In addition, there is a need for efficient collection and use of data. The systematic approach associated with a tiered assessment ensures that data collected and used in a screening-level analysis are consistent with those utilized in the more detailed assessment.

Another primary need for a framework to incorporate multiple disciplines stems from the need to have verifiable modelling protocols. In all of the approaches mentioned so far, individual components (or models) are "hard-wired" into the systems, and to a certain degree, the scientific and quality legacy of the original model that has to be forced into the system is compromised. Any changes to the components will invariably result in changes to the system, because these systems were not designed to accommodate change. If significant modifications are required in these existing systems, the changes tend to be cumbersome, as these models are usually linked to each other in the typical "spider-web" arrangement (i.e., spaghetti code). Experience has clearly demonstrated that modifications within the "spider-web" construct result many times in unnecessary and unexpected changes in other components.

4.3. An Integrated Risk-Assessment Software System

"cleaner" approach for incorporating new models is to reduce the number of variations in the connections so that existing and new attributes maintain their original legacy, realizing that some relatively minor modifications may be necessary. If the interaction and connection of components are focused at the interface between the components, then adding new components or modifying existing ones would not impact the system as a whole. By specifying interface specifications, models can now effectively communicate, as each one will know *a priori* the connection requirements for communication.

A software-based framework for performing environmental risk assessments allows for efficient development and implementation of future environmental simulation software. A framework-based design also allows for individual components (e.g., air transport model) to be developed and inserted directly into a working system that includes all other necessary components. The benefits of this framework-based design include 1) development and testing of new algorithms in the context of a full risk assessment, 2) direct comparison with other algorithms simulating the component processes, and 3) access to standard tools for manipulating and presenting data (e.g., statistical sampling, graphical plotting, and user interfaces).

To allow a suite of users the flexibility and versatility to construct, combine, and couple attributes that meet their specific needs without unnecessarily burdening the user with extraneous capabilities, the development of a computer-based methodology to implement a framework for risk analysis in multiple environmental media was begun in 1994. FRAMES represents a platform that links elements together and yet does not represent the models that are linked to or within it; therefore, changes to elements that are linked to or within FRAMES do not change the framework [9].

FRAMES is an open-architecture, object-oriented framework that

- Interacts with environmental databases
- Helps the user construct a conceptual site model that is real-world based
- Allows the user to choose the most appropriate models to solve simulation requirements
- Presents graphical packages for analysing results.

FRAMES is intended to 1) provide a forum from which various models can interact with each other and 2) facilitate a "plug-and-play" atmosphere for site assessments [9]. FRAMES contains "sockets" for a collection of computer codes that will simulate elements of the transport, exposure, risk assessment, and risk management process, including

- Contaminant source and release to and through overland soils, vadose and saturated zones, air, surface water, and the food supply
- Intake for human health impacts
- Sensitivity/uncertainty analyses
- Ecological impacts
- Geographical Information System (GIS) graphing
- Remediation technology evaluation
- Cost analysis
- Process life-cycle management.

Each of these modules 1) is object oriented, 2) imports the data required for execution, 3) executes the model correctly, 4) correctly exports data to FRAMES data files, and 5) does not have data redundancy. To meet these needs and constraints,

FRAMES structures its data linkages to allow for data files that 1) are used to transfer information between modules; 2) house all user input information, including input from the overall framework user interface (e.g., framework database); and 3) house other information, including output from model simulations, imported data, boundary conditions, and maintained databases.

Figure 14.1 presents a simplified picture of the structure of FRAMES and illustrates file specifications to describe how all information is stored within the framework and passed between modules. Input data are saved to, stored in, and accessed from Global Input Files, which contain all the data required to run the sequence of modules (e.g., source to vadose zone, to saturated zone, to river, to cropland, to exposure, to risk). Only required data are extracted and used by each module. Output data are stored in Global Output Files. These data specifications allow the different modules to communicate and transfer information.

The system helps the user conceptualise the problem by visually expressing the assessment and indicating sources of contamination, contaminant travel pathways through the environment, linkages between contamination and people or wildlife, and impacts associated with the contamination. The framework user interface graphically illustrates this conceptual model and allows the user to see the flow of information and contaminant routing from a source term to releases into the air and subsurface. The interface also illustrates contaminant deposition from the air to agricultural areas as well as contaminated water leaching through vadose zones and local saturated zone to and in nearby rivers. Under this design, a user can choose from a list of models representing different levels of scale and resolution. Scale refers to the physical size and attributes of the problem (e.g., waste unit, watershed, region, or global), and resolution refers to the temporal and spatial resolution of the assessment (for example numerical models have different requirements than analytical models). The appropriate models can be chosen, and the assessment direction can be visually presented, which describes the models and their linkages from source through receptor to the decision-making endpoint. Modules are linked as the direct result of user selection.

14.4. Using an Integrated System for a Complex Analysis

The FRAMES modelling framework was applied by the EPA to a national risk assessment methodology for hazardous waste sites. Typically, screening-level tools (models, databases, methodologies) are used to perform national-scale assessments for the purpose of establishing regulatory thresholds, representing "safe" levels of potential contaminant release to the environment. Screening tools are primarily used because sufficient data on a national scale do not exist, and the computational burden of executing complex models on a national scale is too great, even with advancing technological capabilities.

The assessment associated with the Hazardous Waste Identification Rule (HWIR) [10], however, required a comprehensive environmental transport, exposure, and risk

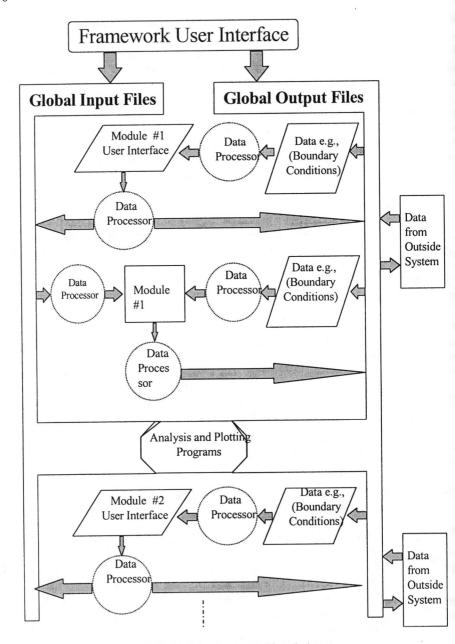

Figure 14.1 Design for an integrated risk analysis system

analysis software system for site-specific, regional, and national application. The
geographic scales associated with most applications range from local (e.g., associated

with a single waste management unit) to urban to watershed. Within an urban area or a watershed there may exist multiple point and nonpoint sources. Regional and national applications reflect scales of regulatory concerns, as opposed to fundamental geographic scales of analysis. For example, a national assessment must include the simulation of exposure and risk at numerous sites located throughout the nation, the intent being to characterize the distribution of exposures and risks as a function of the national collection of individual sources.

Figure 14.2 presents an overall structure of the software framework, which consists of a user interface and a series of data processors. The software structure outlined in this figure is designed to allow for the processing of thousands of individual sites for a national assessment. The framework contains sophisticated screening-level tools/models and allows performance of deterministic and Monte-Carlo-based assessments. Processors (represented by circles in the figure) process information accessed from databases and stored in data files. Some data files (represented by vertically elongated rectangles) are used by succeeding processors. Others (represented by rectangular boxes) are populated externally to supply data to the system.

Each database may contain any or all of the parameters that are needed to conduct a risk assessment and are thus designed to be structurally similar. The distinguishing characteristic of the databases is the source of information and applicability of the data. The collection of databases is utilized in a hierarchical manner. The objective is to build the necessary data files reflecting an individual site by scanning and extracting all available data from the Site-Specific Database. It is typically the case, however, that a complete set of data for a site does not exist. To fill the missing data gaps, the regional database, which is part of the Regional and National Variability Statistics Database, is scanned with a value randomly selected for each parameter missing from the Site-Specific Database. Finally, when regional data do not exist, the national database is scanned with values again randomly selected for data missing from both the site-specific and regional databases. Although Figure 14.2 may imply that the databases and data files might be associated with one processor or another, they actually represent linkages between these processors.

Final risk values are calculated in the Exit Level Processor. This processor actually has two components, as shown in Figure 14.3 (i.e., ELP I and II). The first component retrieves information from the Global Output Files and processes it to produce a Risk Summary Output File. The y axis associated with the table matrix in Figure 14.3 refers to the number of randomly selected sites, indexing from one to "N." The columns (as indexed across the top of the matrix) refer to the number of statistical sampling iterations, implementing a discrete deterministic input file (as extracted from the Global Input Files), indexed from one to "M." Each deterministic run produces a risk that is arrayed to the site and sampling iteration. For each column, the percentages of the risk values that are below a predetermined safe limit (e.g., risk of 10^{-6}) are summed and stored along the bottom of the matrix. This summation at the bottom of each column represents a level of protection. For example, 80% means that 80% of the randomly

252

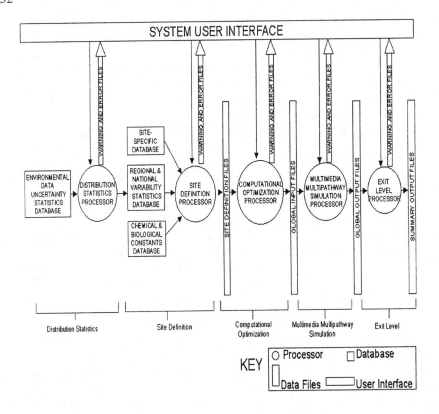

Figure 14.2 Design for a national risk assessment software system

selected sites (i.e., "N" number of sites), which do not have to be similar, contain risks that are below a predetermined safe limit.

The second component of the Exit Level Processor takes the Risk Summary Output File and generates the Protective Summary Output Figure, a series of curves representing percentiles for the percent protection with a given regulatory waste concentration. The percentiles, as indicated in Figure 14.3, include the 95th, 50th, and 5th percentiles of the percent protection. In the case where a regulatory concentration limit is set at "C" (see the x axis of the figure in Figure 14.3), the percent protection for a given degree of uncertainty can be identified. For example, for a regulatory concentration limit of 0.1 mg/L [i.e., Log(1/C)=1] and a probability of protection of 80%, the chance that the

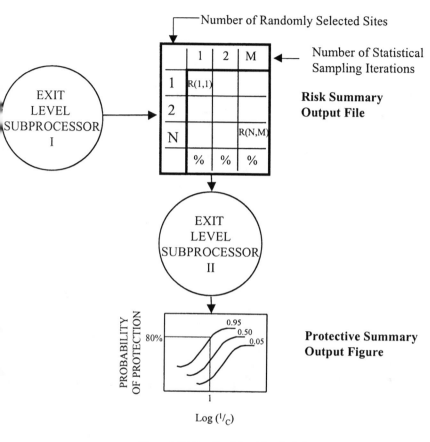

Figure 14.3 Details of the risk calculation

orresponding protection level would be below 80% would be 95%. In other words, 1ere is a 5% probability that the corresponding protection level would be greater than 0%.

This site-based exposure and risk information could be used to establish a national istribution of risks. The national distribution of risks, and all related data, would form 1e technical basis to select chemical-specific exit levels in the waste stream that, if xceeded, define whether the entire waste stream is hazardous or nonhazardous, or /hether it represents a *de minimus* impact. The simplest output of the Exit Level rocessor would be a list of chemical-specific exit levels. However, because so many ıctors influence the actual concentration determined as an exit level, the Exit Level rocessor could output additional information that describes the dimensions of the exit vel.

14.5. Conclusions

The increasing complexity of risk analyses often mandates the integration of national, regional, and site-specific characteristics into a cogent, scientifically defensible, yet cost-effective approach for setting standards and clean-up levels and assessing alternatives to remediate problems. Technology has advanced such that the assessment of human and ecological health risks must include the simultaneous release of contaminants from a waste unit to each environmental medium, the fate and transport of the chemical through a multimedia environment, and the receptor-specific exposures that result. The assessment must also include an estimation of the potential exposures per exposure pathway/receptor, and an estimation of the resulting health impacts/risks. To meet these needs, software systems must incorporate flexibility. The FRAMES platform provides a current example of the type of computerized approach that can be invaluable in managing the risks related to legacy waste.

14.6. References

1. Laniak, G.F., Droppo, J.G. Jr., Faillace, E.R., Gnanapragasam, E.K., Mills, W.B., Strenge, D. L., Whelan, G., and Yu, C. (1997) Overview of the Multimedia Benchmarking Analysis for three risk assessment models: RESRAD, MMSOILS, and MEPAS, *Risk Analysis*, **17**, 203.
2. Mills, W.B., Cheng, J.J., Droppo, J.G. Jr., Faillace, E.R., Gnanapragasam, E.K., Johns, R.A., Laniak, G.F., Lew, C.S., Strenge, D.L., Sutherland, J.F., Whelan, G., and Yu, C., (1997) Multimedia Benchmarking Analysis for three risk assessment models: RESRAD, MMSOILS, and MEPAS, *Risk Analysis*, **17**, 187.
3. Moskowitz, P.D., Pardi, R., Fthenakis, V.M., Holtzman, S., Sun, L.C., and Irla, B. (1996) An evaluation of three representative multimedia models used to support cleanup decision-making at hazardous, mixed and radioactive waste sites, *Risk Analysis*, **16**, 279.
4. Whelan, G., Buck, J.W., Strenge, D.L., Droppo, J.G. Jr., Hoopes, B.L., and Aiken, R.J. (1992) An overview of the multimedia assessment methodology MEPAS, *Haz. Waste Haz. Mat.* **9**, 191.
5. Whelan, G., Strenge, D.L., Droppo, J.G. Jr., and Steelman, B.L. (1986) Overview of the Remedial Action Priority System (RAPS), in Y. Cohen (ed.), *Pollutants in a Multimedia Environment*, Plenum Press, New York, p.338.
6. U.S. Nuclear Regulatory Commission (1994).*Ranking Hazardous-Waste Sites for Remedial Action.* National Research Council., National Academy Press, Washington, DC.
7. U.S. Nuclear Regulatory Commission (1998) Review of Dose Modeling Methods for Demonstration of Compliance with the Radiological Criteria for License Termination, NUREG/CP-0163, U.S. Nuclear Regulatory Commission, Washington, DC.
8. Whelan, G. and Laniak, G.F. (1998) A risk-based approach for a national assessment, in C.H. Benson J.N. Meegoda, R.B. Gilbert, and S.P. Clemence (eds.), *Risk-Based Corrective Action and Brownfields Restorations*, Geotechnical Special Publication Number 82., American Society of Civil Engineers, Reston, Virginia, pp. 55-74.
9. Whelan, G., Castleton, K.J., Buck, J.W., Gelston, G.M., Hoopes, B.L., Pelton, M.A., Strenge, D.L., and Kickert, R.N. (1997) *Concepts of a Framework for Risk Analysis in Multimedia Environmental Systems (FRAMES)*, PNNL-11748, Pacific Northwest National Laboratory, Richland, Washington.
10. 60 FR 66344-469 (1995) Hazardous waste management system: identification and listing of hazardous waste: Hazardous Waste Identification Rule (HWIR), *Federal Register,* Thursday, December 21, 1995.

15. Using Integrated Quantitative Risk Assessment to Optimise Safety in Chemical Installations

Across Cold War facilities, management and organizational factors have often been cited as the cause of accidents. This chapter presents a methodology to estimate release frequencies of hazardous substances in chemical installations. The methodology extends accident risk analysis into an examination of safety management systems. Organizational models describing the characteristics of the management system are employed with the ultimate objective of developing these models along with the technical ones to a level where the effect of the management systems on the parameters of the model will be clear and quantifiable.

Past accidents in industrial plants involving hazardous processes indicate that management factors are a frequent underlying cause[1-4]. Therefore, management and organization play an important role in achieving and maintaining a high level of safety. Most chemical and petrochemical companies have already adopted Safety Management Systems (SMS). Furthermore, such systems are required by the Seseso II Directive (96/82/EC) in the European Community for certain establishments storing hazardous substances. The need to evaluate SMS on site is well established, and great effort has gone into analysing the elements of SMS that might affect system safety[5]. This analysis is usually achieved through some kind of audit[6,7].

On the other hand, quantitative risk assessment (QRA) provides quantified risk indices characterizing the level of safety in the plant and takes into account hardware failures and human actions[8]. It identifies causes of accidents and ways to reduce the likelihood of accidents. If the results of the evaluation of the SMS could be linked to a QRA, quantitative indices taking into account the elements of the SMS could be derived. Various attempts to include organizational and management effects into QRAs of chemical installations were confined into direct judgmental modifications of the frequencies of releases according to the results of audits of the safety management systems of an installation[9,10], or to sensitivity analysis of the management factors affecting risk[11].

The chapter presents a methodology to estimate release frequencies of hazardous substances in chemical installations; this methodology is able to incorporate the effects of a particular SMS employed in the installation. Such SMS can affect systems and actions involved in mitigating consequences, but they are not considered in this risk assessment. The work reported is part of an overall project to quantify the effects of management and organizational factors and incorporate them in QRA in chemical industries[12]. Project I-RISK[12] aims at advancing the state-of-the art by employing detailed "technical" models to estimate the release frequency of hazardous materials in terms of parameters that characterize the stochastic aspects of performance of hardware and humans. Next "organizational" models describing the characteristics of the management system are employed with the ultimate objective to develop these models along with the "technical" ones down to a level where the effect of the former on the parameters of the latter will be clear and quantifiable.

15.1. Master Logic Diagram

In a QRA, the basic approach to identify events is the Master Logic Diagram (MLD)[13]. This logic diagram resembles a fault tree but without the formal mathematical properties of the latter. It starts with a "Top Event," which is the undesired event (like "Loss of Containment"). It continues decomposing into simpler contributing events. Events of one level will, in some logical combination, cause the events of the level immediately above. The development continues until a level is reached where events are identified that directly challenge the various safety functions of the plant. For chemical installations, such as event could be the potential of release of a hazardous substance to the environment. Loss of containment (LOC), for example, means a discontinuity or loss of the pressure boundary between the hazardous substance and the environment, resulting in a release of hazardous substances.

A generic MLD for LOC in installations handling hazardous substances is shown in Figure 15.1. This diagram is partly based on the "Generic Fault Trees" concept[14]. Most of the events in the last level of development in the tree describe categories of causes that, alone or in some combination, result in a LOC of the hazardous substance. Some of these causes can be further developed into joint events consisting of an initiating event and the failure of one or more safety functions. Examples of such event-trees are these leading to failure from overpressure. Other events, however, require different models (e.g., Multistate Markov model).

Two major categories of events lead to LOC: those resulting in a structural failure of the containment and those resulting in containment bypassing because of an inadvertent opening of an engineered discontinuity in the containment (e.g., valves, hatches).

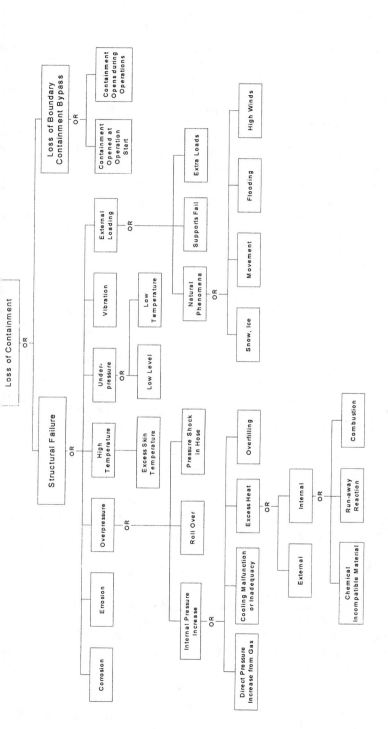

Figure 15.1. Master Logic Diagram for "Loss of Containment" in chemical plants

15.1.1. Loss of Containment as a Result of Structural Failure

Structural failure of the containment can be caused in seven general ways: overpressure, underpressure, corrosion, erosion, external loading, high temperature, and vibration. Each of those fundamental physical processes can induce stresses that will exceed the strength of the containment or can reduce the strength of the containment to levels that cannot withstand normal stresses. Each of these causes of failure can be considered the result of an "initiating event" coupled with the failure of one or more safety functions. The latter are combinations of engineered systems and human actions based on specific procedures to prevent the initiating event from causing containment failure. Note that the frequency of each initiating event can vary from extremely low values (e.g., the frequency of a large earthquake) to very high values (e.g., almost continuous operation in a corroding environment). The following subsections describe those structural events that can be further subdivided into additional detail. Corrosion, erosion, high temperature, and vibration are not considered further.

Overpressure
The second level of decomposition of the MLD in Figure 15.1 follows the possibility of failure from overpressure. In overpressure, the internal pressure increases to such a level that the stresses induced on the containment overcome its strength. Overpressure may be created through an internal pressure increase, rollover, or pressure shock.

An internal pressure increase may occur in four ways: 1) a direct pressure increase from gas material, 2) cooling malfunction, 3) excess heat, or 4) overfilling. A fourth level of decomposition is possible for the cause "Excess heat," which can be decomposed into "internally generated" and "externally generated" excess heat. The former of these two causes can be further decomposed into two contributing causes: "run-away reaction" and "combustion."

Thus, the generic development of the MLD for LOC from overpressure stops after having identified the following subcategories of causes: a) direct pressure increase from gas material, b) cooling malfunction, c) run-away reaction, d) combustion, e) external excess heat production, f) overfilling, g) rollover, and h) pressure shock. Further development of models to identify and quantify which of these causes is possible and in what ways requires more specialized understanding of the particular installation under analysis.

Underpressure
Underpressure, meaning lower internal pressure than external pressure, can lead to containment failure if the induced stress by the pressure difference becomes larger than the strength of the containment material. The result is an implosion. Underpressure can be caused by a low level of liquid in the containment or low temperature in the containment. Further development of the MLD requires more specialized understanding of particular systems.

External Loading

Structural failure of containment from external loading occurs when such external loads induce stresses to the containment exceeding material strength. External loading can be caused by loading form natural phenomena, failure of supports, or external loads on the containment. The first category can be further subdivided into four types of natural phenomena: 1) earthquakes, 2) flooding, 3) high winds, and 4) snow or ice. Further development of the MLD requires a more specific understanding of the system.

15.1.2. Loss of Containment as a Result of Bypassing

Opening of an engineered feature of the containment during operations (e.g., manual opening of a valve or hatch by an operator), or failure to close such a feature when operations start, amounts to LOC. For example, manual or power valves or hatches might be left open for other causes and not closed before operations start.

15.2. Event Tree-Fault Tree Analysis

A number of direct causes of LOC can be further analysed and modelled as a joint event consisting of an "initiating event" and failure of one or more safety functions. Detailed models for this type can be built in terms of event trees and fault trees. To quantify logic models, three major categories of parameters must be estimated: frequencies of initiating events, component unavailability, and probabilities of human actions[8]. Frequencies of initiating events are either estimated directly from historical data or from detailed logic models (e.g., fault trees). This latter approach is necessary when there are dependences among the initiating events and the successful operation of one or more systems.

Component availability is distinguished as continuously monitored and non-continuously monitored. The state of continuously monitored components is always known and their average unavailability is given, as shown in Table 15.1. The state of components that are not continuously monitored can be revealed only through periodic tests. Four conditions contribute to the unavailability of these components: 1) hardware failure between tests, 2) repair of detected failures, 3) routine maintenance, and 4) other maintenance, as shown in Table 15.1, Case A. Case B in Table 15.1 shows the unavailability of untested monitored components. (Cases C and D for non-repairable and repairable components, respectively).

In the logic models, human errors are assumed to occur if 1) an operator does not perform an action (foreseen in the operating procedures) and 2) this error is not detected and recovered by another operator. The probability of this combination is set equal to

$$H = Q_{01}Q_{02} \tag{15.1}$$

where

Q_{01} = probability of not performing the action
Q_{02} = probability of not detecting and recovering the error.

TABLE 15.1 Average unavailability for different types of components

Case A: Non-continuously monitored, periodically tested components
$$\overline{U} = \overline{U}_1 + \overline{U}_2 + \overline{U}_3 + \overline{U}_4$$

1) Hardware failure between tests λ: failure rate T: mean time between tests	$$\overline{U}_1 = 1 - \frac{1 - e^{-\lambda T}}{\lambda T}$$ if $\lambda T \ll 1$ $\overline{U}_1 \cong \frac{1}{2}\lambda T$
2) Repair of detected failures T_R: duration of the repair	$$\overline{U}_2 = \frac{\dfrac{e^{-\lambda(T+T_R)} + \lambda(T + T_R) - 1}{\lambda(T + T_R)}}{1 + (1 - e^{-\lambda T_R})e^{-\lambda T}} + \frac{1 - e^{-\lambda T_R}}{e^{\lambda T} + 1 - e^{-\lambda T_R}}$$ if $\lambda(T_R + T) \ll 1$ $\overline{U}_2 \approx \frac{1}{2}\lambda T + \lambda T_R$
3) Routine maintenance f_m: frequency of maintenance T_m: duration of maintenance	$$\overline{U}_3 = \overline{U}_2 \frac{1}{1 + f_m T_m} + \frac{f_m T_m}{1 + f_m T_m}$$ if $f_m T_m \ll 1$ $\overline{U}_3 = \overline{U}_2 + f_m T_m$
4) Other maintenance Q_{M1}: prob. of committing an error Q_{M2}: prob. of not detecting errors	$$\overline{U}_4 = \overline{U}_3(1 - Q_{M1}Q_{M2}) + Q_{M1}Q_{M2}$$ if $Q_{M1}Q_{M2} \ll 1$ $\overline{U}_4 = \overline{U}_3 + Q_{M1}Q_{M2}$

Case B: Noncontinuously monitored, untested components	
λ: failure rate T_p: fault exposure time	$$\overline{U} = 1 - \frac{1 - e^{-\lambda T_p}}{\lambda T_p}$$

Case C: Monitored, non-repairable components	
λ: failure rate T_M: duration of maintenance	$$\overline{U} = 1 - \exp(\lambda T_M)$$

Case D: Monitored, repairable components	
λ: failure rate μ: repair rate	$$\overline{U} = \frac{\lambda}{\lambda + \mu}$$

15.3. Accident Sequence Quantification

The next major procedural step of quantitative risk assessment includes all the tasks associated with the quantification of accident sequences. This quantification implies manipulation according to the laws of Boolean algebra. The frequency of the accident sequences is then expressed in terms of the number of accident sequences (cut sets) of the form.

$$d = \sum c_i \text{ (rare event approximation) with } c_i = f_i \cdot \prod_{j=1}^{J} U_j \cdot \prod_{k=1}^{K} H_k \qquad (15.2)$$

Each basic cut set can be expressed in terms of parameters comprising the frequency of the initiating event (f_i), failure rate (λ), duration of repair (T_R), mean time between tests (T), frequency of maintenance (f_m), duration of maintenance (T_M), probability of committing an error (Q_{M1}), probability of not detecting errors (Q_{M2}), human error probability of an action (Q_{o1}), and probability of not detecting and recovering the error (Q_{o2}). These parameters are modified according to the quality of the SMS, as discussed in the following section.

15.4. Modification of the Frequency of Loss of Containment According to the Safety Management System

All basic events and their corresponding technical parameters are grouped in such a way that all members of a group are influenced by a common management system (i.e., operation, maintenance, or emergency, and then, for example, maintenance of mechanical components, of electrical components, etc.). Each of the resulting groups is in general influenced by the following eight management delivery systems: 1) availability of personnel, 2) commitment and motivation to carry out the work safely, 3) internal communication and coordination of people, 4) competence of personnel, 5) resolution of conflicting pressures and demands antagonistic to safety, 6) plant interface, 7) plans and procedures, and 8) delivery of correct spares for repairs[15]. The overall influence of the SMS on a technical parameter is quantified by

$$m_j = \sum_{i=1}^{8} y_i w_{ij} \qquad 0 \le mj \le 10 \qquad (15.3)$$

where

m_j = modification factor of the jth technical parameter
y_i = quality of the ith delivery system (i=1,...,8)
w_{ij} = weighting factor assessing the relative importance of the ith management delivery system on the influence of the jth technical parameter
k = an index running over the basic events of the kth group.

The dependence on k is not shown here because the application under discussion assumes that all basic events are influenced by only one management subsystem.

Once the modification factors are assessed, the technical parameters are modified according to:

$$\ln f_j = \ln f_i + \frac{(\ln f_u - \ln f_l)}{10} m_j \qquad (15.4)$$

where

f_j = modified value of the jth technical parameter

f_l = lower value of each parameter, for the installation with the poorest SMS in the industry

f_u = upper value of each parameter, for the installation with the best SMS in the industry.

The weighting factors w_{ij} are based on expert judgement; the values shown in Table 15.2 have been used for the application at hand[16]. The quality of the delivery systems y_i are assessed according to an audit[8]. Lower and upper values of the technical parameters are based on expert judgement (see [16]. Once the modified technical parameters are obtained, the frequency of LOC is obtained through classical quantification.

TABLE 15.2 Weighting factors of the delivery systems affecting basic event parameters

Function	Availability	Commitment	Communications	Competence	Conflict resolution	Interface	Rules and Procedures	Spares and Tools
Q_{O1}	0.06	0.15	0.07	0.16	0.18	0.2	0.18	0
Q_{O2}	0.05	0.14	0.05	0.21	0.21	0.2	0.14	0
Q_{M1}	0.08	0.19	0.06	0.14	0.14	0.08	0.17	0.14
Q_{M2}	0.05	0.13	0.05	0.22	0.18	0.18	0.15	0.04
f_i	0.1	0.2	0.1	0.1	0.1	0	0.4	0
λ	0.08	0.12	0.12	0.08	0.08	0.08	0.16	0.28
T	0.05	0.24	0.14	0	0.28	0.05	0.19	0.05
f_m	0.05	0.21	0.16	0	0.32	0.05	0.16	0.05
T_R	0.12	0.07	0.21	0.09	0.1	0.19	0.1	0.12
T_M	0.12	0.08	0.21	0.08	0.12	0.17	0.08	0.14

15.5. Case Study

The methodology described so far is exemplified through its application to the risk assessment of a liquefied petroleum gas (LPG) scrubbing tower of an oil-refinery. A detailed technical model simulating the response of the system to various initiating events can be developed along with a detailed model simulating the influence of the plant-specific management and organizational practices. The overall effect is quantified through the frequency of release of LPG as a result of a LOC in scrubbing towers of the refinery.

In such a case study, the unit in question contains three towers (T6654, T6655, and T6656) where LPG is scrubbed to remove hydrogen sulphide. In the first tower (T6654) (see Figure 15.2) methylethylamine (MEA) absorbs most of the hydrogen sulphide contained in the LPG stream. When the stream enters the second tower (T6655), hydrogen sulphide is further scrubbed by caustic (NaOH). The LPG stream leaves this tower and enters the last tower (T6656), where water washes any entrained caustic. This last tower is equipped with two safety valves, which open in case of high pressure (see [17] for details).

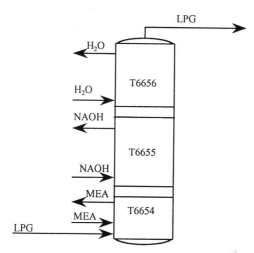

Figure 15.2 Simplified diagram of a liquefied petroleum gas scrubber

15.5.1. Designing the Master Logic Diagram of the Situation

Following the MLD methodology, the following direct causes for LOC can be identified for this case study:

Tower failure from material aging or corrosion
Tower failure from overpressure through pressure increase caused by heat flux from an external heat source
Tower failure from overpressure as a result of overfilling
Tower failure from freezing
Extra loads from a road accident.

5.5.2. Combining Initiating Events and Safety Functions/Systems Failures

Each "direct cause of LOC" in the MLD can be considered a joint event consisting of one initiating event and the failure of one or more safety functions that are served by either systems (hardware) and/or operator procedures. In certain circumstances, no safety functions are present, and the direct cause will be the initiating event itself. Two direct causes (tower failure from material aging and tower failure from freezing) are actually combinations because the safety function in this case is the structural strength of the tower material that by definition is exceeded by stress. Two other direct causes (tower failure from overpressure through pressure increase caused by heat flux from an external heat source and tower failure from overpressure as a result of overfilling) are considered as joint events consisting of one initiating event and failure of one or more safety systems. These two events are further analysed to identify all the initiating events of this system, which are presented in Table 15.3. The safety systems required to prevent the occurrence of LPG release for all initiating events are presented in Table 15.4.

TABLE 15.3 Initiating events

1. Operating conditions off specifications
2. External fire
3. High inlet of MEA from valve failure
4. No outlet of MEA
5. High inlet of caustic (NaOH)
6. No outlet of caustic (NaOH)
7. High inlet of water from valve failure
8. No outlet of water
9. High inlet of LPG
10. No outlet of LPG

TABLE 15.4 Safety systems

1. Pressure detection system
2. Fire suppression system
3. Pressure safety valves
4. Low-level protection system in Tower T6654
5. High-level protection system in Tower T6654
6. Low-level protection system in Tower T6655
7. High-level protection system in Tower T6655
8. Low-level protection system in Tower T6656
9. High-level protection system in Tower T6656
10. Tower integrity

Next, event trees can be constructed for all initiating events presented in Table 15.3, defining the response of the plant and the spectrum of the resulting damage states. A typical event tree constructed for one such initiating event is presented in Figure 15.3. This tree is for the scenario in which a high inlet of MEA from valve failure causes a LOC. The first two tree paths (#1, #2) lead to a safe state; the third leads to tower rupture from overpressure. A total of 10 event trees corresponding to 10 initiating events could be developed. Failures of systems are modelled through the Fault Tree technique. Nine Fault Trees could be constructed for the first nine safety systems presented in Table 15.4.

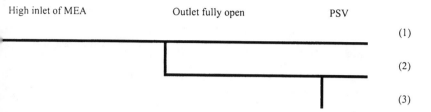

High inlet of MEA   Outlet fully open   PSV

(1)

(2)

(3)

Figure 15.3 Event tree for initiating event "High inlet of MEA from valve failure"

15.5.3. Modifying Data According to the Plant Management Model

The frequency of LOC in any of the three scrubbing towers is expressed in terms of 41 basic events. The safety management audit team would determine that all these events are affected by a single management system. As a result, only eight management delivery systems need to be assessed and quantified (for details, see Hale et al.[15]). The qualities of the eight outputs y_i ($i=1,...,8$) when combined with the weighting factors w_{ij} given in Table 15.2 provide the modification factors given in Table 15.5.

TABLE 15.5 Modification factors of technical parameters

Technical Parameter	Description	Modification Factor
Q_{o1}	Probability of not performing an action	9.1
Q_{o2}	Probability of not detecting and recovering from an error	9.0
Q_{M1}	Probability of committing an error during maintenance	9.3
Q_{M2}	Probability of not detecting an error during maintenance	9.0
f_i	Frequency of initiating event	9.5
Λ	Failure rate	9.3
T	Mean time between tests	9.4
f_m	Frequency of maintenance	9.3
T_R	Duration of repair	9.1
T_M	Duration of maintenance	9.2

5.5.4. Quantifying Accident Sequence

Three large fault trees with top events "Tower T6654 Failure," "Tower T6655 Failure," and "Tower T6656 Failure" could be created, each consisting of an "OR" gate with accident sequences leading to the corresponding top event as inputs. Each accident sequence would then be developed in terms of an "AND" gate with system failures and the initiating event of each accident sequence as inputs. Quantification would be performed for three cases according to the specific management system of the installation. The first case would use the values of the parameters corresponding to the best management system or f_l. The second case would use the values of the parameters corresponding to the worst management system or f_u. The third case would use the modified values of the parameters f_j according to Equation (2) and the values of Table 15.5. Lower and upper values of the technical parameters for the equipment of this tower are presented in Table 15.6. The results of the three sets of calculations providing the frequencies of failure of Towers T6654, T6655, and T6656 are presented in Table 15.7.

15.5.5. Assessing Consequence

The three LPG towers (T6654, T6655 and T6656) contain flammable LPG, which will be released to the environment in case of a LOC. If LPG is ignited immediately, a

TABLE 15.6 Lower and upper values of technical parameters

Equipment	Parameter	Lower	Upper
Safety valves, remote control valves	T_r, T_m (hr)	24	8,760
All equipment	T	Plant data x 0.9	Plant data x 100
Safety valves, remote control valves	λ	1.71×10^{-6}	3.15×10^{-5}
All equipment	Q_{m1}	1.00×10^{-4}	0.5
All equipment	Q_{m2}	5.00×10^{-2}	1
Safety valves fail in open position	λ	8.50×10^{-7}	3.40×10^{-5}
Manual valves	λ	2.74×10^{-7}	5.04×10^{-6}
Manual valves	T_r, T_m, T (hr)	Plant data x 0.9	Plant data x 100
Flow instruments	λ	8.30×10^{-7}	5.59×10^{-6}
Flow instruments	T_r, T_m (hr)	24	336
Instruments where equipment has to be taken apart for repair	T_r, T_m (hr)	24	8,760
Level instrument	λ	2.50×10^{-6}	1.10×10^{-5}
Pressure instrument	λ	2.50×10^{-7}	2.94×10^{-6}
Temperature instrument	λ	3.00×10^{-8}	2.97×10^{-5}
Process pump	λ	4.50×10^{-5}	2.28×10^{-4}
Process pump	T_r, T_m (hr)	24	8,760
Human error	Q_{o1}	1.00×10^{-4}	5.00×10^{-1}
Human error	Q_{o2}	5.00×10^{-2}	1.00
Corrosion	Λ	5.00×10^{-8}	5.00×10^{-6}
Corrosion	T_r (hr)	24	8,760
Water of fire fighting system	λ	5.00×10^{-8}	5.00×10^{-6}
Water of fire fighting system	T_r (hr)	2	331
Heater	λ	5.00×10^{-8}	5.00×10^{-6}
Heater	T_r (hr)	24	8,760
External fire	f_i	1.00×10^{-6}	1.00×10^{-5}
High inlet of NaOH in tower	f_i	1.00×10^{-6}	1.00×10^{-5}
High inlet of LPG in tower	f_i	1.00×10^{-6}	1.00×10^{-5}
Ambient temperature very low	f_i	1.00×10^{-6}	1.00×10^{-5}
Conditions off specifications	f_i	1.00×10^{-9}	1.00×10^{-8}
Impact from road	f_i	1.00×10^{-12}	1.00×10^{-11}

boiling liquid expanding vapour explosion will occur. Otherwise, LPG will disperse to the atmosphere as a dense cloud and either a flash fire or an explosion will occur. It is assumed that, in case of delayed ignition, there is a probability of 1/3 for flash fire and 2/3 for explosion[18]. All the possible sequences that might occur in case of tower failure are presented in Table 15.8. In all cases, individual conditional risk of death does not exceed 1×10^{-4}/yr at a distance of 300 m away from the towers.

15.5.6. Risk Integration

The area above certain risk levels can be calculated for three management assessments: the specific system of this installation and the two bounding values for the best and worst possible systems. This level of area versus risk is presented in Figure 15.4. Interpreting this figure, if the best management system were used, the area with unconditional individual risk higher than 10^{-8}/year would be equal to 0.143 km². If the worst management system were used, this area would be equal to 3.36 km². In the actual case with a management system judged to be very good, this area is equal to 0.231 km².

TABLE 15.7 Frequencies of failure

Catastrophic Failure of Tower T6654
a) Best possible case 1.1 x 10⁻¹⁰/hr
b) Worst possible case 1.2 x 10⁻⁴/hr
c) Plant as assessed 4.7 x 10⁻¹⁰/hr
Catastrophic Failure of Tower T6655
a) Best case 1.3 x 10⁻¹⁰/hr
b) Worst case 2.6 x 10⁻⁴/hr
c) Specific model 5.3 x 10⁻¹⁰/hr
Catastrophic Failure of Tower T6656
a) Best case 1.1 x 10⁻¹⁰/hr
b) Worst case 3.1 x 10⁻⁴/hr
c) Specific model 3.9 x 10⁻¹⁰/hr

TABLE 15.8 Accident sequences of plant

Catastrophic Failure of Tower T6654
1. Boiling liquid expanding vapour explosion (2,700 Kg LPG)
2. Flash fire (2,700 Kg LPG)
3. Explosion (2,700 Kg LPG)
Catastrophic Failure of Tower T6654
4. Boiling liquid expanding vapour explosion (1,200 Kg LPG)
5. Flash fire (1,200 Kg LPG)
6. Explosion (1,200 Kg LPG)
Catastrophic Failure of Tower T6654
7. Boiling liquid expanding vapour explosion (800 Kg LPG)
8. Flash fire (800 Kg LPG)
9. Explosion (800 Kg LPG)

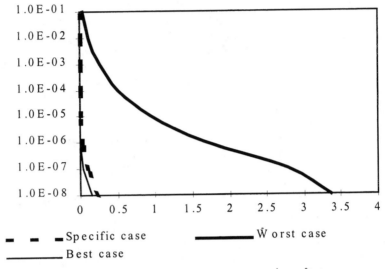

Figure 15.4 Area (km²) above certain risk levels (10⁻¹ to 10⁻⁸/yr)

15.6. Conclusions

The I-RISK methodology provides a step-by-step approach for integrating the effects of SMS into the quantification of risk for an installation handling hazardous materials. The methodology consists of a technical model and a management model linked at a point where specific managerial tasks influence specific parameters determining the probability of occurrence of specific basic events affecting the quantified risk indices. The ability to integrate these management aspects into risk quantification provides some guidance into determining the importance of maintaining or improving safety on certain systems.

15.7. References

1. Rasmussen, J. (1995). "The experience with the major accident reporting system from 1984 to 1993." EUR 16341 EN, Joint Research Centre of the European Commission, Ispra, Italy.
2. Hurst, N.W., Bellamy, L.J., Geyer, T.A., and Astley, J.A. (1991). A classification scheme for pipework failures to include human and sociotechnical errors and their contribution to pipework failure frequencies. *Journal of Hazardous Materials*, **80** (11) 66-69.
3. Pate-Cornell, M.E., and Bea, R.G. (1992) Management errors and system reliability: a probabilistic approach and application to offshore platforms. *Risk Analysis*, **40** 239-257.
4. Pate-Cornell, M.E., and Firchbeck, P.S. (1993) PRA as a management tool: organizational tools and risk based priorities for the maintenance of the tiles of the space shuttle orbiter. *Reliability Engineering and System Safety*, **40**, 239-257.
5. Mitchison N., and Papadakis, G.A. (1999) Safety management systems under Seveso II: implementation and assessment. *Journal of Loss Prevention in the Process Industries*, **12** (1), 43-51.

6. Bellamy, L.J., Wright, M.S., and Hurst, N.W. (1993) History and development of a safety management system audit for incorporation into quantitative risk assessment. *International Process Safety Management Workshop*, AIChE/CCPS, 22-24 September 1993, American Institute of Chemical Engineers, Washington, D.C.

7. Hurst, N.W. (1993) Auditing and safety management. CEC DGXII/ESReDA Conference, *Occupational Safety Seminar*, Lyon, France, 14-15 October 1993.

8. Papazoglou, I.A., Nivolianitou, Z., Aneziris, O., and Christou, M. (1992). Probabilistic safety analysis in chemical installations, *Journal of Loss Prevention in the Process Industries*, **5** (33), 181-191.

9. Hurst, N.W., Young, S., Donald, I., Dibson, H., and Muyselaar, A. (1996) Measurers of safety management performance and attitudes to safety at major hazard sites. *Journal of Loss Prevention in the Process Industries*, **9** (2), 161-172.

10. Pitaldo, R., Williams, J.C., and Slater, D.H. (1990) Quantitative assessment of process safety programs. *Plant/Operations Progress*, **9** (3), 169-175.

11. Papazoglou, I.A., and Aneziris, O. (1999) On the quantification of the effects of organizational and management factors in chemical installations. *Reliability Engineering and System Safety*, **63** 33-45.

12. Oh, J.I.H., Brouwer, W.G.J., Bellamy, L.J., Hale, A.R., Ale, B.J.M., and Papazoglou, I.A. (1998). The IRISK Project: development of an integrated technical and management risk control and monitoring methodology for managing and quantifying on-site and off-site risks, in A. Mosleh and R.A. Bari (eds.), *Proceedings of the 4th International Conference on Probabilistic Safety Assessment and Management*, Vol 1, p2485, New York.

13. Kaplam, S., Apostolakis, G., Garrick, B.J., Bley, D., and Woodard, D. (1981) *Methodology for Probabilistic Risk Assessment of Nuclear Power Plants*. PLG-0209, Pickard, Lowe and Garrick, Inc., Irvine, California.

4. Van der Mark, R. (1996) *Generic Fault Trees and the Modelling of Management and Organisation: Final Year Report*. Delft University of Technology, Department of Statistics, Probability and Operations Research, Delft.

5. Hale, A.R., Guldenmund, F., and Bellamy, L. (1998) An audit method for the modification of technical risk assessment with management weighting factors, in A. Mosleh and R.A. Bari (eds.), *Proceedings of the 4th International Conference on Probabilistic Safety Assessment and Management*, Vol 3, 2093, New York.

6. Bellamy, L., Papazoglou, I.A., Aneziris, O., Ale, B.J.M., Hale, A.R., Post, J.G., Morris, M, and Oh, J.I.H. (1999). I-RISK. Development of an integrated technical and management risk control and monitoring methodology for managing and quantifying on-site and off-site risks, Contract ENVA-CT-96-0243, Rpt 2tEU, Ministry of Social Affairs and Employment, Den Haag, Netherlands.

7. Papazoglou, I.A., and Aneziris, O. (1998) System performance modeling for quantification of organizational factors in chemical installations, in A. Mosleh and R.A. Bari (eds.), *Proceedings of the 4th International Conference on Probabilistic Safety Assessment and Management*, Vol 3, 2081, New York.

8. --- (1992) Methods for the Calculation of Physical Effects Resulting from Releases of Hazardous Materials (Liquids and Gases) (Yellow Book), TNO, Committee for the Prevention of Disasters, Voorburg, The Netherlands.

16. Site-Specific Modification of Ground-Water Generic Criteria as Applied to a Contaminated Site

In Western countries in particular, and in the growing Green movement in the Eastern countries as well, how risks are regulated is a key component to the management of Cold War legacies. Regulatory approaches that allow site-specific risk analysis must be considered because regulations and associated guidance are often the basis for defining acceptable cleanup levels. This chapter provides an example of how one regulatory group has approached this issue. The example shows how generic regulatory guidelines can be implemented in a manner that allows for site-specific risk-based evaluations of cleanup levels.

The definition of what is "clean" or "acceptable" in terms of residual environmental concentrations after remediation is a formable task facing those undertaking the remediation of sites with radioactive or hazardous chemicals stemming from Cold War activities. The cost of cleanup is normally directly related the selected target cleanup level. Experience in the United States has been that very stringent cleanup levels can often result in prohibitive remediation costs, even for relatively small sites.

National or regional (state, province, etc.) regulations and associated guidance have historically been based largely on defining a single standard or norm for acceptable environmental concentrations. Although often risk-based in their origin, the "one-size-fits-all-sites" approach generally means these standards or norms are conservative values designed to protect in extreme situations. When applied generically at all sites, many sites have to clean to unnecessarily restrictive levels.

This chapter provides an example of how one regulatory group has approached this issue. The example shows how generic regulatory guidelines can be implemented in a manner that allows for site-specific risk-based evaluations of cleanup levels. Although a number of federal and state agencies in the United States have, or are working towards, such an approach, an example from Canada was selected to illustrate a state-of-the-art implementation. The provincial government of Ontario, Canada, has implemented relatively unique generic guidance. Although most of their sites are not the result of Cold War activities, the principles are applicable to other such contaminated sites. Their approach allows the use of site-specific risk information, which can potentially help focus remediation efforts on sites with the highest actual or potential risk to people and the environment.

Considerable time and effort can go into determining the particular contaminants of concern and establishing appropriate restoration levels. Wrestling with streamlining this process, the Ontario Ministry of the Environment and Energy published guidelines[1] in 1996. These guidelines are applicable to any contaminated site in the province, except Potentially Sensitive Sites. Generic criteria for 115 organic and inorganic chemicals were derived and listed in four tables, depending on intended land use, type of required ground water restoration, and depth of required soil restoration. In addition, different values for each soil criterion were provided depending on the soil texture (i.e., coarse and medium/fine soil texture). The generic criteria development process was described in detail in a separate document[2]. Additional documents[3,4] were also published to establish the cleanup procedures to be followed in the province.

The generic criteria are based on conservative assumptions and models and provide protection for all contaminated sites in the province. However, remediation activities up to these levels often result in overprotection and may require extensive effort, time, and funding. To avoid unnecessary remediation and/or restrictive risk management measures, the Ministry of Environment and Energy allowed proponents to develop site-specific criteria by following the procedure applied in the derivation of generic criteria and by substituting the default conservative parameter values with site-specific ones. Also, the proponent was given the opportunity to apply a relevant risk management decision, especially in cases where contamination exceeded the developed site-specific criteria.

In the discussions of this presentation at the NATO Advanced Study Institute (that provides the basis for this textbook), significant misunderstandings occurred based on judging this contribution as a comprehensive research approach rather than a resource-limited screening application. Several participants clearly did not appreciate the innovative and useful aspects of the regulatory guidance approach–and as result were highly critical. Although such regulations need to be based in "sound science," the constraints of such actions are often misconstrued as not having such a basis. These reactions are similar to those discussed in Chapter 6 relative to the early attempts to adopt risk based approaches in United States, showing this problem of misunderstanding by "scientific experts" of risk-based approaches occurs in other countries as well.

The consideration of regulatory approaches that allow site-specific risk analysis is critical because the regulations and associated guidance are often the primary basis for defining acceptable cleanup levels. The following regulatory guidance example illustrates of the type of approach that can allow derivation of site-specific criteria for defining acceptable site-specific cleanup levels. This particular regulatory guidance approach is relatively unique because it includes several media-to-media linkages such as is discussed in terms of multimedia risk-based modelling in Chapter 14.

16.1. Development of Generic Criteria for Ground Water—Component Selection

A modification of criteria starts with the development of generic criteria. This development is based on a component approach. The Ministry of Environment and Energy assumed three principal pathways in the derivation of generic criteria for ground water:

- Use of ground water as drinking water
- Migration of ground water vapour to indoor air
- Migration of ground water to surface water.

Accordingly, three components have been designed. Figure 16.1 demonstrates the derivation.

The derivation process consists of three steps. In the first step, all existing values for **Criterion Components** must be collected and entered into the process. The parameters used to estimate each criterion component are as follows:

- GW1 (Drinking Water Quality) Component is represented by the Ontario Drinking Water Objective or, if that objective does not exist, by the health-based and odour/taste value for the chemical of interest
- GW2 (Migration: ground water vapour to indoor air) Component is based on the background indoor air value, the health-based indoor air value, and the odour recognition value in air
- GW3 (Migration: ground water to surface water) Component is based on the lowest toxicity value for freshwater species.

During the second step, called the **Value Selection Process**, the values of the criterion components are identified and the lowest value is selected. In the third step, **Risk Management Decisions**, this value is compared to several numbers to arrive at the final criterion value:

The lower of half of the solubility and the ceiling value (selected to minimize the potential for continuous degradation of the ground water in the province)
The higher of the method of detection limit (MDL, selected on the basis of enforceability)
The ground-water background concentration for the contaminant of interest (selected on the basis of enforceability).

16.2. Site-Specific Modification Of Generic Criteria—Value Selection

Site-specific criteria provide a tool that assists risk managers in evaluating and comparing alternatives for site development and remediation. The following case study illustrates the suggested procedure of site assessment and risk modelling.

274

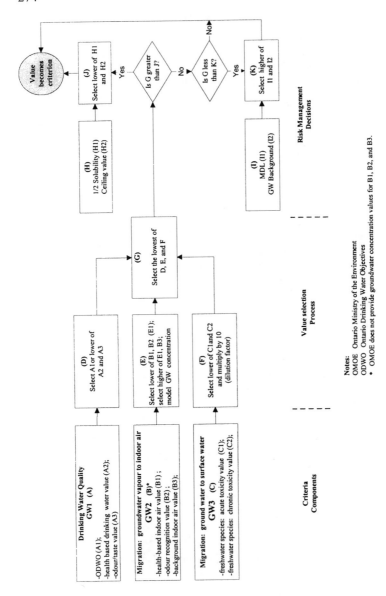

Figure 16.1 Ground-water generic criterion development flowchart (based on [4])

Drinking Water Quality GW1 (A)

-ODWO (A1);
-health based drinking water value (A2);
-odour/taste value (A3)

Migration: groundwater vapour to indoor air GW2 (B)*

-health-based indoor air value (B1);
-odour recognition value (B2);
-background indoor air value (B3);

Migration: ground water to surface water GW3 (C)

-freshwater species: acute toxicity value (C1);
-freshwater species: chronic toxicity value (C2);

(D) Select A1 or lower of A2 and A3

(E) Select lower of B1, B2 (E1); select higher of E1, B3; model GW concentration

(F) Select lower of C1 and C2 and multiply by 10 (dilution factor)

(G) Select the lowest of D, E, and F

(H) 1/2 Solubility (H1) Ceiling value (H2)

(J) Select lower of H1 and H2

Is G greater than J?

Yes

No

Is G less than K?

Yes

No

(K) Select higher of I1 and I2

(I) MDL (I1) GW Background (I2)

Value becomes criterion

Criteria Components

Value selection Process

Risk Management Decisions

Notes:
OMOE   Ontario Ministry of the Environment
ODWO   Ontario Drinking Water Objectives
*   OMOE does not provide groundwater concentration values for B1, B2, and B3.

16.2.1. Site Assessment

The site used for the case study (Figure 16.2) is located in a mixed residential, commercial, and industrial area of a big city. The nearest ecological receptor is a creek located 1.5 km southeast of the site. The site was developed as an industrial property in 1957; before that, the land was agricultural. The facility built on this site manufactured electric heaters and controls. The facility building was demolished in the later half of 1996 to prepare for future residential development. The site plan included 129 lots, 115 of which were proposed for townhouses/linkhomes and the remainder for various other residential, commercial, and public uses.

The environmental and geo-technical findings allowed classifying the site as non-sensitive. The soil consisted of silty clay and sand and was classified to be coarse. The ground water appeared to be heavily contaminated. Different ground-water remedial options were evaluated at the site. However, the analytical results from many rounds of sampling demonstrated that ground water could not be remediated to acceptable levels of contamination (assuring protection of human health and ecological receptors) in a cost-efficient way. As a result, it was decided to evaluate the site contamination risk and to apply risk management controls if necessary to redevelop the site. The ground water satisfied the requirements for non-potable use because 1) the area was serviced by municipal water, 2) present or future surface or ground-water sources would not be affected, and 3) the municipality was notified about the proposal to restore the site to non-potable levels[1].

Conceptual Model

A limited area of soil polychlorinated biphenyl (PCB) contamination was discovered and later excavated and treated *ex situ*. The ground water was found to be located in an upper and lower aquifer. The ground-water flow was determined to be in a southeastern direction toward the creek; flow velocity was calculated to be approximately 9 m/yr. Volatile organic compound (VOC) contamination was identified in ground water in the upper aquifer beneath a former chemical storage area. In addition, a discontinuous perched water table was discovered in a sand unit found above the upper aquifer unit. The lateral extent of the ground-water contamination is shown on Figure 16.2

Contaminants of Interest

Ground-water samples were collected and analysed from the monitoring wells installed in the area of contamination. The measured concentrations were compared to the corresponding generic criteria (Table 16.1). Any substances that exceeded their generic criteria in any sample were selected for detailed assessment. In addition, all non-detected chemicals (i.e., vinyl chloride) with detection limits higher than the generic criteria were also chosen for detailed analysis. The analysis of the data allowed the following conclusions:

Generic criteria for all chemicals of interest exist

Five chemicals (trichloroethylene, trichloroethane, and related degradation
   products) were found above their generic criteria in ground water.
These five chemicals were selected for further specific evaluation.

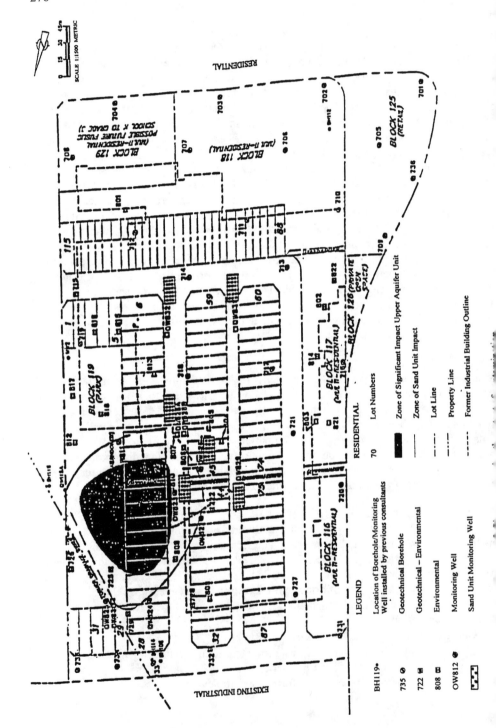

TABLE 16.1 Chemicals that exceeded the generic ground-water criteria, in μg/L

Chemical → Well and Unit ↓	TCE 50.0*	1,1,1-TCA 200.0*	1,1-DCE 0.66*	cis-1,2-DCE 70.0*	VC 0.5*
OW827 , UA	2,300.0	594.0	27.0	--	ND, 5.0
OW826, SU	4,920.0	3,260.0	212.0	ND, 100.0	ND, 5.0
OW825, SU	--	346.0	22.4	--	--
OW824, UA	--	--	1.4	--	--
OW823, SU	4,150.0	3,230.0	178.0	--	ND, 5.0
OW833, UA	3,930.0	4,330.0	194.0	82.0	ND, 5.0
BH103, UA	1,300.0	2,130.0	76.0	--	ND, 8.0
BH106, UA	--	--	0.8	--	ND, 1.3
OW115A, UA	--	--	--	--	ND, 0.5
MW1, UA	2,030.0	2,710.0	208.0	ND, 200.0	ND, 500.0

Notes: TCE = trichloroethylene; * = generic criteria for non-potable ground water; 1,1,1-TCA = 1,1,1-trichloroethane; 1,1-DCE = 1,1-dichloroethylene; cis-1,2-DCE = cis-1,2-dichloroethylene; VC = vinyl chloride; OW, BH, MW = monitoring wells; UA = upper aquifer; ND = not detected, less than generic criteria;. SU = Sand Unit.

Pathways and Receptors

No soil pathways were identified at this site since the contaminated soil was excavated and removed. Because ground water would not be used as a source of potable water, all pathways relevant to that use were excluded. Ground-water pathways of concern included migration of ground-water vapour to indoor air and migration of ground water to surface water.

Appropriate receptors on this site were selected based on their exposure potential. Child and adult individuals living onsite were selected as representative human receptors. The exposure duration is 30 years, everyday, for human receptors.[5]

All ecological receptors were excluded based on the following reasons:

The site is a former industrial facility undergoing redevelopment for residential/commercial uses. There are no sensitive ecological receptors within or near the site.

The limited area of contamination, habitat and lifestyle characteristics, duration of potential exposure, and cement/asphalt cover over the contaminated area negate the exposure of plants, animals, and birds.

The distance between the source and fish and other water organisms in the nearest creek (nearest potential receptors, located 1.5 km southeast of the site), ground-water velocity of 9 m/yr and contaminants' biodegradation half-lives (maximum half-life of less than 8 years[6] made their exposure unlikely.

6.2.2. Site Restoration Approach

The framework shown on Figure 16.3 was developed to demonstrate the most appropriate restoration approach for this case study and the rationale followed to identify . There are three options for restoration of a site for a specific use. One approach is to restore the site up to the generic criteria levels. Another approach is to restore the site up to background levels. However, in some situations, both of these approaches can be costly, not feasible, and/or not permissible (for example, the generic criteria approach

278

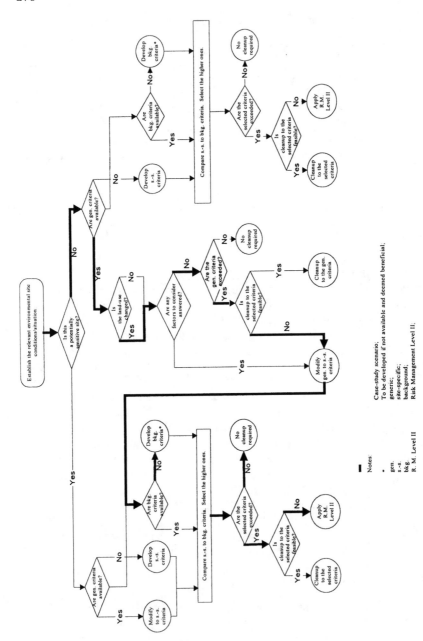

Figure 16.3 Framework for selection of site restoration approach

Notes:

Case-study scenario:

* To be developed if not available and deemed beneficial;

gen.   generic;
s.-s.   site-specific;
bkg.   background;
R. M. Level II Risk Management Level II.

can not be used for sensitive sites). In such cases, a site-specific risk assessment should be performed and appropriate engineered risk management techniques may be applied, where necessary. This third approach allows the development of a site in a manner that is protective of human and environmental health, and which considers aesthetic factors such as odour, without requiring the site to be cleaned up to the generic or background levels. In this way, it is also cost efficient.

A summary of all possible scenarios, steps, and outcomes with their corresponding risk management decisions is shown on Figure 16.3. The case study restoration pathway is shown in bolded lines. It starts with evaluation of the site conditions, as was described in the preceding sections of the chapter. The evaluation is followed by a determination of site sensitivity. The site can be identified as not potentially sensitive[1] for three reasons:

- It does not include or have an effect on a nature reserve, environmentally sensitive area, habitat of endangered species, park, etc.
- More than 2 m of overburden overly the bedrock
- The background concentrations for inorganic parameters were not exceeded, and the soil pH is between 5.0 and 9.0.

The next step determines whether generic criteria are available. Generic criteria are available for all chemicals of interest, as shown in Table 16.1. The next determination involves land use. The land use will be changed from industrial to residential type. Next comes the determination about the resolution of factors. No factors differ from those considered during the development of the generic criteria. Factors to be considered include presence or likelihood of adverse effect (on and off property), receptors and pathways different from those considered in the development of generic criteria, quantitative dose-response relationships for sensitive receptors from all exposure pathways different from that used in the development of generic criteria, and site conditions impacting the contaminant migration different from those considered in the development of generic criteria[1].

The next determination involves exceeding criteria. The generic ground-water criteria are exceeded at different locations, as shown in the Table 16.1. In answering the next question, it was determined that the analytical results from many rounds of sampling demonstrated that the ground water could not be remediated to acceptable levels of contamination (assuring protection of human health and ecological receptors) in a cost-efficient way.

Following the shown procedure, a decision was made to modify the existing generic to site-specific criteria to evaluate the existing risk in a more realistic way. Unfortunately, no ground-water background criteria are available for the Province of Ontario. The development of background criteria is optional and was not deemed to be beneficial. As a result, the site-specific criteria were followed in the procedure.

Comparing the Maximum Observed Concentrations (at the site) and the corresponding site-specific criteria (described later in this chapter) showed that criteria were not exceeded for a number of chemicals at specific areas of the site. For such chemicals/areas, no cleanup was deemed necessary. However, chemicals did exceed criteria at other site areas. Because it had already been determined not to be feasible to

280

cleanup the ground water under the site, a Level II risk management approach was applied to all areas where unacceptable risk was identified.

16.2.3. Selection of Criterion Component(s) for Modification

Figure 16.4 presents the process of selecting criteria component(s) for modification. This process consists of comparing the component value to the Maximum Observed Concentration. All exceeded criteria components are eligible for site-specific modification.

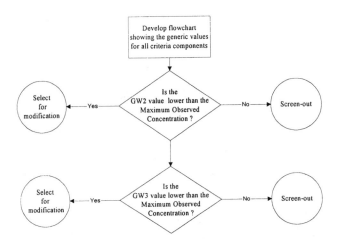

Figure 16.4 Selection of non-potable ground-water criteria components for modification

Exceeding a criteria for a certain component means that a risk of the corresponding effects exists. Because conservative "worst-case" values were used in the generic criteria development (possibly overestimating the risk), the component can be modified by applying the more realistic site-specific parameter values. If the component value is not exceeded, no risk is expected and no modification is necessary. Table 16.2 compares the Maximum Observed Concentrations and the values for GW2 and GW3 components for the five chemicals of interest. The analysis of the table data allowed the following conclusions:

- The GW2 Component can be modified (for all chemicals of interest)
- The GW3 Component must rely upon criteria component values established in the development of the corresponding generic criteria (for all chemicals of interest).

TABLE 16.2 Components of the non-potable ground-water criteria and maximum observed ground-water concentrations at the site (in μg/L)

Chemicals of Interest	TCE	1,1,1-TCA	1,1-DCE	cis-1,2-DCE	VC
Maximum Observed Concentration or highest detection limit for non-detects	4,920	4,330	212.0	82.0	ND, 500.0
GW2* (ground-water concentration corresponding to the allowable indoor air concentration; basis is volatility risk)	30.0	4,200	0.66	N/A	0.01
GW3 (ground-water concentration corresponding to the allowable surface water concentration; basis is AWQC)	220,000	180,000	120,000	120,000	3,600,000
Criterion component selected for modification	GW2	GW2	GW2	GW2	GW2

Notes: TCE = trichloroethylene; 1,1,1-TCA = 1,1,1-trichloroethane; 1,1-DCE = 1,1-dichloroethylene; cis-1,2-DCE = cis-1,2-dichloroethylene; VC = vinyl chloride; ND = not detected ; * = U.S. Environmental Protection Agency data; N/A = Not Available; AWQC = U.S. Environmental Protection Agency fresh water aquatic criteria.

Figure 16.5 shows a flowchart of the generic values for all criterion components for 1,1- dichloroethylene. The procedure for development of site-specific criteria may be summarized as presented on Figure 16.6. It consists of replacing the generic component values with modelled site-specific ones followed by application of the Value Selection and Risk Management Decisions steps.

16.2.4. Site-Specific Modelling of Vapour Migration from Ground Water to Indoor Air

The site-specific GW2 concentration was calculated from the allowable indoor air concentration using the formula shown in the guideline[1]:

$$OHM_{gw} = \frac{OHM_{air}}{\alpha x d x H x C} \tag{16.1}$$

where

OHM_{gw} = Calculated ground-water concentration that would not result in an indoor air concentration greater than OHM_{air} ($\mu g/L$)

OHM_{air} = Allowable target indoor air concentration ($\mu g/m^3$)

α = Calculated soil gas attenuation factor that relates the indoor air concentration to the concentration in soil gas directly above the ground water source based on the heuristic model[7]

d = Modification factor to convert the theoretical equilibrium concentration of ground water to soil gas to a realistic environmental concentration (equilibrium conditions are assumed to be unlikely)

282

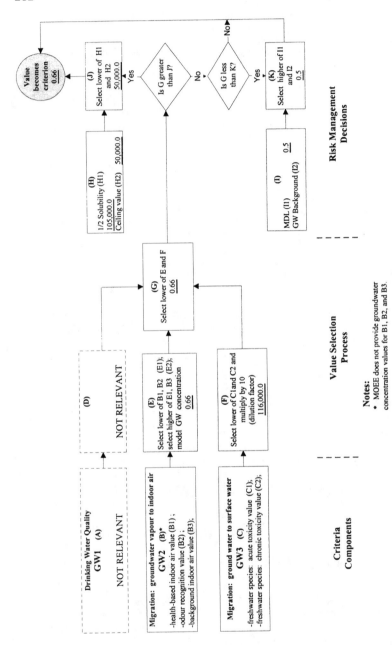

Figure 16.5 Generic criterion development flowchart for 1,1-dichloroethylene in ground water (µg/L[1])

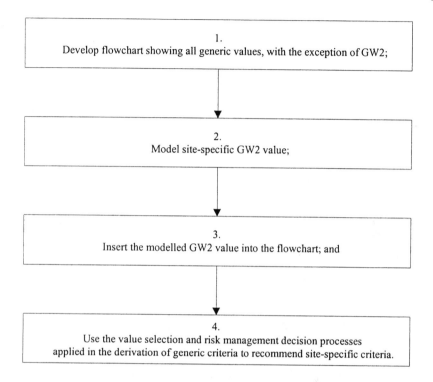

Figure 16.6 Summary of site-specific criteria deviation process

$\mathcal{H}$ = Chemical-specific, dimensionless Henry's Law constant

$\mathcal{C}$ = Unit conversion factor of 10^3 L/m^3.

The calculation of α relies on calculation of several equations providing values for its parameters. Those equations rely in turn on a number of site- and building-specific parameters. The application of site- and building-specific values allowed the derivation of site-specific criteria.

$$\chi = \frac{\left(\dfrac{D_T^{eff} A_B}{Q_{building} L_T} \right) x \exp\left(\dfrac{Q_{soil} L_{crack}}{D^{crack} A_{crack}} \right)}{\exp\left(\dfrac{Q_{soil} L_{crack}}{D^{crack} A_{crack}} \right) + \left(\dfrac{D_T^{eff} A_B}{Q_{building} L_T} \right) + \left(\dfrac{D_T^{eff} A_B}{Q_{soil} L_T} \right)\left[\exp\left(\dfrac{Q_{soil} L_{crack}}{D^{crack} A_{crack}} \right) - 1 \right]}$$

$$(16.2)$$

where

D_T^{eff} = Overall effective porous media diffusion coefficient for the contaminant between the contamination source and the building foundation (cm^2/s)

A_B = Cross-sectional area of the building, equivalent to the total below-grade area (floor and walls) of the building (cm^2)

$Q_{building}$ = Building volumetric ventilation rate (cm^3/s)

L_T = Distance between the contamination source and the building foundation (cm)

Q_{soil} = Volumetric flow rate of soil gas into the building (cm^3/s)

L_{crack} = Length of cracks/openings in the foundation through which contaminant vapours enter the building, equal to the foundation thickness (cm)

D^{crack} = Effective vapour-pressure diffusion coefficient for the contaminant through the cracks/openings (cm^2/s); D^{crack} is assumed to equal D_T^{eff}

A_{crack} = Area of cracks/openings through which contaminant vapours enter the building (cm^2).

Table 16.3 shows the site- and building-specific parameters are applied in the derivation of the soil gas attenuation factor α.

TABLE 16.3 Parameters applied in the derivation of the soil gas attenuation factor

Site-Specific Parameters		Building-Specific Parameters	
ε_T	Total porosity	A_B	Cross-sectional area of the building (floor and walls)
θ_m	Soil moisture content	L_{crack}	Length of cracks/openings in the building foundation
ρ_b	Bulk soil density	ACH	Building air exchange rate
K	Saturated hydraulic conductivity	V	Building volume
L_T	Distance between the contamination source and the building foundation	X_{crack}	Total floor/wall seam perimeter
P_1	Pressure in vadose zone	Z_{crack}	Depth of crack below ground surface
T	Absolute temperature of the vadose zone		

Figure 16.7 demonstrates the site-specifically modified ground-water criterion for 1,1-dichloroethylene. The generic GW2 Component value of 0.66 was replaced by the modelled site-specific value of 481.0. This value became a site-specific criterion as a result of following the Value Selection and Risk Management decisions steps. Similarly, site-specific criteria were derived for all substances of interest at all locations of concern.

16.2.5. Dealing with Uncertainty

Uncertainty must be factored into any risk calculation. The most important potential sources of uncertainty are as follows:

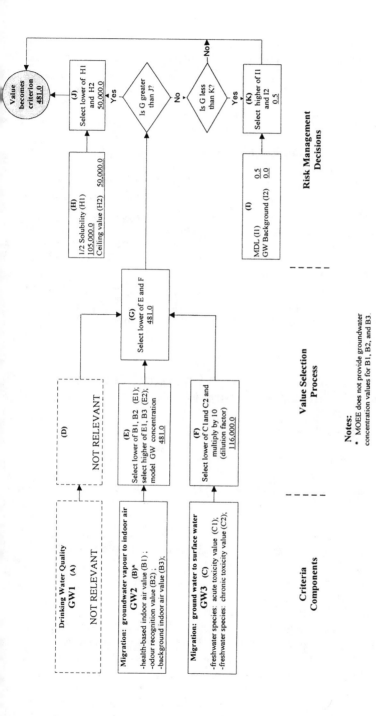

Figure 16.7 Site-specific ground-water criterion development for 1,1 dichloroethylene (in μg/L) for Lots 45, 46, and 47[1]

- **Accuracy of Supporting Site-Specific Information**. All required procedures for quality assurance/quality control were followed during the collection, transportation, and analysis of samples.
- **Toxicity Information**. Toxicity information provided in this chapter was derived from credible sources. In addition, the derivation process assumed identical end points and effects levels as those assumed when establishing the corresponding generic criteria.
- **Modelling of Vapour Migration into Basements**. Model and parameter values used in establishing generic criteria are inherently conservative and tend to overestimate the potential for exposures via inhalation of indoor air. The same model was employed with site-specific data. The use of site-specific data reduced the magnitude of uncertainty in the model predictions.
- **Short-Term Exposures During Construction Activities**. Risks to construction workers and nearby off-site receptors during construction were not assessed because there was insufficient information concerning timing, methods, and other relevant factors. However, it was not anticipated that exposures to receptors associated with construction would be significant. This assumption was supported by the fact that all substances in soil were at levels below the corresponding generic criteria. Furthermore, exposures during construction, if any, were expected to be relatively brief and intermittent.

16.3. Risk Management Decisions Based on the Site-Specific Modified Ground-Water Criteria

The exposure potential depends on three factors, namely:
1. **Intended land use** (home or open space, including park, walkway, roadway, etc.). The open space is expected to have lower air concentrations because of the contaminant dispersion in the air.
2. **Magnitude of contaminant concentration.** The higher the ground-water concentration, the higher the indoor air concentration (i.e., the higher the risk level all other conditions being equal).
3. **Soil properties** conditioning the vapour migration potential. Soil properties with a higher migration potential are expected to generate higher indoor air concentrations and higher risk.

Considering these three factors, three different areas of the site were identified (Figure 16.8), as follows:
- Lots 45, 46, and 47 located above the Sand Unit.
- All other residential lots excluding Lots 20 through 33. Lots 20 through 33 are subject to a development freeze (because of other than environmental reasons).
- Parkland, walkways, and roadways. Because no buildings will be constructed on these areas, there is no potential for the substances present in the ground water to gain entry into the indoor air.

Accordingly, different sets of site-specific ground-water criteria were derived for each area of the site (Tables 16.4 to 16.6). Table 16.4 shows that the site-specific

287

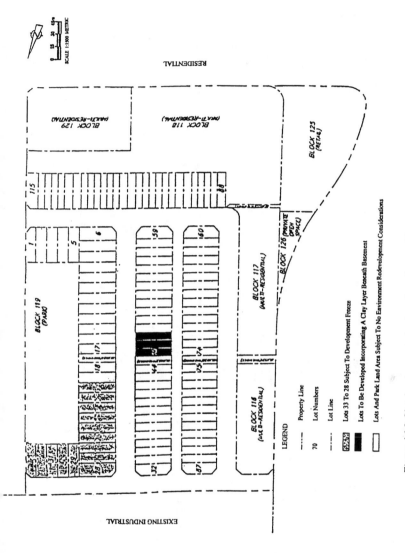

Figure 16.8 Site development plan based on site-specific ground-water criteria

TABLE 16.4 Recommended site-specific criteria and risk management level at the sand unit

Chemicals of Interest	Maximum Analysed Concentrations, in µg/L	Site-Specific Criteria, in µg/L	Recommended Site-Specific Criteria with Clay Layer, in µg/L	Recommended Risk Management Level
TCE	4,920.0	50,000.0	50,000.0	
1,1,1-TCA	3,260.0	50,000.0	50,000.0	
1,1-DCE	212.0	118.0	505.0	Level II (clay layer installation)
cis-1,2-DCE	ND, 100.0	424.0	2,046.0	
VC	ND, 5.0	57.0	218.0	

Notes: TCE = trichloroethylene; 1,1,1-TCA = 1,1,1-trichloroethane; 1,1-DCE = 1,1-dichloroethylene; cis-1,2-DCE = cis-1,2-dichloroethylene; ND = not detected; VC = vinyl chloride

TABLE 16.5 Recommended site-specific criteria and risk management level at the upper aquifer

Chemicals of Interest	Maximum Analysed Concentrations, in µg/L	Recommended Site-Specific Criteria, in µg/L	Recommended Risk Management Level
TCE	3,930.0	50,000.0	
1,1,1-TCA	4,330.0	50,000.0	
1,1-DCE	194.0	393.0	Level I (application of site-specific criteria)
cis-1,2-DCE	82.0	1,564.0	
VC	ND, 5.0	174.0	

Notes: TCE = trichloroethylene; 1,1,1-TCA = 1,1,1-trichloroethane; 1,1-DCE = 1,1-dichloroethylene; cis-1,2-DCE = cis-1,2-dichloroethylene; VC = vinyl chloride; ND = not detected.

TABLE 16.6 Recommended site-specific criteria and risk management levels at parkland, walkways, and roadways

Chemicals of Interest	Maximum Analysed Concentrations, in µg/L	Recommended Site-Specific Criteria, in µg/L	Recommended Risk Management Level
TCE	3,930.0	50,000.0	
1,1,1-TCA	4,330.0	50,000.0	
1,1-DCE	194.0	419.0	Level I (application of site-specific criteria)
cis-1,2-DCE	82.0	1,927.0	
VC	ND, 5.0	550.0	

Notes: TCE = trichloroethylene; 1,1,1-TCA = 1,1,1-trichloroethane; 1,1-DCE = 1,1-dichloroethylene; cis-1,2-DCE = cis-1,2-dichloroethylene; VC = vinyl chloride; ND = not detected

criterion for 1,1-dichloroethylene is exceeded. Because the criterion for this chemical was exceeded, and because of the potential for a future increase of vinyl chloride (which is a degradation product), risk managers decided to apply Risk Management Level II by installation of an artificial clay layer beneath the basements of Lots 45 through 47. The recalculated site-specific criteria based on the installation proved that this clay layer would provide satisfactory protection against vapour migration.

Table 16.5 shows no calculated site-specific criterion was exceeded. As a result, no engineering or administrative controls in these areas were needed. For the parkland, walkways, and roadways, monitoring of the migration from ground water to the ambient air demonstrated that none of the corresponding Ontario ambient air standards were exceeded, ensuring lack of outdoor inhalation risk. Site-specific criteria were developed to assess the vapour migration potential if houses were later built on these areas . As shown on Table 16.6, the derived site-specific criteria are not exceeded. Once again, remediation was not needed for these areas.

16.4. Conclusions

The procedure presented in this chapter:

- Affords protection to human health, ecological receptors, and the environment at a level equivalent to that provided by the corresponding generic criteria
- Provides a fast and cost-efficient alternative to site cleanup (saving money from cleanup, additional site investigations, etc.)
- Can be applied wherever generic or background criteria are used and a different solution is sought, because it relies on risk assessment principles and chemical-specific information.

In addition, the procedure can be easily adapted or improved. The exposure models (equations) can be replaced by others that better fit the specific needs of the site. The toxicity data should be updated as soon as newer valid data become available. Also, different risk management decisions can be applied under different conditions (ceiling, Maximum Detection Level, background, 1/2 solubility, or others). Similar procedures can be developed for other media (e.g., soil, surface water, and air).

16.5. References

Ontario Ministry of the Environment and Energy. (1996) Guideline for Use at Contaminated Sites in Ontario. Ontario Ministry of the Environment and Energy, Standards Development Branch, Calgary.

Ontario Ministry of the Environment and Energy. (1996) Rationale for the Development and Application of Generic Soil, Groundwater and Sediment Criteria for Use at Contaminated Sites in Ontario. Ontario Ministry of the Environment and Energy, Standards Development Branch, Calgary.

Ontario Ministry of the Environment and Energy. (1996) Guidance on Site-Specific Risk Assessment for Use at Contaminated Sites in Ontario. Ontario Ministry of the Environment and Energy, Standards Development Branch, Calgary.

Ontario Ministry of the Environment and Energy. (1996) Guidance for Sampling and Analytical Methods for Use at Contaminated Sites in Ontario. Ontario Ministry of the Environment and Energy, Standards Development Branch, Calgary.

U.S. Environmental Protection Agency. (1989) Risk Assessment Guidance for Superfund: Volume I – Human Health Evaluation Manual (Part A). Interim. EPA/540/1-89/002, U.S. Environmental Protection Agency, Washington, D.C.

Mackay, D., Shiu, W.Y. and Ma, K.C. (1993) Illustrated Handbook of Physical-Chemical Properties and Environmental Fate for Organic Chemical: Volume III–Volatile Organic Compounds. Lewis Publishers, Chelsea, Michigan.

Johnson, P.C., and Ettinger, RA. (1991) Heuristic model for predicting the intrusion rate of contaminated vapors into buildings, Environmental Science and Technology 25 1445-1452.

PART IV: FUTURE DIRECTIONS

Men reject their prophets and slay them, but they love their martyrs and honour those whom they have slain.
Fyodor Dostoyevsky

If we open a quarrel between the past and the present, we shall find that we have lost the future.
Winston Churchill

Many are focused on the contamination that is the legacy of the Cold War. In the United States, Canada, and Western Europe, approaches are being developed to return the land to other uses and ensure the protection of human life. Eastern Europe is just beginning the process of addressing their legacies. Analysts in both the East and the West who are new to the risk assessment process can find guidance on how to proceed effectively with limited budgets in the methods expounded in the NATO Advanced Study Institute. The recommendation from the institute on which this book is based was to use a holistic multimedia, multi-hazard approach that considers uncertainties. This approach can be used to set priorities with respect to accidents, contaminants, locations, and time frames. This understanding is the only way to be confident that limited resources can provide maximum health benefits to populations and the environment.

17. East Meets West: Teaming on Risk Assessment

One of the outcomes of this NATO Advanced Study Institute was the recognition that those charged with assessing, managing, and communicating risks in the East and West have unique challenges and approaches worthy of sharing. This chapter describes one mechanism being developed to facilitate this sharing, an International Risk Network.

The North Atlantic Treaty Organization (NATO) Advanced Studies Institute on Risk Assessment of Cold War Facilities and Environmental Legacies, held in Bourgas, Bulgaria, 2 to 11 May 2000, brought together risk assessment, management, and perception experts from eastern and western countries. It should come as no surprise that when so many educated, experienced scientists meet in one place, additional ideas are born. One such idea was proposed by representatives from the United States (Dr. Alvin L. Young, Center for Risk Excellence, Department of Energy) and Russia (Dr. Vitaly Eremenko, Department of the International Chair Network on Transfer Technologies for Sustainable Development under the United Nations Education, Science and Culture Organization, and Educational Centre TRAOMD, Moscow). They saw the need to establish an international risk assessment network for Cold War facilities and environmental legacies. This chapter discusses the need for such a network, the goals of the network, its organization, its initial successes, and future directions.

7.1. Why Here, Why Now?

As mentioned elsewhere in this book, the United States and eastern block countries have been faced with remediating hundreds of facilities and thousands of square kilometres of land associated with Cold War activities. This remediation effort includes decommissioning, decontaminating, and dismantling facilities used for military purposes (such as producers of nuclear, chemical, and bacteriological weapons); disposing of hazardous waste; and either remediating or providing stewardship for contaminated areas that may not be clean enough for even industrial use in our lifetimes. In the eastern countries in particular, remediation also carries with it a social cost, with a deteriorated standard of living for former government and industry specialists, retired servicemen and women, and their families.

This Cold War legacy poses a number of risks to both the countries in which it is found as well as other countries around the world. The Chernobyl disaster proved that environmental contamination does not respect national boundaries. A disaster of similar scale involving legacy waste could easily result in contamination elsewhere. Even if contamination were confined to the host country, the subsequent cleanup efforts could ultimately result in an international SOS to secure citizen safety and a heavy price tag for countries responding.

On the other hand, advances in risk analysis methodologies could help identify and resolve contamination problems before they spread. Some of these advances were shared at the NATO Advanced Study Institute in Bourgas, but conversations with participants indicated an even richer field of risk techniques waiting to be shared. Such an exchange could

- Assist with planning and prioritising environmental and Cold War facility protection activities
- Establish a baseline for determining the residual risks present from Cold War facilities and for measuring the progress of the cleanup efforts
- Determine level of risk and/or hazard reduction appropriate for different materials and settings
- Enhance the leveraging of funds and the focusing of current and new international efforts that support these objectives.

Such an exchange could be developed through an international network of risk analysis experts from academia, industry, and government.

This concept was presented to western and eastern government agencies and other interested organizations by the two project directors, Dr. Alvin L. Young and Dr. Vitaly Eremenko. Dr. Young briefed leadership in both the U.S. Department of Energy's Environmental Management Office and the Department of Defense's Office of the Deputy Undersecretary of Defense for Environmental Security. In addition, Dr. Young gathered support from existing risk programs at three of the Department of Energy national laboratories (Argonne National Laboratory, Illinois; Pacific Northwest National Laboratory, Washington; and Brookhaven National Laboratory, New York), the Medical University of South Carolina, and a consortium of six universities (Environmental Risk Management Alliance- ERMA). The ERMA institutions include Purdue University, New Mexico University, Colorado State University, Carnegie Mellon University, University of Virginia, and Harvard University. Together, these federal agencies, laboratories, and universities provide tremendous expertise and capabilities essential to the success of the network.

Simultaneous with the efforts in the United States, Dr. Vitaly Ermenko briefed Russian leadership and began to establish ties with coordinators in the initial interested countries for implementing the network. Many of these countries are mobilising their own pools of academic, industrial, and governmental resources (see insert on Armenia).

17.2. Network Goals and Objectives

The goal of the Risk Assessment Network is to establish a functional, self-sustaining, risk assessment capability within each participating country to support national decision making and to secure international financial contributions for Cold War environmental legacies of national and international significance. Cold War facilities may include, but not be limited to, former military-industrial complexes producing weapons of mass destruction or sites at which the by-products from these activities were disposed. Environmental legacies of national and international concern may include facilities presenting transboundary pollution problems. Facility and legacy issues to be examined within each country will not encroach on the national security of the participating countries and will not require the use or release of classified or country-sensitive information.

The multilateral objectives of this program include:

- Cooperatively engage governmental and nongovernmental entities responsible for management of Cold War legacy sites and facilities
- Forge relationships among multidisciplinary and multi-agency teams within network countries
- Identify risks within the participating countries, especially transboundary

---

The International Risk Assessment Network—an Armenian Perspective

In August 2000, the National Group of Risk Assessment Experts of Armenia was established. It comprises 16 specialists from the Ministries of Energy, Education, and Agriculture; Departments of Nuclear Control and Emergency Situations; Engineering Academy; and National Academy of Sciences of Armenia. Items of highest priority include science, information support, and personnel training.

The first step was to compile as quickly as possible information on scientific developments and risk assessment methodologies in Armenia. About 30 topics representing important areas for exchange of information with other network countries were selected and approved, including developments of the assessment of economical and social, technological, and medico-hygienic risks and defence profiles, as well as risks relevant to natural catastrophes (earthquakes), nuclear energy, and agriculture.

The work performed by experts and analysts has enabled Armenia to choose the areas and ways for the realization as follows:

- Select the best developments, analysis, and risk management methodologies on the level of national and international models
- Organize scientifically justified support of risk assessment in the governmental decision-making system in natural resource management and environmental conservation
- Establish an information service for risk assessment and a system of personnel training.

The environmental aspect of education is very important in Armenia. One of the first interchanges with the United States brought muc needed literature, methodological material, practical recommendations, and audio- and videocassettes on environmental risk assessment from the Medical University of South Carolina.

issues, and benefits of alternative management strategies

- Identify areas of priority concern where protective or remediation efforts are needed to reduce both domestic and transboundary hazards
- Focus international attention and assistance on Cold War legacy sites in areas with the highest risks threatening health of present and future generations, and threatening international relationships
- Provide training through classroom, web-based courses, workshops, and internships for current and future (e.g., student interns) decision makers
- Share decision tools (e.g., risk management software) and best management practices needed to conduct risk assessments.
- Ensure that the assessment process is done in an open and lawful manner so that national security of the participating countries is not threatened.

17.3. Network Organisation

The initial countries within the network include Armenia, Bulgaria, Czech Republic, Greece, Hungary, Kyrgystan, Lithuania, Republic of Georgia, Romania, Russia, Turkey, Ukraine, and the United States. In addition, in a NATO meeting of the Science Committee on 20 October 2000, additional countries were given the opportunity to join. Kazakhstan, Albania, and Macedonia expressed initial interest. Each country provides a focal point or country coordinator, which in turn are led by the two project directors. Table 17.1 lists the country coordinator for the initial countries participating.

These coordinators will form country teams, which will familiarize themselves with software, technologies, and methodologies that have been applied across the network to deal with risk issues associated with Cold War legacy sites.

TABLE 17.1 Participants in the International Risk Assessment Network

Country	Coordinator
Armenia	Olga A. Juharyan, Group Manager, Ecocenter, Academy of Sciences
Bulgaria	Marusja Ljubcheva, Associate Professor, Scientific Advisory
Czech Republic	Jaroslav Volf, Director, Health Officer
Greece	Olga N. Aneziris, Group Manager, National Centre for Scientific Research
Hungary	Tamas Madarasz, Group Manager, University of Miskoic
Kyrgystan	Azamat Tynybekov, Head, International Science Centre
Lithuania	Kestutis Kadunas, Head, Hydrological Division
Republic of Georgia	Petr I. Metreveli, Group Manager, Ministry of Foreign Affairs
Romania	Florin Glodeauu, Head, Health Physics and Environmental Protection Department
Russia	Petr L. Gusika, President, International Association of Ecological Safety
Turkey	Ah Esat Karakaya, President, Turkish Society of Toxicology
Ukraine	Georgy V. Lysychenko, Director, Ukrainian Society for Sustainable Development
United States	James G. Droppo, Group Manager, Multimedia Environmental Assessment, Pacific Northwest National Laboratory

17.4. Initial Successes

Since the Advanced Studies Institute and the birth of the network, the individual countries have participated in a number of information exchanges. The first exchange occurred on 17 May 2000, less than a week following the initial conference. Additional information exchanges occurred on 21 June, 24 August, and 5 October 2000. More than 20 publications on all aspects of risk assessment, risk management, and risk communication have already been provided through these exchanges.

In addition, the bi-monthly newsletter, *Risk Excellence Notes*, published by the Center for Risk Excellence, was a vehicle for initiating and maintaining contact with this network of nations. The June/July 2000 issue of the newsletter contained the *International Memorandum of Agreement* and was published in Russian and English. Each subsequent edition included a section published in Russian and English of activities submitted by scientists in the various nations.

17.5. The Future

Building on its successes and goals, the network hopes to move forward both organizationally and scientifically. As mentioned, the first step will be to develop country teams within each participating nation. Each country team will be multidisciplinary in composition and include representatives from governmental and nongovernmental entities engaged in the management of Cold War facilities. These representatives will be principally applied scientists who have technical skills and interest in the overall risk assessment process. Potential customers (e.g., regulators or funding organizations/investors) may also be invited to participate in the teams.

Participating countries nominated sites to the project directors for preliminary risk characterization studies. Country teams presented information on the nominated sites at the May 2001 ECO-INFORMA Conference held in the United States in Chicago, Illinois.

The network also plans to establish an intern program to support training and student exchange programs. Skilled technicians are needed to perform the basic requirements of environmental compliance such as workplace safety and health monitoring, daily sampling, analysis, manifesting, training, and record keeping associated with environmental compliance. Trained management personnel are also needed to oversee and develop policies and procedures as part of environmental management systems. Lastly, students who have completed their Bachelors Degree, for example in risk management or environmental science or engineering, need the opportunity to visit first hand the challenges involved in the cleanup and management of the legacies of the Cold War. The design of this internship program must allow students to have appropriate three-month projects with "hands on" experience that will be relevant to their country's needs as identified by their country team. To ensure this relevance, students must be carefully selected, given an opportunity to experience risk assessment in a real-world situation, and then returned to their respectively country to continue serving the activities of the network. This transmittal of information will ensure that these young people take responsibility for continuing these programs into the next generation.

In addition, the network plans to establish a virtual network to enhance communication; consolidate and integrate efforts; minimize travel; organize and disseminate risk management methodologies and lessons learned; provide access to top-level expertise and resources among the network countries; and facilitate training and technology/information transfer. The last point will be very critical to the success of the network. Thus, large efforts will focus on providing interactive training for tools, models, and methods.

Finally, the network will focus on the need to resolve risk-based scientific issues that are common to the participating nations. These issues include the following:

- Impact of transportation and storage of spent nuclear fuel and nuclear waste
- Magnitude and severity of risks and the level of cleanup needed at former military sites engaged in nuclear, chemical, or biological weapons research and production
- Risks associated with transboundary consequences from natural accidents, (e.g., floods) and human-related events (e.g., operator accidents or terrorism) involving nuclear, chemical, or biological materials stored at former military installations
- Risks associated with deteriorating structures (science issues of concern to multiple countries will become points of more detailed study for network countries and participants)
- Applied methods for rapid assessment of risks and for making operative remediation-based decisions.

Breakthroughs in any of these areas may serve as subjects of annual meetings in Eastern Europe of the country teams from each of the participating countries.

The project directors are currently seeking to leverage funding to support these and other activities. International interest is high, and the need is pressing. It remains to be seen whether the subject nations can rise to this important challenge.

18.   Where Are We Going?

Another outcome if the NATO Advanced Study Institute was the identification of a number of areas in which risk assessment, risk management, and the understanding of risk perceptions are in their infancy. This chapter suggests possible directions for the future of these disciplines in meeting the challenges of managing Cold War legacies.

Decision makers in all countries with hazardous facilities, stored waste, environmental contamination from the Cold War and similar challenges should find this book helpful in understanding the potential use of different applied risk methodologies to protect people and the environment. The information in the book was provided by experts in collecting and analysing data for making decisions related to industrial safety, environmental protection, and public health and safety, who gathered together at the Advanced Study Institute (ASI) on which this book is based. They came from Eastern and Western countries, including Armenia, Georgia, the Kyrgys Republic, Lithuania, Russia, the Ukraine, and other countries of the former Soviet Union (FSU), as well as Bulgaria, the Czech Republic, Hungary, Romania, Greece, Turkey, the U.S., and Canada. They discussed the possible benefits to applying Western experience in using risk methodologies to ensure safe management of waste storage and environmental contamination and agreed that major risks exist associated with the management of Cold War legacies.

In the case of environmental contamination, participants from the Eastern countries felt that they were already suffering exposure and health effects, with efforts just starting to inventory and characterize the risks. Participants from the West felt that risks in their countries are largely characterized, and efforts are now moving forward with activities to reduce these risks. However, in both the East and the West, large risks remain, and much work remains to be done to ensure the public's safety. In this chapter, the material in previous chapters is used to project what can be done to realise a wider use of applied risk assessment and management to address the legacy issues.

Two views of the future are given – one from the Western viewpoint of evolving methodologies and their appropriate applications and the other from the Eastern viewpoint of implementation in countries of the FSU and Eastern Europe.

18.1. View from the West: Methodologies and Applications

To set the direction for future applications, this book provided a unified view of risk methodologies for decision makers and their experts. The recent Western experience is that often some combination of approaches is necessary to meet the needs of decision makers. The effective and appropriate use of these methodologies has been referred to as a Complementary Risk Management approach. The following sections lay out additional requirements for implementing Complementary Risk Management.

18.1.1. Conduct Flexible Application-Specific Approaches

A Complementary Risk Management approach balances flexibility within specific applications. Various chapters in this book provided examples of successfully applied risk assessment and management efforts with very different endpoints—each of which is appropriate for the particular issue being addressed in that specific case. These case studies show the flexibility of using the risk analysis methodologies to address different situations in quite different, but appropriate ways.

For such a complementary approach to be effective, decision makers must clearly define in advance exactly what issues are being addressed. Experience has shown that clear definitions of the products and their application are essential before starting to conduct an applied risk analyses, if results are to be meaningful in the context of the decisions to be made.

18.1.2. Consider Many Aspects of Risk

The proactive consideration of the many aspects of risk is a relatively new development. During the Cold War, the emphasis was on production; risks to people and the environment were at best secondary considerations. Only since the latter part of the 1970s has risk become widely recognized as a major concern. The *Reactor Safety Study* (Rasmussen Report)[1], a landmark document for conducting probabilistic risk assessments (PRA), was published in 1975. The Three-Mile Island Accident in the United States in 1978 further stimulated application of the *Reactor Safety Study* and PRA methodology.

Chapter 6 of this book provides insights into the early efforts and the institutional barriers and challenges to implementing Cold War waste management and remediation policies based on computed risk. One of the important outcomes of this period was the realisation that decisions makers must understand the many facets of risk; they cannot rely on a single risk number as was once proposed.[2] Since the early 1980s, a single risk number was proposed for accident analyses related to production reactors and waste storage and processing facilities. As early as the 1970s, such a number was proposed, based on the expected number of deaths (or injuries, etc.) for assessing nuclear reactor accidents. In this same period, the U.S. Environmental Protection Agency developed and promulgated the idea of risk analyses that assume the release occurs (i.e., release has a probability of 1.0) and that evaluate the risk to individuals. While some proponents still argue for the use of the single number for accident risk assessment, most have moved to a more holistic approach.

An even broader outcome of these early years of risk analysis was the realisation that computed risks are only one of many factors which decision makers must balance. This realisation also led to a more holistic approach. For example, Chapter 4 describes a risk profile information system used by the U.S. Department of Energy to consider major waste management and contamination sites at Cold War facilities. In addition, Chapter 16 provides insights into how risk criteria have been implemented in West.

The trend for the future is clearly away from using single measures of risk and simple upper bounds as input to decision makers. As much as a single number is an appealingly simple approach, decision makers must consider many aspects of risk – and make decisions as a balance of the different types of risk. Furthermore, a single number can, at best, offer a vague comfort, if the number is low. It provides no understanding of the causes of risk, the uncertainty in the results, or what can be done to control the risk.

18.1.3. Broaden the Applications of Accident Risk Analysis

Accident risks from the storage and destruction of chemical weapons have been studied since the early 1990s, but, as shown in Chapter 8, those characterization analyses are just being completed. Risk analyses at contaminated sites in the FSU are just beginning and can benefit from the risk methodologies developed previously. However, as shown in Chapter 10, the application of accident risk modelling techniques to weapons handling is relatively new, even in the West.

In the world of commercial nuclear power (and soon in the nuclear reactor world of the U.S. Department of Energy as licensing by the Nuclear Regulatory Commission begins again), regulations are being changed to be "risk-informed." Not previously an explicit requirement, risk analysis now becomes one aspect of the licensing decision process. Many regulatory guides and Standard Review Plans have been issued by the U.S. Nuclear Regulatory Commission through the U.S. Code of Federal Regulations. Of these, Part 50 is considered "risk-informed."

18.1.4. Provide a Better Balance of Risks, Cost, and Technical Factors

In the West, the cost of remediation and long-term management of legacy wastes has proved to be very high. Countries of the FSU cannot afford the magnitudes of costs being experienced in the West, and thus must carefully invest what resources they can in keeping risks to a minimum. The Western approach using a balance of risk management, risk analysis, and risk perception is seen as a means of effectively directing priorities for management and cleanup efforts based on maximizing potential population safety.

The U.S. remediation efforts are being variously conducted under rules and guidelines for the Resource Conservation and Recovery Act (current operations) and Comprehensive Environmental Response, Compensation, and Liability Act (past operations). Although the former is largely a process control regulation that protects the environment based on process-specific emission limits, it does include provisions to consider risk in certain land disposal options. The latter, on the other hand, has a clearly defined process for handling remediation, which calls for a balance of cost, risk, and technical factors. As a result of the widespread implementation of remediation efforts at facilities such as mines, mills, weapons factories, test areas, research areas, reactors, waste tanks, and military bases, a great wealth of experience and knowledge exists in addressing such issues in the United States.

18.2. View from the East: Implementation

Risk methodologies are currently seldom used in decision-making processes in the countries of the FSU. In addition to the normal lag in implementation of new methods, there are reasons specific to these countries for this situation:

1. **Unfamiliar Concepts and Approaches**: The applied risk methodologies discussed in this book were developed for conditions peculiar to the Western world infrastructure and cultural outlook. Scientists in the FSU and Eastern Europe have put considerable effort into developing theoretical approaches. Many of the principles and potential advantages of the Western applied risk methodologies are unclear to the managers of the FSU and Eastern Europe. One of the objectives of this book is to communicate information on the Western risk analysis methodologies and recommend how they might be adapted to the conditions of Eastern Europe.

2. **Economic Pressures**: There has been little demand for risk methodology applications in the FSU and Eastern Europe with these counties facing economic problems of survival. Decisions related to increasing population safety or improving the environment are postponed. A second objective of this book is to show that these problems need to addressed and understood. Good decisions can be made with or even in spite of an obvious absence of economic resources. The most cost-effective time to understand these problems is now so that even meagre resources can be most effectively used.

3. **Local Infrastructure**: The use of the proven Western applied risk methodologies in countries of Eastern Europe is limited by insufficiently formed democratic procedures to address the most important social problems. Although the situation is improving, everywhere risk methodologies are used to a lesser degree, democratic relations between authorities and the population are more primitive. This book is not meant to influence local political situations. However, the opinion was expressed at the ASI that a wider use of applied risk methodologies is seen in these Eastern countries as a natural part of the development of the democratic process.

The editors of this book understood and carefully considered these difficulties in applying risk methodologies. The promotion, communication, and implementation of the methodologies contained in this book will require consideration of and respect for political and national preferences in decision-making processes. The materials in this book were jointly prepared by applied scientists from both the West and East. All share the concerns for environmental legacies of the Cold War in their and neighboring counties. Each, however, developed the risk approach within certain political and military situations that will limit what problems can be addressed.

The recommendation from the ASI is that proven applied risk methodologies can be used in these East European countries for Cold War legacies that do not involve major national political or military restrictions/secrets. Such test subjects are not easy to identify, and it is often harder still to get the required approval of national authorities to go forward. The location of appropriate sites may well be the most difficult problem for implementing Western risk methodologies.

The discussions at the ASI suggested five possible location types:

1. Storage for irradiated and spent nuclear fuel as unwanted by-products of the Cold War.
2. Transport of the irradiated and spent nuclear fuel to and from storage.
3. Other types of long-term nuclear and chemical waste storage sites. These sites are of special concern because many are old, in need of maintenance, and located in residential areas.
4. Legacy sites and military-industrial activities resulting in trans-boundary hazard transport.
5. Contaminated zones outside borders of closed administrative territories related to numerous Cold War activities and facilities. The Russian dose reconstruction study described in Chapter 9 is an excellent example of such an application.

The challenge is for risk analysts from the FSU and East European countries to carry out analyses relative to risks to workers and the population. These applications should satisfy any established national norms and standards of an acceptable risk. Successful applications will result in optimal risk-based decisions, taking into account available domestic resources and social factors.

Risk analysis results have been proposed to provide a basis for defining protective safety, remedial, or alternative actions. One of the most important proposals is to estimate incremental health treatment costs for populations as well as the size of appropriate insurance guarantees for those living in these zones. Such a use of risk analysis would be a departure from the Western view that the only acceptable risks are those with trivial risk levels. Another important proposal that does have an analogue in U.S. air emissions management is to use risk results to define optimal measures to protect the population--even if that new protection is not directly connected with the proposed new activity or facility. The idea is to reward the region that agreed to accept the new hazardous activity by reducing large existent current risks produced by other sources.

These types of risk analyses, recommended for decision makers of Eastern Europe, cannot be limited to a single risk analysis methodology. The ASI identified a general set of methodologies that must be considered as part of a Complementary Risk Management Effort. These methodologies share many of the same factors as Western types of analysis but differ in the purpose for the effort:

- Facility-Centred Risk Analysis is used mainly to define or demonstrate acceptable risk-based operating parameters for facilities.
- Human-Centred Risk Analysis is used mainly to study and understand the human exposures and risk for environmental contamination. These studies normally are based on fixed operating parameters for hazardous facilities.
- Risk Perception Assessment is the analysis of the perception of risks by the involved parties (decision makers and local populations).

Misunderstandings about the roles of these analyses can lead to apparent inappropriate competitive views. In fact, all three are needed as part of a Complementary Risk Management approach.

18.2.1. Support for Complementary Risk Management

As this book was completed, the first application of Complementary Risk Management was begun in Russia. Local decision makers and the population see a high perceived risk (Risk Perception Assessment) in the future operations of the facility in question. The design studies for the proposed facility (Facility-Centred Risk Assessment) indicate that operations should be relatively safe for those working and living around the site. The risk assessment for the local populations (Human-Centred Risk Assessment) confirms that expected emissions will have trivial impacts on the surrounding populations. This assessment also indicates that there are likely significant impacts to surrounding populations from the current ambient environmental quality. With these complementary results, the challenge to the decision makers is to define solutions that will be acceptable to the involved parties. The path forward is to implement the applied risk methodologies, adapted for local conditions, to protect populations and the environment for the initial set of five types of Cold War legacy sites listed above. Some materials and tools are available to support the implementation of these methodologies. English-to-Russian and English-to-Bulgarian glossaries of general terminology related to risk assessment and cleanup efforts were provided to the participants at the ASI. A version of the English-to-Russian glossary is included as an appendix to this book to provide Russian language readers with extended explanations of some of the new concepts presented.

For Facility-Centred Risk Assessment, several tools are available:

- *Severe Accident Risks for VVER Reactors: The Kalinin PRA Program*[3], adapted into the Russian language by Vitaly A. Eremenko (1994)
- Set of computer programs SAVE I, II, and III (*System for Quick Calculation of Physical Effects and Risks*, The Netherlands), designed by Esko Blokker for Russian users (1995)
- *Categorizations and Priorities of Risk from Severe Accidents in Non-Nuclear Technological Processes*[4], adapted into the Russian-language by Vitaly A. Eremenko for Russian users (1996)
- Software and documentation for an emergency preparedness model RASCAL by the International Atomic Energy Agency and U.S. Nuclear Regulatory Commission (1997).

For Human-Centred Risk Assessment, the following tools are available:

- Multimedia Environmental Pollutant Assessment System (the software MEPAS, described further in Chapters 6 and 14 (MEPAS is available as a 1998 Russian language version.)
- *Digest of RAAS' English-Russian Glossaries*, an unpublished document based on an American software system for conducting remediation efforts, contain unique Russian language explanations of American technologies used for Human-Centred Risk Assessment (prepared by Vitaly A. Eremenko, 1997).
- Other relevant software packages, or newer versions of the above packages, available in their original languages or in semi-adapted packages (e.g., COSIMA 1994)

The most important role of groups such as the International Risk Network created as a result of the NATO ASI will be in the circulation of risk analysis and application support materials during seminars and training sessions. The concepts for the methodologies need to be understood before national risk analysts can expect any movement toward concept acceptance and implementation.

18.3. As Challenges Evolve

At the time of publication of this book, many of the efforts to address technological legacies of the Cold War can be best characterized as "underway." In the United States, the characterization and cleanup efforts have been underway for more than a decade. A number of legacy studies have attempted to define the nature and challenges of the legacies (see, for example, the Department of Energy Environmental Survey[5] and the Baseline Environmental Management Reports[6]). Cleanup and weapons destruction have started at major U.S. Department of Defense and U.S. Department of Energy facilities.

It is also encouraging that a number of activities have been conducted, or are underway, in the counties of the FSU and Eastern Europe that are significant steps in implementing a wider use of risk analysis. This book provides examples of such activities. Chapter 7 describes a cleanup effort that has been successfully completed. Chapter 11 discuses developing institutional structures. Chapter 15 addresses widening the concern from nuclear to other types of hazardous facilities. Efforts are also underway to understand perceived risks (see Chapters 5 and 13). In addition, initial assessments have been conducted, such as the radiation factor assessment described in Chapter 12.

The U.S. agencies have cleaned up and restored an impressive volume of waste and area of land (see Chapter 4). These, however, largely represent the "easy" issues remaining from the Cold War. The efforts for sites with high-level radioactive wastes and the destruction of weapons are proving to be much slower and much more expensive programs. The factor driving these high costs and slow progress is the desire to conduct these operations with minimal risks to the workers and surrounding populations.

In counties where risks from Cold War legacies are just beginning to be considered, new unexpected pathways will likely be found for health impacts. The researchers in the

306

United States, Russia, Ukraine, and other affected countries continue to recognize new issues and pathways (such as new areas of contamination or new contaminants). Good examples are the expanding list of contamination at Department of Energy sites and the recognition of the potential impact of beryllium.

The overall risks in the counties of the FSU and Eastern Europe may be potentially greater than that found in the United States. There are two main differences: 1) many of the Russian-built facilities were located within or nearby population centres, and 2) at the more remote facilities in the FSU, the waste management policies allowed much more material to be discharged to the local environments.

Future efforts will likely employ a multimedia approach that involves multi-contaminant, multi-pathway multi-effect integral analysis of risks and other indexes of hazard. First-generation versions of software models to support this approach[2] have been available and used for more than a decade. Second-generation models have been developed and applied in the past few years by the Environmental Protection Agency. In 2001, the Environmental Protection Agency, Department of Defense, Department of Energy, and others signed a memorandum of understanding that calls for multi-agency collaboration and cooperation in the development of the next generations of these "multimedia models."

An important advance in these multimedia models is the consideration of contaminant concentrations, doses, hazard indicators, and risks from both radioactive and hazardous chemicals in a single analysis system. As seen in the chapters in this book, the current risk analysis largely emphasizes nuclear materials in the countries of the FSU. Multimedia analyses have proven to be valuable in highlighting the importance of previously overlooked contamination of hazardous chemicals from nuclear operations. Based on the United States experience, many of these chemical materials typically move faster in the environment than radionuclides and can pose major health and environmental effects much sooner than the radionuclides.

Similarly, the application of a multimedia environmental assessment approach will result in the understanding of the importance of alternative exposure pathways and the consideration of linkages between pathways (for example, see Chapter 14).

Based on current trends, the future will see a wider use of the different aspects of risk analysis for important applications. The NATO ASI and this book are important steps in describing contemporary risk methodologies. Through this information, the potential application of these methodologies to Cold War legacy sites should be better understood, allowing decision makers in the East and West to make more optimal use of the limited resources available to address the important population safety issues.

18.4. References

1. U.S. Nuclear Regulatory Commission. (1975) Reactor Safety Study. WASH-1400, NUREG-75/014, U.S. Nuclear Regulatory Commission, Washington, D.C.
2. Droppo, J.D., Jr., Strenge, D.L., Buck, J.W., Hoopes, B.L., Brockhaus, R.D., Walter, M.B., and Whelan G. (1989) Multimedia Environmental Pollutant Assessment System (MEPAS) Application Guidance, Volumes 1 and 2. PNL-7216, Pacific Northwest National Laboratory, Richland, Washington.

3. --- (1994) Severe Accident Risks for VVER Reactors: The Kalinin PRA Program, Volume 3: Procedure Guides. NUREG/CR-6572, Vol. 3, Part 1, BNL-NUREG-52534, prepared by Brookhaven National Laboratory for the U.S. Nuclear Regulatory Commission, Washington, D.C.
4. --- (1994) Categorization and Priorities of Risk from Severe Accidents in Non-Nuclear Technology Processes. IAEA-TECDC-727, International Atomic Energy Agency, Vienna.
5. U.S. Department of Energy. (1988) Environmental Survey Preliminary Summary Report of Defense Production Facilities. DOE/EH-0072, U.S. Department of Energy, Washington, D.C.
6. U.S. Department of Energy. (1995) Estimating the Cold War Mortgage, The 1995 Baseline Environmental Management Report, Volumes I and II. U.S. Department of Energy Office of Environmental Management, Washington, D.C.

Appendix A
Programme from NATO Advanced Study Institute, Risk Assessment Activities for the Cold War Facilities and Environmental Legacies

Hotel Bulgaria, Bourgas, Bulgaria, 2 to 11 May 2000

Tuesday, 2 May 2000

Breakfast: 8:00-9:00

Registration Desk in Lobby: 8:00-18:30

Morning Lecture Hall Programme: Formal Opening of Institute 11:00-12:00

11:00-12:00 Formal Meeting Opening
Opening:
Dr. Dennis C. Bley, Buttonwood Consulting, Inc., USA
Director, NATO ASI (5 minutes)
English, translated into Russian and Bulgarian

Welcome Speeches:
Mr. Ioan Kostadinov, Mayor of Bourgas (10 minutes)
Bulgarian translated into English and Russian

Mr. Ivan G. Karapenev, Bulgarian Ministry of Defense (5 minutes)
Bulgarian translated into English and Russian

Dr. Simeon Simeonov, Representative of Bulgarian Ministry of Environment (5 minutes)
Bulgarian translated into English and Russian

Dr. Alvin Young, US DOE, Center for Risk Excellence (5 minutes)
English, translated into Russian and Bulgarian

Dr. Georgy Lisitchenko, The Ukraine State Scientific Center of Environmental Radiogeochemistry, (5 minutes)
Russian, translated into English and Bulgarian

Mrs. Beyza Űntuna, General Consulate of Turkey in Bourgas

12:00-13:00 Break

Lunch: 13:00-14:00 (at hotel)

Afternoon Lecture Hall Programme: Keynote Lectures 14:30-17:00

14:30-15:30 Advanced Study Institute Introductions
Directors: Drs. Dennis Bley, Vitaly Eremenko, and Jim Droppo
Review Lecture Programme and Introduce Lecturers

15:30-16:00 Coffee/tea

16:00-17:00 Session
History of risk assessment methods
Dr. Vitaly Eremenko, International Center of Educational Systems, RF

Open time: 17:00-18:30

Dinner: 18:30-19:30 (at hotel)

Wednesday, 3 May 2000

Breakfast: 8.00-9.00 (at hotel)

Registration Desk in Lobby: 8:00-10:00

Morning Lecture Hall Programme: Integrated Risk Assessments 9:00-12:30

9:00-10:00 Session
    Facing the Risk Issues of the Cold War Legacy: A U.S. View
    Dr. Alvin Young, US DOE, Center for Risk Excellence, USA
    (45 minute lecture, 15 minute discussion).

10:00-10:30 Coffee/tea

10:30-11:30 Session
    Risk Assessment – European View of Key Principles
    Judith Lowe, CLARINET, UK
    (45 minute lecture, 15 minute discussion)

11:30-12:30 Session
    Integrated Risk Assessment – Technologies for Risk Assessment for
    Optimization of Management Decisions
    Dr. Vitaly Eremenko, ICES, RF
    (45 minute lecture, 15 minute discussion)

Lunch: 12:30-13:30 (at hotel)

Afternoon Lecture Hall Programme: Integrated Risk Assessments (cont.) 14:00-17:30

14:00-15:00 Session
    Management of Risk Portfolios for Weapon Site Cleanup
    William Andrews, PNNL, USA

15:00-15:30 Coffee/tea

15:30-17:30 Session
    Round Table #1. Challenges of Risk Assessments in East and West
    Each participant will be asked to give their opinion on the challenges.
    Led by Directors and Lecturers

Open time: 17:30-18:00

Reception with Cocktails, Dinner, and Entertainment (at hotel) 18:00-20:00

Thursday, 4 May 2000

Breakfast: 8.00-9.00 (at hotel)

Morning Lecture Hall Programme: 9:00-12:30

9:00-10:15 Session
Accident Risk Assessment
Dr. Dennis Bley, Buttonwood Consulting, Inc., USA

10.15-10:45 Coffee/tea

10:45-12:30 Session
Accident Risk Analyses and Applications for the Disposal of Chemical
Agents and Munitions
Susan Bayley, SAIC, USA

Lunch: 12:30-13:30 (at hotel)

Afternoon Lecture Hall Programme: 14:00-17:30

14:00-15:00 Session
Integrating Management Effects into Quantitative Risk Assessment
Dr. Olga Aneziris, Demokritos, Greece

15:00-15:30 Coffee/tea

15:30-17:30 Session
Round Table #2. Facility-Centered Risk Assessment. Small groups.
Facilitators (Directors/Lecturers) ask students to propose problems from
their home regions. Groups will define issues and discuss solution options.
Led by Directors and Lecturers

Open time: 17:30-18:30

Dinner: 18:30-19:30 (at hotel)

Friday, 5 May 2000

Breakfast: 8:00-9:00 (at hotel)

Morning Lecture Hall Programme: 9:00-12:30

9:00-10:00 Session
Programmatic Risk Assessment
Dr. James Droppo, PNNL, USA

10:00-10:30 Coffee/tea

10:30-11:30 Session
Site-Specific Modification of Ground Water Generic Criteria as Applied
to a Contaminated Site – a Canadian Approach
Dr. Hristo Hristov, Ontario Ministry of the Environment, Canada

11:30-12:30 Session
Comprehensive Risk Management Programs for the Disposal of Chemical
Agent and Munitions
Susan Bayley, SAIC, USA

Lunch: 12:30-13:30 (at Hotel)

Afternoon Lecture Hall Programme: 14:00-17:30

14:00-15:00 Session
Risk Perception : The Psychological Aspects
Dr. Vladilena Abramova, Obninsk Institute of Nuclear Power
Engineering, RF

15:00-15:30 Coffee/tea

15:30-16:30 Session
Environmental Pollution and Environmental Health: Dynamic of Risk
Perception and Risk Communication over the Last 15 Years
Daniela Kolarova, Sofia University, Bulgaria

16:30-17:30 Session
Facilitated Exercises in Risk Perception
Dr. Vladilena Abramova, RF and Daniela Kolarova, Bulgaria

Open time: 17:30-18:30

Dinner 18:30-19:30 (at hotel)

Saturday, 6 May 2000

Breakfast: 8:00-9:00 (in hotel)

Morning Lecture Hall Programme: 9:30-12:30

9:30-11:00 Session
  Approaches through Multimedia Assessment [Facilities-Centered Health
  Risk Assessment
  Dr. Gene Whelan, PNNL, USA

11:00-11:30 Coffee/tea

11:30-12:30 Session
  Methods and Tools in the Management of Technological Risk
  Assoc. Prof. Dr. Lyubcho Lyubchev, University "Prof. Dr. Assen
  Zlatarov," Bourgas

Lunch: 12:30-13:30 (at hotel)

Local sight-seeing; excursion to the old city Nesebár: 14:00-18:30

Dinner 18:30-19:30 (at hotel)

Sunday, 7 May 2000

Breakfast: 8:00-10:00 (at hotel)

Free time for study, informal discussions, and other activities.

12:30-13:30 Lunch (at hotel)

Afternoon Special Event, 14:00 to 18:00

  International Gymnastics Competition, Bourgas

18:30-19:30 Dinner (at hotel)

Monday, 8 May 2000

Breakfast: 8:00-9:00 (at hotel)

Morning Lecture Hall Programme: 9:30-12:30

  9:00-10:00 Session

Cleanup of Vromos Bay
Dr. Simeon Simeonov, Director, Regional Inspectorate to Environmental Protection and Water of Ministry of Environment in Bourgas
Dr. Eng. Ilko Bonev, Director, Bourgas Copper Mines, Bulgaria

10:00-13:00 Site Visit
Vromos Bay, Drs. Simeonov and Bonev
Excursion to old city Sozopol

Lunch: 13:00-14:00 (at hotel)

Afternoon Lecture Hall Programme: 14:00-17:30

14:00-15:00 Computer lab
Risk assessment software workshop
Jim Droppo, Coordinator

15:00-15:30 Coffee/tea

15:30-17:30 Software Demonstrations

Open time: 17:30-18:30

Dinner 18:30-19:30 (at hotel)

Tuesday, 9 May 2000

Breakfast: 8:00-9:00 (in hotel)

Morning Lecture Hall Programme: 9:00-12:30

9:00-10:30 Session
Potential for Risk Assessment Research in the Chernobyl Exclusion Zone
Aleksey Ryabuskin, International Radiology Lab, Ukraine

10:30-11:00 Coffee/tea

10:30-11:30 Session
Modular Risk Assessment–Hanford Example
Dr. Gene Whelan, PNNL, USA

11:30-12:30 Session

Chernobyl Catastrophe Problems: Radiation Factor Risk Assessment
within the Exclusion ChNPP Zone
Dr. Georgy Lisitchenko, State Scientific Center of Environmental
Radiogeochemistry, Ukraine

Lunch: 12:30-14:00 (at hotel)

Afternoon Lecture Hall Programme: 14:00-17:30

14:00-15:30 Session
Quantitative Risk Assessment of Peace-Time Activities
Associated with Complex Weapon Systems in Military Installations
Steve Fogarty, ARES Corporation, USA

15:30-16:00 Coffee/tea

16:00-17:30 Session
Model selection
Dr. Jim Droppo, PNNL, USA

Open time: 17:30-18:30

Dinner: 18:30-19:30 (at hotel)

Wednesday, 10 May 2000

Breakfast: 8:00-9:00 (at hotel)

Morning Lecture Hall Programme: 8:30-12:30

9:00-10:30 Session
Radiation Legacy Of The Former Soviet Union (Weapon Complex)
Yuri Gorlinsky, Science and Technology Association "Computer
Technologies and Information Systems for Science and Research
Development," RF

10:30-11:00 Coffee/tea

11:00-12:00 Session
US Weapons Production–Hanford Dose Reconstruction
Bruce Napier, PNNL, USA

12:00-13:00 Session
 Collaborative Risk Assessment in the Russian Federation
 Bruce Napier, PNNL, USA

Lunch: 13:00-14:00 (at hotel)

Afternoon Lecture Hall Programme: 14:00-17:30

 14:00-15:00 Session
  Risk Assessment and Risk Management in the Netherlands
  Dr. Esko F. Blokker, DCMR Environmental Protection Agency, The
  Netherlands

 15:00-15:30 Coffee/tea

 15:30-17:00 Session
  Regulatory Rule Making Example – US EPA Rule Making
  Dr. Gene Whelan, PNNL, USA

Open time: 17:30-18:30

Dinner: 18:30-19:30 (at hotel)

Thursday, 11 May 2000

Breakfast: 8:00-9:00 (in hotel)

Morning Lecture Hall Programme: 9:00-12:30

 9:00-10:00 Session
  Civil Protection in Bulgaria
  Dipl. Eng. Svetoslav Andonov, Deputy Director of Civil Protection,
  Bulgaria

 10:00-10:30 Coffee/tea

 10:30-12:00 Session
  Round Table #3. Risk Assessment Problems and Programs in Home
  Countries.
  Facilitator (Directors/Lecturers) asks students to propose problems from
  their home regions. Groups will define issues and discuss solution
  options.

12:00-13:00 Closing Panel
    Organizers – Drs. Dennis Bley, Vitaly Eremenko, Jim Droppo, and Esko Blokker
        Summary Statements
        Lecturer and Participant Input and Recommendations/Statements
        Closing

Lunch: 13:00-14:00 (at hotel)

Appendix B
Acronyms and Abbreviations Used in Text

The following acronyms and abbreviations are used in the text, figures, and tables of this book. Acronyms and abbreviations used in equations are described for each equation and are not included here.

Nonletters

%	percent
/	per
μR	microRoengten(s)
μSv	microSievert(s)

A

APET	accident progression event tree
ASI	Advanced Study Institute

B

BD	Bryansk District
BEMR	Baseline Environmental Management Report
Bq	Bequerel

C

CDC	Centers for Disease Control and Prevention
CERCLA	Comprehensive Environmental Response, Compensation, and Liability Act
ChEZ	Chornobyl Exclusion Zone
ChNPP	Chornobyl Nuclear Power Plant
Ci	curie(s)
cm	centimeter(s)

CMF	common mode failures
CNPP	Chernobyl Nuclear Power Plant
Cs-137	cesium-137
CSv	xSievert(s)

D

DESCARTES	Dynamic Estimates of Concentrations and Accumulated Radionuclides in Terrestrial Environments
DM's	decision makers
DOD	U.S. Department of Defense
DOE	U.S. Department of Energy
DPM	Defense Priority Model
Dr.	Doctor of Philosophy (U.S) or Science (Europe)
Drs.	Doctors of Philosophy (U.S) or Science (Europe)

E

EDE	estimated dose equivalent
e.g.	for example
EM	U.S. Department of Energy Office of Environmental Management
Eng.	Engineering
EPA	U.S. Environmental Protection Agency
ERMA	Environmental Risk Management Alliance
ETRC	Extended Techa River Cohort

F

FCRA	facility-centred risk assessment
FRAMES	Framework for Risk Assessment in Multimedia Environmental Systems
FS	feasibility study
FSU	Former Soviet Union

G

g	gram(s)
GBq	gigaBequerel(s)
GIS	geographic information system
GW	gigawatt(s)
Gy	Gray(s)--check

H

ha	hectare(s)
HCRA	human-centred risk assessment
HEDR	Hanford Environmental Dose Reconstruction (Project)
hr	hour(s)
HRS	Hazard Ranking System

HVAC	heating, ventilating, and air conditioning
HWIR	Hazardous Waste Identification Rule

I

IA	Industrial Association
IAEA	International Atomic Energy Agency
IIASA	International Institute for Applied Systems Analysis
ICRP	International Commission on Radiation Protection
i.e.	in essence
IE	initiating events
Inc.	Incorporated

J

K

KA	Kaluga District
kBq	kiloBequerel(s)
K_d	distribution coefficient
KD	Kurst District
keV	kilo-electron-volt(s)
kg	kilogram(s)
km	kilometre(s)
km^2	square kilometre(s)
km^3	cubic kilometre(s)
kt	kilotonne(s)
kW	kilowatt(s)

L

l	litre(s)
LOC	loss of containment
LPG	liquefied petroleum gas

M

m	meter(s)
m^3	cubic meter(s)
Max	maximum
mBq	millibequerel(s)
MCi	megaCurie
MDL	method of detection limit
MEA	methylethylamine
MEPAS	Multimedia Environmental Pollutant Assessment System
MeV	mega-electron-volt(s)
Min	minimum

Minatom	Ministry of the Russian Federation for Atomic Energy
mL	millilitre(s)
MLD	Master Logic Diagram
mm	millimeter(s)
mSv	milliSievert(s)
MW	megawatt(s)

N

NATO	North Atlantic Treaty Organization
NEPA	National Environmental Policy Act
NM	State of New Mexico
No.	Number
NPL	National Priorities List
NPP	nuclear power plant
NRC	U.S. Nuclear Regulatory Commission
NV	State of Nevada

O

P

PBq	petaBequerel (page 198—check)
PCB	polychlorinated biphenyls
PEIS	Programmatic Environmental Impact Statement
PL	Public Law
PMCD	Program Manager for Chemical Demilitarization
PNNL	Pacific Northwest National Laboratory
POD	process operations diagrams
PRA	probabilistic risk assessment

Q

QRA	qualitative risk assessment

R

^{226}Ra	radium-226
R&D	research and development
RA	risk assessment
RAAS	Remedial Action Assessment System
RAMEH	Risk Analysis Methodology for Environment and Health
RATCHET	Regional Atmospheric Transport Code for Hanford Emission Tracking
RCRA	Resource Conservation and Recovery Act
ReOpt	Remediation Option System
RfD	reference dose
RH	relative hazard

RI	remedial investigation
RI/FS	remedial investigation/feasibility study
RM	risk measure
RR	relative risk

S

SAIC	Science Applications International Corporation (check)
SARA	Superfund Amendment and Reauthorization Act
SC	State of South Carolina
SD	sustainable development
SMS	safety management system
SNF	spent nuclear fuel
SPD	scenario progression diagrams
Sv	Sievert(s)

T

TN	State of Tennessee
TRC	Techa River Cohort
TRDS	Techa River Dosimetry System
TSP	Technical Steering Panel

U

UN	United Nations
URF	unit risk factor
U.S.	United States of America
USA	United States of America
USSR	United States of the Soviet Republic
UTF	unit-transfer factor

V

| VOC | volatile organic compounds |

W

| WA | State of Washington |
| WBA | whole-body counter |

X

--

Y

yr   year(s)

Z

Appendix C
Cross-Cultural Guide to the Book

This appendix is derived from an English-Russian cross-culture terminology guide that was provided to the native Russian-speaking participants at the Advanced Study Institute.

In the last decades of the twentieth century in United States, risk analysis methodologies started being used in waste management and environmental remediation efforts. The chapters in this book describe some of these applications for facility operations and environmental restoration. The former includes the design of storage, handling, and treatment/destruction facilities related to both weapons and environmental efforts. Relative to the latter cleanup of sites contaminated by nuclear and chemical materials, laws have been passed, regulations written, and guidance prepared for both waste practices (CERCLA) and current operations (RCRA). The result of the implementation of these actions is that there is a large amount of experience in the West relative to the East on how to approach and accomplish site cleanups. One of the objectives of the Institute was to provide decisions makers in the East access to that experience base relative to the use of risk methodologies.

To understand the motivations and drivers for inclusion of risk in the American environmental remediation culture, one needs to understand the Remedial Investigation/ Feasibility (RI/FS) processes specified by the U.S. Environmental Protection Agency. Similar regulatory-based approaches have not been implemented in the regions of the former Soviet Union. To help bridge this information gap, the Russian-speaking participants were given supplemental background material to help them understand the American culture for conducting cleanup efforts.

The following English-Russian cross-culture terminology guide is adapted from the ASI handout prepared by Vitaly A. Eremenko for the Russian-speaking participants attending the Institute. The English-Russian cross-culture terminology guide is provided to define selected American concepts for native speaking Russians that are unfamiliar with the American environmental remediation culture. Selected environmental remediation concepts are included that require more than a simple word-to-word translation to be understood. These extended Russian-language descriptions provide native-language background material for understanding the various risk-based applications described in this book.

Вспомогательные рубрики, шрифты и сокращения. Для расширения творческой активности и самостоятельности читателей, наиболее важные понятия и определения в Глоссарии сопровождаются дополнительной информацией. Чаще всего используются следующие вспомогательные рубрики: [...] - источник сведений по данному термину, взятый из прилагаемого списка публикаций; может располагаться, либо непосредственно после первого короткого перевода на русский язык, либо после комментария или дополнительных данных; там, где источником является личное мнение Еременко, используется сокращение в виде; "**Генезис**" обозначает происхождение и развитие определяемого термина или физического объекта; "**Математика**"- означает, что определения термина имеет формальную математическую основу; "**Комментарий**" открывает последующие авторские рекомендация по использованию этого термина; "**Совет**" или "**Замечание**" - под этой рубрикой приводятся замечания и советы в тех случаях, когда мнение автора словаря не совпадает с трактовкой данного термина в оригинале; "**Цифры**"- цифровые данные, нормы и др. факты.

В глоссарии используются следующие типы шрифтов: "**жирный**"- для выделения терминов и специальных слов на английском языке, при условии, что их полные переводы и определения можно найти хотя бы в одной из рубрик данного Глоссария; "*курсив*" применяется для выделения в кириллице, наиболее значимых - ключевых терминов, и когда, хотя бы в одной из рубрик данного словаря, имеется толкование термина на русском языке.

American RI/FS process (выбор *реабилитационных и восстановительных мер*, и анализ их оптимальности по стоимости и эффективности [10]) - проведение комплекса исследований по уровням и видам *реабилитационных и восстановительных мер*, уменьшающих опасности и понижающих *риск* для населения и персонала на изучаемом опасном предприятии или загрязненной территории, с параллельным анализом их осуществимости и экономической целесообразности, - то есть, в проведением *модернизированного технико-экономического обоснования (МТЭО)* проектов их реализации.

Комментарий: Смысл и содержание каждой такой процедуры (**RI/FS**), предваряющей практическое решение проблемы защиты *Окр.Ср.* и здоровья людей, состоит в следующем:
- Во-первых, - дать количественную характеристику проблемы в целом. Например, количественно оценить степень загрязнения окружающей среды и опасность накопившихся отходов. При этом принимаются во внимание все имеющиеся на данной территории опасные предприятия и др. *источники* риска, конкретные геологические, метереологические и др. условия, а также текущее состояние окружающей среды, и пр., и пр. Таким способом

составляется количественная модель "*состояния сложного многомерного объекта управлени.*"

- Во - вторых, - расчленить проблему на отдельные сегменты по типу и виду таких *реабилитационных и восстановительных мероприятий,* в рамках которых было бы возможно - практически осуществимо организовать отдельные, конкретно ориентированные действия по улучшению состояния территорий и здоровья, проживающих там людей до некоторого приемлемого уровня, с учетом специфики конкретных геосред, свойств конкретных загрязняющих веществ, и конкретных предприятий и загрязненных территорий, - *источников* радиационной и токсической опасности. Здесь, по сути, формулируется задача управления состоянием сложного многомерного объекта.

- Наконец, в-третьих, - основываясь на предыдущем анализе, выбрать цели и установить соответствующие задачи по реализации оптимальных *реабилитационных и восстановительных мероприятий,* а также определить, на основе *технико-экономического анализа - МТЭО*, какие конкретные их типы и виды будут в наибольшей степени удовлетворять поставленным целям и задачам. На этой фазе **RI/FS process** аналитически решается задача оптимального управления состоянием сложного многомерного объекта.

American RI/FS process analysis framework (анализ, проводимый в рамках *RI/FS процесса* [10,13]) - предусматривает выполнение сложной комбинации вычислительных операций с использованием пяти оценочных, аналитических и оптимизационных модулей: **site conceptual model; contaminant transport; technology selection and performance; human health effects; restoration alternative effects.**

Carcinogenic risk (*риск канцерогенных* заболеваний [5,8,18]) - в методологии **ICRA** определяется двумя составляющими: *индивидуальным риском канцерогенного* заболевания от радиации (см. понятие **Risk Analysis**); *индивидуальным риском канцерогенного* заболевания от воздействия химических канцерогенов.

Carcinogenic risk: Математика [5,18] При определении величины первой составляющей, используют формулу **Rr,rp = H * Drp * 2.555*10E+4,** рекомендованную Национальной Академией Наук США (1992г). Здесь: * - знак произведения; **Rr,rp** - *риск* в течение жизни при *мощности ежедневной экспозиционной дозы* радиоактивного облучения - **Drp** (рентг/день); **H** - *онверсионный фактор риск*а, а цифра 2.555*10E+4 определяет число дней в 70 одах (дн). Вторая составляющая, по рекомендациям EPA 1982г, рассчитывается осредством введения *фактора потенциальной возможности заболевания раком*: **Rc,rp = 1 - exp(- D,rp * q)**, где: * - знак произведения; **Rc,rp** - верхняя граница иска при ежедневной экспозиции человека *дозой* мощностью **D,rp** (безразмерная еличина), в течение всей его 70-летней жизни, а **q** - значение фактора **CPF** для данного химического соединения в (кг*день/мг). Сравнение потенциальных опасностей раковых заболеваний от радионуклидов и канцерогенных химикатов производят на основе усредненной величины *индивидуального риска канцерогенного* заболевания человека в течение его жизни (**R,rp**), которую выражают величинами - *индивидуального риска канцерогенного* заболевания под воздействием *экспозиционных доз* ионизирующей радиации (**Rr,rp**), либо - *индивидуального риска канцерогенного* заболевания под воздействием химических канцерогенов (**Rc,rp**) соответственно. Для практического использования приведенных выше формул необходимо оценить величины факторов **H** и **CPF**. Формулы для их расчетов можно найти в справочниках под рубриками **risk conversion factor - H**, и **risk cancer potential factor - q**.

Chernobyl NPP, ChNPP, Chernobyl Nuclear Power Plant (Чернобыльская АЭС, ЧАЭС, Чернобыльская атомная электростанция).

Competitive Risk Management - CRM (*конкурентное управление риском, КУР* [1]) - см. также многочисленные определения и дополнения к термину.

Risk Containment technologies (технологические *процессы купирования* химических *загрязнений* и радиоактивности, *источников* облучения [5,6,7,10,11, 12,13]).

Contaminant effectiveness parameters (параметры пригодности *загрязнителей* к обработке - ПЗО [10,13])

Contaminant screening (альтернативы технологий [7,14]) - имеется в виду выбор, в *процессе RI/FS*, технологий, ориентированных на тип и свойства радиоактивных и химических *загрязнителей.*

Contaminant transport (*распространение/миграция* химических загрязнений и радиоактивности [7]) - удаление от исходного местоположения источника *сбросов* или *выбросов* опасного радиоактивного или химического вещества, которое вызвано процессами диффузии в *Окр.Ср.* растворенных или взвешенных в геосредах примесей, либо перенос примесей движущими геосредами *Окр.Ср.*, либо, наконец, перемещение опасных веществ транспортными средствами и людьми, работающими в контакте с ними. Термин "*миграция*" введен стандартом 6107/6-86 Международной Организации по Стандартам (ИСО) как "самопроизвольное или принудительное перемещение растворенных или взвешенных веществ или организмов в водном объекте." Параметры *распространения/миграции* оцениваются с помощью моделей распространения (переноса, диффузии) химических *загрязнений и радиоактивности* [2].

Contaminants (*загрязнители* [3])

Daily exposure rate - D,rp (*ежедневная мощность экспозиционной дозы*) - получается среднестатистическим представителем данной группы населения по каждому *дозовому маршруту*, и оценивается в рамках *СО-АР* как: **D,rp = U,r*C,rp*F,r**, где: * - знак умножения; запятая, предваряет индекс (здесь, **rp, r**); **r** - индекс *дозового маршрута* (**I, O, D, E** - их всего четыре); **p** - индекс плотности населения оцениваемого региона или административной территории (**p = 1,2,..P**); **D,rp** - *ежедневная мощность экспозиционной дозы* (мг/кг*день или rem/день); **U,r** - обычная средняя ежедневная скорость поступления в организм, или поглощения, или облучения по *дозовому маршруту* "r" для данного *загрязнителя* (в кг/день - для орального поступления, м3/день - для ингаляционного поступления, час/день - время экспонирования для внешнего облучения); **C,rp** - концентрация *загрязнителя* в среде, контактируемой с человеком (в пище, воде, и пр), или продолжительность *экспонирования* (в мг/кг или пкКи/кг - для перорального поступления, мг/м+3 или пкКи/м+3 - при ингаляционном поступлении, пкКи/м+2 - при внешнем облучении от земли, пкКи/л - при внешнем облучении от воды, пкКи/м+3 - при внешнем облучении из атмосферного воздуха); **x+2, y+3, кг-1** - обозначает возведение "**x**" в степень +2, "**y**" в степень +3, а "**кг**" в степень -1; **F,r** - дозовый *конверсионный фактор* для *дозового маршрута* "r" (в кг-1 или rem/пкКи - для ингаляционного и перорального поступления; rem/час /на пкКи/м2 - при внешнем облучении от земли; rem/час /на пкКи/л - при внешнем облучении от воды, и rem/час/на пкКи/м3 - при внешнем облучении из атмосферного воздуха). В практике часто используется более полное понятие **Daily exposure rate to an average member of the population for each exposure route - D,rp** (*ежедневная мощность экспозиционной дозы для среднего представителя данной популяции и по каждому дозовому маршруту*) - в том числе, конечно, и для расчета *экспозиционной дозы* от радиоактивного облучения.

Emission rate (скорость, расход или мощность генерации опасных материалов и/или энергии и/или отходов, при соблюдении регламента эксплуатации производственных объектов **[2]**) - имеется в виду показатели поступления загрязнителей в среду обитания человека, - то есть, в *Окр.Ср.*, при штатном состоянии и/или эксплуатации их *источников*.

Environment (*окружающая среда - Окр.Ср.* **[12]**, или, *внешняя среда - Вне.Ср.* **[2]**) - совокупность физических, химических, биологических характеристик, а также социальных факторов, способных оказывать прямое или косвенное, немедленное или отдаленное воздействие на живые существа и деятельность человека; или среда, окружающая биоценозы, физическая основа их биоцениотической среды - атмосфера и ее циркуляция, солнечный свет и теплота, материнская порода почвы, ее химические вещества, газы и растворы, вода и влажность атмосферы и подпочвы, общий климат территорий и акваторий.

Environmental Media (природная среда **[5,13,18]**)

Environmental Restoration processes - (*способы восстановления Окр.Ср.* [5,18]): процедура, установленная федеральным законодательством США (**CERCLA**), и определяющая последовательность оценок и действий при решении данной проблемы.

Ex situ media (*удаленные первичного источника* загрязненные и излучающие среды и материалы [7,13])- загрязненные или излучающие геосреды и материалы, находящиеся вне местоположения *источников* первичных *выбросов, сбросов* и облучений, а также первичного поступления отходов.

Ex situ media: Комментарий: Предполагается, что *загрязнители* уже *мигрировали*, и распространились за пределы эпицентра *первичных источников*. Таким образом, **RI/FS process** и **HCRA** ориентируют исследователей риска на распределенные *источники* опасности. Расчеты, выполняемые на **MEPAS,** позволяют оценить распространение загрязняющих примесей по геосредам на расстоянии до 80 кмот эпицентра первичной опасности в случае воздушных переносов, и на неограниченные расстояния при переносе *загрязнений* поверхностными и подземными водами. Все *мигрирующие загрязнители*, а также перемещающиеся загрязненные среды и материалы обязательно *учитываются* также и в сценариях системы **RAAS**. Понятие *"учет"* (**contaminant inventory**) подразумевает регистрацию запасов *загрязнителей*, а также физических и химических их свойств (радиоактивности, токсичности) и характеристик (**media properties**).

Exposure (*экспонирование* [3]) - в общем смысле - *облучение*, как необходимое условие для получения *экспозиционной дозы.*

Exposure pathways (пути распространения *источников экспонирования*/облучения [5,8,18]) - включают природные среды: подземные и поверхностные воды, почва и воздух; *дозообразующие среды* (включают, кроме природных, загрязненные продукты потребления, в частности - питания); загрязненный грузовой и пассажирский транспорт, материалы; и т.п.- все, что может быть средой для распространения химических загрязнений и радиоактивности, и, в конце концов, служить источником *экспонирования* людей.

Exposure point analysis (анализ локальных *экспозиционных доз* [7,18])

Exposure route (*дозовые маршруты* [7,18]) - представляют возможные варианты контактов человека и *загрязнений* при моделировании *риска*. К *дозовым маршрутам* могут быть отнесены, например, пищевые цепи, внешнее облучение (только для радионуклидов), а также различные виды физиологических контактов человека, таких как ингаляционный, пероральный и кожный.

Exposure scenario (*дозовый сценарий* [7,18]) - составляется из цепочки нескольких *дозовых маршрутов,* которые выражают движение загрязнения через пищевые цепи в человеческий организм, и описывает случайные эффекты их воздействия на людей. Пример написания *дозового сценария*: "От загрязненных поверхностных вод -> во время купания -> через пероральное/и ингаляционное случайные поступления -> внутрь организма."

Facility (промышленный или энергетический объект[5,7,18]) - типичный техногенный источник химической или радиационной опасности и *риска;* обобщенное наименование потенциально опасного производственного комплекса.

Facility-Centered Risk Assessment (*Объектно - Ориентированный Анализ Риска, - ОО-АР* **[1,10,14]***)*– первый и старейший из *трех классов прикладных оценок риска,* объединяющий методы и компьютерные системы оценок, анализа и оптимизации результатов, позволяющие подготовить обоснованные решения по мероприятиям и действиям, направленным на совершенствование безопасности ядерных, радиационных и химических объектов и загрязненных территорий. Используется как один из базовых методов при решении проблем ликвидации наследия прошлой холодной войны (**past Cold War Legacy).**

Facility-Centered Risk Assessment: Комментарий *Первый из методов класса ОО-АР,* названный *Объектно - Ориентированным Вероятностным Анализом Риска/Безопасности, (или ОО - ВАР или ВАБ),* был разработан для атомных электростанций в 70-х годах прошлого века, но широко применялся вплоть до настоящего времени, и не только для АЭС, но и для многих других опасных производств. Второй метод этого класса, так называемый *Объектно - Ориентированный Анализ Риска для Здоровья (или ОО - АРЗ),* изучает *риски как* вероятностные последствия рутинной или штатной эксплуатации одиночных технологических объектов. Этим же методом оцениваются риски от незначительных аварийных ситуаций, связанных с отказами оборудования или систем безопасности на отдельных объектах. В США эта область использования риска является прерогативой Агентства по Защите Окружающей Среды (EPA). Наиболее распространенные задачи для методов ОО-АР - *оценки потенциального риска для здоровья и Окр.Ср. от аварий на единичных объектах; от эмиссий радиационных и химических веществ при штатной эксплуатации единичных объектов; а также разработка превентивных мер безопасности для единичных объектов.* Используется как один из базовых методов при решении проблем ликвидации наследия прошлой холодной войны (**past Cold War Legacy).**

Facility-Centered Risk Assessment: Комментарий [1,10] Основными целями методов ОО-АР являются обеспечение оптимизации решений *превентивного* характера *для единичных объектов.* Такие решения направлены на предотвращение *потенциальных аварий* или ограничение *рутинных выбросов или сбросов. Методология ОО-АР* используется также для *планирования действий* на аварийном объекте или прилежащей территории при возникновении *чрезвычайных нештатных ситуаций.* Решения, разработанные на основе методов *ОО-АР,* реализуются путем *совершенствования систем безопасности* анализируемых объектов, а также посредством *поддержания предварительной готовности* чрезвычайных служб в зонах их потенциального воздействия.

Facility-Centered Risk Assessment: Замечание [1]. Если на интересующей заказчика территории расположен не один, -единичный, а несколько потенциально опасных объектов, то анализ методами *ОО-АР* производится отдельно для каждого из таких объектов. Процедура *оптимизации управленческих решений* по *превентивным* мерам и действиям становится значительно более эффективной, если первыми использовать возможности методов *СО-АР для всех источников опасности*, а затем уже инструмент *ОО-АР* для *особенно значимых единичных* объектов.

Facility-Centered Probabilistic Risk Assessment (*Объектно - Ориентированный Вероятностный Анализ Риска/Безопасности, ОО-ВАР/ВАБ* [1,17]) - Старейший из методов оценки риска этого класса был разработан для атомных электростанций в 1974. Начиная с ядерной аварии на АЭС США ТМА в 1978, *ВАБ* широко применялся и для многих других технологий в течение последующих 25 лет. *ВАБ (ВАР)* был разработан для оценки рисков потенциальных аварий на одиночных технологических объектах, причинами, которых могут быть как внутренние, так внешние события. Результаты таких оценок используются при *планировании мер*, уменьшающих потенциальные риски посредством *увеличения надежности и безопасности технологических объектов*. Удельная стоимость таких мер для больших ядерных, химических и других промышленных предприятий часто оценивается в миллионах долларов. Используется как один из базовых методов при решении проблем ликвидации наследия прошлой холодной войны (**past Cold War Legacy**).

Facility unit (разнотипные объекты, имеющие сходные регламенты обеспечения безопасности [5,8,18]) - объединяемые, при использовании методологии *СО-АР,* в одну группу объекты по тому признаку, что *реабилитационные меры* для них близки по целям способам и средствам реализации. Как следствие, в компьютерных системах типа **MEPAS/RAAS**, где используется эта классификация, предполагается, что *реабилитационные мероприятия* для группы таких объектов будут осуществляться по единому плану. Требуется, однако, чтобы такие объекты, представляющие разнообразные *источники выбросов/сбросов* или *вторичные потоки отходов*, тем не менее, находились в пределах одной *ПНТ*, промплощадки или промзоны (**installation, site, waste site**).

Facility-Centered Health Risk Assessment (*Объектно - Ориентированный Анализ Риск для Здоровья, - ОО - АРЗ* [1,10,14,17]) - Второй из *методов класса ОО-АР*, задача которого состоит в изучении и оценке рисков от рутинной (штатной) эксплуатации одиночных технологических объектов. Результаты оценок и анализа на основе *ОО-АРЗ* используются при установлении стандартов и норм для *предельно – допустимых выбросов или сбросов* во внешнюю среду, а также при назначении *предельно- допустимых концентраций* загрязняющих веществ в *окружающей среде.* Используется как один из базовых методов при решении проблем ликвидации наследия прошлой холодной войны (**past Cold War Legacy**).

Facility-Centered Health Risk Assessment: Комментарий. В США, РФ, Украине, например, эта область использования риска является прерогативами соответственно Агентства по Защите Окружающей Среды, Госкомитета РФ и Министерства Украины. Этот метод по сравнению с *OO-АР* типа *ВАБ/ВАР* допускает ограничиться более простым анализом. Результаты оценок и такого анализа используются при установлении норм и стандартов для предельно – допустимых выбросов или сбросов от опасных объектов во *внешнюю среду*, а также для предельно- допустимых концентраций загрязняющих веществ в окружающей среде. Такие оценки также лежат в основе определения штрафов для объектов – нарушителей норм. Не редки случаи, когда удельные годовые штрафы, установленные на основе *OO-АРЗ*, для некоторых больших ядерных, радиационных и химических объектов – нарушителей достигают сотен тысяч долларов. В результате сложилась парадоксальная ситуация: в убытке находятся как собственники опасных предприятий (громадные штрафы !), так и близлежащее население, поскольку, согласно существующему законодательству, предприятия "купили индульгенции" на отравление людей и природы.

Feasibility study - FS (анализ осуществимости предлагаемых проектов [7,8,18]) - известная процедура технико-экономического обоснования, наиболее близка по содержанию к **FS**. Далее **FS** понимается как американский вариант модернизированного *ТЭО - МТЭО*. Включает: постановку целей; оценку стоимости; а также предполагаемого выход продукции, здесь - эффективности снижения риска; принимается во внимание конкретный потребитель проекта и спрос на проекты такого типа; оцениваются способы отступлений в случае неудачи. Используется как одна из базовых процедур при решении проблем ликвидации наследия прошлой холодной войны (**past Cold War Legacy**).

Features of American RI/FS process, characterizing an order of a technology selection (особенности выбора технологий *в рамках RI/FS процесса*) - в этой части *RI/FS процесса* выбираются технологии *реабилитационных мер* и оценивается их эффективность. Наиболее значимыми составляющими этой части процесса, разработанными в **US DOE PNNL**, являются [7]: База Знаний по возможным *реабилитационным мерам*, содержащая сведения по более чем 100 специальным технологиям, ориентированным на более чем 400 различных *загрязнителя* и содержащая описания опыта в 14 различных случаях их применения. БЗ является частью компьютерных систем **ReOpt и RAAS.** Рекомендуемые БЗ технологии *реабилитационных мероприятий* могут быть реализованы как в эпицентре *выбросов* или *сбросов загрязнений,* так и на значительном расстоянии от них, а также при разнообразных сочетаниях *загрязнителей, загрязненных сред,* и *загрязненных* материалов (**technology applicability for 400 contaminants**). В процессе предварительного рассмотрения технологий для конкретных *реабилитационных мер* производится оценка мощности *вторичных потоков* *отходов (*secondary waste stream*)* и методов *обработки и переработки* последних.

Features of American RI/FS process, characterizing module of evaluation of potential efficiency chosen technologies (особенности оценки потенциальной эффективности выбранных технологий в *процессе RI/FS*) - заключаются в проведении двух этапов специальной процедуры оценки. Наиболее значимыми функциями данной части *процесса RI/FS* являются: 1). **Alternative effectiveness -** оценка эффективности анализируемых вариантов, и значений **residual contaminant inventory/mobility/human health effects**; и 2).**Alternative performance-** сопоставление анализируемых вариантов, включающее определение и сравнение не менее шести наиболее важных показателей, характеризующих результативность выбранной технологии. Основные из них - стоимость, затраты времени, объем обрабатываемых *загрязненных* сред, и проч. (**cost, time, permanence, volume, toxicity, mobility reduction**).

Features of HCRA concerning assessment of human health (особенности *оценки риска для здоровья* в рамках *RI/FS процесса* [7]) - учитываются как результат проведения вычислительных операций на "модуле *ущерба для здоровья человека*" (**human health effects**). Наиболее значимыми составляющими оценки в этом случае являются:1). *риск канцерогенных* заболеваний (**carcinogenic risk**), 2). *квоты опасности* (**hazard quotient**), 3). характеристики *рецептора* (**variable receptor characteristics**), 4). *дозы*, получаемые при пребывании на *загрязненных* землях (**land use variable dose**).

Ground-water exposure routes (*дозовые маршруты* в подземных водах [7])- здесь оперируют с тремя *дозовыми маршрутами* (r=1,2,3), каждый из которых заканчивается тремя видами физиологических контактов *рецептора - человека* (**ingestion, dermal, inhalation**), а также с пятью *дозообразующими* средами (**3 through Ingestion; 1 through Dermal; 1 through Inhalation**).

Hazard quotients (*квоты опасности* [5,8,18]) - безразмерные величины, определяющие меру *ущерба для здоровья человека*. В документах на английском языке обозначается как **HQ**, в мат. формулах - как **In,rp**. **Hazard quotients: Математика [5, 8, 18]**: Величина **HQ** определяется как частное, - отношением *мощности ежедневной экспозиционной дозы* **D,rp** к *минимально - опасной дозе токсического* вещества **RfD,k**, то есть: **In,rp = D,rp/RfD,k**.

Hazard quotients: Комментарий: В методологии **HCRA** этот подход используется для учета влияния не - канцерогенных химических токсикантов на здоровье человека. Для практического использования приведенной формулы необходимо знать величины **RfD**, k. Источником таких сведений для пользователей в США является федеральная директива Агентства Защиты *Окр.Ср.* США (ЕРА, 1989г.).

Human factor a source of risk (*"человеческий фактор"* как источник *риска* [6,9,16]) - эта составляющая *полного риска*, во-первых, может быть обусловлена нерациональностью или иррациональностью действий специалистов, работа которых связанна с той или иной стадией жизненного цикла опасных технологий (см. **Stages of "life cycle" of hazardous technologies**). Во-вторых, генераторами таких же действий, приводящих к негативным последствиям, и потому с полным правом относимых к источникам *риска*, могут быть и *лица, принимающие решения, - ЛПР* (**Decision Makers, DM's**). Вообще же. не только технические специалисты и управленцы разного уровня, но и любой человек по разным причинам (в том числе, например, в силу "комплекса Герострата") может быть источником самых изощренных исходных замыслов, которые, при определенных условиях, перерастают в тяжкие последствия. Именно в таком контексте потенциального носителя антропогенного *риска "человеческий фактор"* изучается, а соответствующий *риск* оценивается в методологии *КУР* (**CRM, - Competitive Risk Management**), см. в [1]: **Analysis of "Perception" of the objective Risk by DM'S - PRA-DM'S**. Метод, позволяющий выполнить необходимые оценки на качественном уровне, изложен в известном документе МАГАТЭ "**ASCOT Guidelines**"[16]. Используя этот документ, возможно получить экспертную оценку "уровня культуры безопасности" специалистов и *ЛПР*. Общепризнанного метода численной оценки *риска* от *"человеческого фактора"* пока не существует. Большинство из созданных в 80-90-тые годы методов численного анализа предназначались, за небольшим исключением, для оценки надежности операторов опасных производств [APJ(Absolute Probability Judgement,1983); TESEO (Tecnica Empirica Stima Errori Operatori,1982); THERRP (Technique for Human Error Rate Prediction,1984); HCR (Human Cognitive Reliability, 1983); Facility- Centered Risk Assessment from Human Factor, 1995]. Исключение представляет последний из перечисленных выше методов - **Facility- Centered Risk Assessment from Human Factor, - FCRA-HF** (*Объектно - Ориентированный Анализ Риска от Человеческого Фактора, - ОО-АР-ЧФ*). *Суть* этого метода *кратко представлена в публикациях* [1,4,6,17,18,19]. Однако они еще не опробованы в практике, так как исследования по ним прекращены в 1995г.,из-за отсутствия финансирования.

Human health effect (*ущерб здоровью человека* [7,18]) - нарушение здоровья отдельного человека или групп людей от воздействий техногенных *источников опасности* при регламентной эксплуатации последних, то есть, вне острой фазы аварий. Выражается в методологии **HCRA** двумя показателями: 1). как *индивидуальный риск канцерогенного* заболевания в течение его жизни (70 лет) под *воздействием радионуклидов и/или химических канцерогенов*, и 2). как *индивидуальный ущерб* для здоровья от не канцерогенных химикатов, выражаемый *через квоты опасности*.

Human health effect: Комментарий: Посредством этих двух прямых количественных показателей на терминалах системы **MEPAS** представляются, и далее, в системах **ReOpt** и **RAAS** используются: во-первых, результаты оценки *ущерба здоровью человека*, и, во-вторых, - интегральные показатели текущего состояния территорий, технологических установок, отдельных производств и их комплексов.

Human-Centered Risk Assessment/Analysis (*Субъектно-Ориентированный Анализ Риска, - СО-АР* [1,10,14,17]) - методология *СО-АР* предназначена для поддержки *лиц, принимающих решения, - ЛПР* (**DM's**): 1) данными по оценке текущего и будущего состояния исследуемых промзон и/или населенных территорий по показателям *загрязненности* компонентов *окружающей среды* и *опасности/риска для здоровья людей*, 2) отборочными сценариями всевозможных последствий хозяйственной активности/ бездействия в условиях применения/отказа от применения *реабилитационных мер*, 3) данными и оптимизирующими процедурами для комплекса *реабилитационных мер* и действий по *восстановлению окружающей среды*, и т.д. Предполагается, что, в результате выполнения *СО-АР*, кроме того, будет подготовлены для *ЛПР* (**DM's**) рекомендации, воплощение которых позволит: 4) снизить поступление в среду обитания человека опасных отходов современной индустрии, а также материалов и/или опасных излучений, способных нарушить функции окружающей среды; 5) уменьшить *опасности и риски* от поступающих в среду обитания человека материалов и/или потоков излучений, посредством снижения их концентрации в *окружающей среды*; 6) удалять опасные материалы от мест плотного проживания людей, фауны и флоры, а также ослаблять потоки опасных излучений, перемещая материалы их в места управляемого хранения и создавая *зонтики безопасности* (**Umbrella of Safety**) для ослабления воздействия опасных материалов и/или излучений; 7) формулировать требования к местоположению и техническим характеристикам оптимальных по стоимости и эффективности хранилищ и *зонтиков безопасности*. Естественно, что такой перечень возможностей методологии *СО-АР* позволяет эффективно участвовать в решении проблем ликвидации наследия прошлой холодной войны (**past Cold War Legacy**).

Human-Centered Risk Assessment/Analysis: Генезис Идея этого подхода возникла в США еще в 1986 г. Тогда, в процессе первого практического осуществления **American RI/FS process**, предусматриваемого американским законодательством по защите окружающей среды **CERCLA -1980** и **SARA - 1986,** выяснилось, что выполнение только начальной фазы **RI/FS process (Risk Assessment)** требует значительных затрат финансово-материальных средств и времени даже в простейших случаях: в среднем от 36 до 40 чел*мес. Вследствие ограниченности средств на эти цели, правительством США (DOE, EPA) была поставлена задача замены рутинной измерительной и исследовательской работы путем создания специальных методик и поддерживающих систем для ускоренной оценки, анализа и оптимизации подготавливаемых в **RI/FS process** решений. К таким методикам и поддерживающим системам предъявлялись требования удобства, простоты и дешевизны в обращении при достаточной научной обоснованности. В соответствии с этими требованиями в PNNL в период 1993-1998, было создано семейство компьютерных поддерживающих систем MEPAS - ReOpt - RAAS, реализующих следующий перечень основных задач: 1) максимально достоверной и объективной оценки состояния *загрязненности* всех компонентов окружающей среды в исследуемом регионе (воздуха, почвы, поверхностных и подземных вод, и т.д.), и, как конечный результат, - оценки показателей *опасности* и *риска для здоровья людей*, проживающих и/или работающих на конкретной территории (в понятие *"риск для здоровья"* вкладывался смысл канцерогенного исхода от всех воздействий, в понятие же *"опасности для здоровья"* - неканцерогенного эффекта только от биологических и химических токсикантов); 2) выбора оптимальных по стоимости и эффективности мер, снижающих *опасности* и *риск* от всех *источников* расположенных в границах исследуемой территории (до 10000 *источников*); 3) анализа практической осуществимости и экономической целесообразности намечаемых мероприятий (в нашей терминологии - *технико-экономическое обоснование* проекта, - **ТЭО**), и, наконец, 4) оптимизации адресности и последовательности действий по восстановлению качества окружающей среды.

Human-Centered Risk Assessment/Analysis: Комментарий. Ядром комплексной системы компьютерной поддержки *СО-АР* являются физически обоснованные вычислительные программы для персонального компьютера. На платформе **INTEL WINDOWS**. Нормативная часть программ базируется, в основном, на стандартах **EPA**. Моделирующая часть построена на относительно стандартных подходах в расчетах переноса и воздействия. Подавляющее большинство используемых физико-химических моделей не оригинальны, и, напротив, многократно проверены в прошлом, и согласованы с соответствующими моделями из архивов Международных Агентств ООН (WHO, UNEP, IAEA). Пожалуй, единственной отличительной чертой моделирующей части комплексной системы является то, что разнообразные модели состыкованы и объединены в единую систему. Разработчики комплексной системы компьютерной поддержки *СО-АР* нашли свой вариант решения в среде специализированных аналитических моделей. Этот вариант хорошо понятен не только пользователю с профессиональным образованием, но и, что принципиально важно, - *ЛПР*. В результате созданная в **PNNL** система поддержки *СО-АР* стала уникальным и популярным помощником *ЛПР* (**DM's**) в решении проблем ликвидации наследия прошлой холодной войны (**past Cold War Legacy**). Кроме того, внедрение и использование информационно - оптимизирующих технологий типа *ОО-АР и СО-АР* позволяет свести до минимума возможности лоббирования государственных бюджетов, а также уменьшить болезненность перераспределения ограниченных личных и местных бюджетов, обусловленную,увы, неизбежным дальнейшим развитием цивилизации по пути промышленного и энергетического развития. Наконец, внедрение и использование технологий *ОО-АР и СО-АР* позволяет начать действительное, а не декларативное, выполнение рекомендаций "Повестки дня XXI века" Рио 1992, и предстоящей конференции "10 лет после Рио," на уровне местных и/или частных бюджетов, в том числе и странах бывшего СССР и Восточной Европы.

In situ media (загрязненные и излучающие среды *"на месте"* [7])- загрязненные и излучающие среды и материалы, находящиеся в непосредственной близости к источнику первичной опасности. Предполагается, что *миграция* загрязнителей не произошла. В **American RI/FS process** это понятие введено для ориентации пользователей при определении места внедрения избираемых технологий.

In situ treatment technologies (технологические *процессы переработки или обработки "на месте" загрязненных сред,* материалов и *источников* облучения [7]) - часть из перечня технологий в рамках **RI/FS** процесса, содержащихся в Базе Знаний компьютерной системы **ReOpt**, и рекомендуемых для применения *"на месте,"* то есть, в непосредственной близости от *источников первичных выбросов или сбросов* радиоактивных или токсически опасных веществ.

Initial technology screening criteria (*критерии для выбора стратегий* в рамках **RI/FS** процесса [7,13]), - оцениваются варианты предпочтительных технологий для реализации *реабилитационных мер.*

Institutional control technologies (нормируемые операции контроля и обслуживания [7]) - установленные нормативными документами операции контроля и обслуживания загрязненных площадок, промзон, т.д.

Land use variable dose (*дозы,* получаемые при пребывании на загрязненных землях [,5,7, 8, 18]) - в **RI/FS process** вычисляются по каждому *загрязнителю,* и на каждый *дозовый маршрут,* по формуле: **Drp** (мг/кг*день или rem/день) = Ur*Crp*Fr, где: * знак умножения; **r** - индекс дозового маршрута (r = 1,2 3,4); p - индекс технопатогеной зоны (p = 1,2,..P); Ur - средняя ежедневная скорость поступления в организм - поглощения данного загрязнителя, или облучения по дозовому маршруту r (кг/день - для орального поступления, м3/день - для ингаляционного, в час/день - для внешнего облучения); **Crp** - концентрация загрязнителя в контактируемой с человеком среде (здесь в почве), Fr - *дозовый конверсионный фактор* (**dose conversion factor**). Значения **D,rp** оцениваются для каждого загрязнителя, способного потенциально экспонировать человека или животного, и для каждого дозового маршрута.

Levels of Protection from Non-rational & Irrational Deeds of Specialists (уровни защиты от нерациональной и иррациональности деятельности специалистов - УЗНИДС [4, 6, 16]), - *источники* опасности, содержатся в трех основных областях человеческой деятельности: профессиональной; психологической мотивационной, а также в рамках определенных структур организации деятельности. Вычислительные алгоритмы УЗНИДС, разработанные в лаборатории В. Еременко (РФ), предусматривает пошаговый анализ и выявление *источников* опасности на всех стадиях жизненного цикла опасных технологий (**R&D, Project making, Manufacturing & Construction, Assembling; Installing, Operation, Decommissioning, Preserved**). Выходными показателями анализа являются вероятности появления отказов по общей причине (**common mode failures - CMF**) и исходных событий (**Initial Events - IE**)., и, как следствие, - вероятность и масштаб аварий, то есть искомую оценку риска от так называемого человеческого фактора (**HF**).

Maximum exposed individual risk estimate (наибольший *риск* для индивида из данной группы *экспонированных/облученных* людей, оцениваемая в рамках *RI/FS процесса* [5, 8, 18]) - наибольшая вероятность канцерогенного заболевания для индивида из данной группы *экспонированных*, в частности - облученных людей, которая оценивается с учетом сведений о каждом субъекте - рецепторе, подверженном воздействию радиационной и токсической опасности, координатах и характеристиках места его расположения, а также о характере и времени пребывания в данном месте- поведении рецептора.

Maximum exposed individual risk: Цифры [15]: Современными нормативными документами России установлены пороговые значения *индивидуального риска*: *предел индивидуального радиационного риска* - $1*10E-3$ /год*чел и $5*10E-5$ год*чел - для персонала и населения соответственно. При рассмотрении этих величин следует иметь в виду, что в РФ нормируется *полный риск*, включающий *риски* от смертельного рака, серьезных наследственных эффектов и не смертельного рака, приведенного по вреду к последствиям от смертельного рака. То есть, *канцерогенный риск* составляет только часть от полного.

Media/location (загрязненная среда,продукт, материал/ее местоположение [7,18])

Media properties (свойства *загрязненных сред* и материалов [7, 13,18]) - один из четырех основных факторов в перечне ограничений и выходных характеристик, которые играют главную роль в принятии решения на применение предварительно выбранных технологий *реабилитационных мер для СО-АР*.

Multimedia Environmental Pollutant Assessment System,-MEPAS (Система комплексной Оценки Состояния Загрязненных Территорий, - МИПАС [5,8,18])

Multi-Media System (компьютерная программа для комплексных оценок *Окр. Ср.* [4,8]) - типа **MEPAS**, используемая при решении проблем восстановления окружающей среды, покрывающая весь спектр природных сред, являющихся потенциальными пунктами аккумуляции, *путями распространения и дозообразующими средами* для радиационных и токсических химических загрязнений. На терминалах системы отображаются количественные данные по состоянию *Окр.Ср.* и здоровью людей до и после реализации *реабилитационных мер* в рамках RI/FS процесса, по относительной опасности радиационных и химических предприятий и загрязненных территорий, по *эффекту применения* восстановительных технологий, и пр., используемые при анализе проблемы и подготовке рекомендаций по оптимальным действиям.

Multimedia transport (*распространение/миграция* химических загрязнений и радиоактивности и радиоактивности в геосредах **[7]**) - распространение химических загрязнений и радиоактивности и радиоактивности в атмосферном воздухе, поверхностных и подземных водах, а также по суше. Понятие является производным от ранее введенных понятий - **transport of contaminants, pathways, transport pathways,** и **release pathways.**

Normal operating procedures (*руководства для нормальной или штатной эксплуатации*). В заголовках проектных, наладочных и эксплуатационных документов встречаются такие разновидности наименований различных *руководств,* как: *Р. по функциональным испытаниям (***Functional test procedures***); Р. проектированием (***Design management***); Р., официально утвержденное (***Procedure, formally approved***).*

Operable unit (однотипные эксплуатируемые объекты, имеющие сходные регламенты обеспечения безопасности **[5,8,18]**) - промышленные объекты, для которых мероприятия, реализующие *реабилитационные меры для СО-АР,* предусмотрены их эксплуатационными инструкциями. При оценках уровня загрязненности и *рисков* для здоровья людей на основе **RI/FS process** рассматриваются как одиночные, так и группы таких объектов, при условии, что *СО-АР* будет использоваться применительно к однотипным источникам *выбросов* и *сбросов,* либо к однотипному технологическому оборудованию, либо к подобным источникам *вторичных потоков отходов,* и включать однотипные операции, например, контроля над характеристиками источника *выбросов* и *сбросов,* или мониторинга состояния обслуживаемых хранилищ, либо контролируемых подземных вод, геологических пластов и проч.

Pathways (*пути* **[5,8,18]**) - термин для обозначения *путей* распространения химических *загрязнений* и радиоактивности применяемый в методологии **HCRA**

Population exposure estimate (*экспонированное/облученное* население **[3,7]**) - условие для получения коллективных *экспозиционных доз.*

Practicality limits (*границы полезности* **[7,13]**)

Probabilistic Risk Assessment (*Вероятностный анализ риска, ВАР* **[10,17]**)

Probabilistic Safety Assessment (*Вероятностный анализ безопасности,-ВАБ* **[10,17]**)- компьютеризированные методы расчета и анализа *аварийного риска.*

Probability Risk Assessment: Математика [10,17]. В самом грубом приближении вычислительный алгоритм составляется из следующих соображений: 1. Если знак "si" будет выражать определение или описание некоторого сценария развития аварии; а под знаком "фi" будут помещены случайные величины вероятностей аварии по данному сценарию (в практике используется не вероятность, а частота реализации данного сценария), и этому сценарию соответствуют некоторые отрицательные последствия, например, количественные выражения ущерба "x," то из численных значений вышеперечисленных величин можно составить такую Таблицу:

Сценарий	Частота	Последствие	Совокупная частота
S1	ф1	x1	$\Phi1=\Phi2+\phi1$
S2	ф2	x2	$\Phi2=\Phi3+\phi2$
.	.	.	.
Si	фi	xi	$\Phi i=\Phi i+1+\phi i$
.		.	.
S,N-1	ф,N-1	x,N-1	$\Phi,N-1=\Phi,N+\phi,N-1$
S,N	ф, N	x,N	$\Phi,N=\phi,N$

Теперь, каждая i-тая строка данной таблицы может быть представлена в виде комбинации трех ее основных величин. То есть, для каждой i-той строки можно записать комбинацию <si,фi,xi>, смысл которой, по сути, уже и будет количественным выражением аварийного риска для i-того набора значений соответствующих величин: R = <si,фi,xi>, i=1, 2,....N. 2. Вместе с тем, предыдущее выражение не полно по семантическому определению. Согласно последнему, показатель риска должен выражать не только практическую вероятность появления данного сценария, и также вероятность конкретных отрицательных последствий - потерь, сопровождающих данный сценарий. Принципиальным считается необходимость учета неопределенностей таких определений, так как семантическое выражение риска есть: R = <"неопределенность" и "потери">. Поэтому работу с таблицей продолжают в направлении учета неопределенностей. 3. Для этого: сценарии аварий располагают в таблице в порядке возрастания серьезности потерь, то есть: x1<x2<x3<xN; плотность вероятности для "фi" в i-том сценарии выражают функцией "Pi(фi)" (как правило, частоты реализации той или иной категории сценариев "Si" неизвестны), а неопределенность в определении размера ущерба обозначают "qi(xi)." Теперь значение *аварийного риска*, уже с учетом "неопределенности" в определении частоты и ущерба, ищут в виде: R = <si,Pi(фi),qi(xi)>. Такое выражение для расчета *аварийного риска* относят *в ВАР и ВАБ* к первому уровню приближений.

Realise unit (*источник выбросов /сбросов* [5,18]).

Receptor characteristics (характеристики *рецептора* [5,7,18]) - понятие, производное от определений **receptor locations**, **remediation strategy**, и **receptor exposures.**

Receptor characteristics: Комментарий. Термин **receptors** используется в литературе и руководствах по **RI/FS process** неоднозначно. Иногда им обозначают

непосредственные *источники* опасности, прямо воздействующие на человека
[5,18].

Receptor exposures (*экспонирование*/облучение *рецептора* - человека, растения, животного, и т.д. [18]) - необходимое условие получения *экспозиционной дозы* в практической жизни, и достаточное - для выполнения соответствующих расчетов в рамках *RI/FS процесса* (например, на базе компьютерных систем **MEPAS, RAAS**).

Reduced human health effects (снижение *ущерба для здоровья* человека) - производное от определения **human health effects (см.).**

Release (*выброс/сброс* [5, 8,18]) - выход, или, по терминологии, принятой в расчетах по промышленной безопасности - *"выброс"* в атмосферу, или *"сброс"* в водоемы, а также на поверхность суши, или в подземные водоносные горизонты опасного токсичного (химического) вещества или радиоактивности в газообразном, паровом, жидком, пылевом или твердом состояниях за первоначальные границы любого источника опасности, в частности - выход химических *загрязнений и радиоактивности* за пределы санитарной зоны технологических установок или аппаратов (иногда добавляют, - *первичный выброс или сброс* - **primary release**), как при регламентном, штатном, то есть, безаварийном использовании (**release, routine**), так и, в особенности, при нарушении их герметичности (**release, accidental**).

Release Mechanisms (*механизмы выброса/сброса* загрязнений [13,18]).

Release pathways (пути распространения *выбросов /сбросов* [7,18]) - составное понятие; определения составляющих приведены в **release,** и **pathways.**

Release site ("пункт" [5,7]) - наиболее обобщенное наименование потенциально опасного компонента источника опасности; чаще всего этот термин используют для обозначения промплощадки, загрязненного участка производственной территории, в том числе сельскохозяйственного назначения.

Release site assessment (загрязненный *пункт*, в сложных случаях - *ПНТ*, рассматриваемые как источник *выбросов* или *сбросов* в рамках *RI/FS процесса* [5,8]) - учитывается весь спектр возможных *выбросов* и *сбросов* от идентифицированных опасных *источников* и их сочетаний, размещенных в границах данных промплощадки, промзоны, в общем случае - *ПНТ* (промышленной населенной территории), или вне их, но создающих угрозу для субъектов - *рецепторов*, находящихся внутри *ПНТ*, на ее границах или в непосредственной близости от них.

Release site assessment: Комментарий [7]: Понятие "всего спектра" *выбросов* и *сбросов ПНТ* включает *выброс* в атмосферу, *сбросы* в водоемы, на поверхность суши, или в подземные водоносные горизонты одного из более чем 600 опасных токсичных - химических или радиоактивных веществ - элементов, изотопов, которые содержатся в Базах Данных и Знаний компьютерных систем **MEPAS и ReOpt**. При моделировании распространения *загрязнений*, а также дозовых нагрузок, *канцерогенных рисков* и *квот токсичности*, опасные примеси могут рассматриваться в газообразном, паровом, жидком, пылевом или твердом состояниях. В таком виде они находятся на месте их зарождения (**in situ**), либо выходят за первоначальные границы любого источника потенциальной опасности (**hazardous sourses**), в частности - за пределы санитарной зоны (**ex situ**) энергетических, промышленных или транспортных установок (**operable/facility units**), хранилищ отходов (**waste storages**), либо просто загрязненных территорий (**release site**).

Remedial Action Assessment System - RAAS (методология оценки *реабилитационных мер* в рамках *CO-AP* [7]) - программное обеспечение компьютерной системы поддержки принятия решений по проблемам загрязненной окружающей среды, отходов и безопасности населения, разработанное в Национальной Тихоокеанской Северо - Западной Лаборатории Минэнерго США (**PNNL DOE, Battelle**) для оказания профессионалам из данного ведомства США в обоснованном выборе механизмов специальных технических мероприятий - *реабилитационных мер* (**remedial action**), которые бы соответствовали наиболее полному решению проблемы, что обеспечивается оптимизацией их показателей по технологической и экологической эффективности и стоимости, выполняемых согласно установленной законодательством США (**CERCLA, SARA**) специальной процедуры анализа (**RI/FS**).

Remedial Action Assessment System: Генезис термина: Минэнерго США в течение уже более десяти лет серьезно озабочено необходимостью перманентного решения дорогостоящей и неотложной проблемы, связанной с очисткой и восстановлением приемлемого состояния сотен загрязненных опасными веществами территорий и объектов - предприятий, разбросанных по всей стране. Решение этой проблемы в условиях ограниченного времени и относительно скромного финансирования оказалось возможным при выполнении многочисленных и серьезных исследований по *риску* воздействий химических загрязнений и радиоактивности на здоровье людей и обоснований целесообразной степени его снижения, которые позволяют оптимизировать по стоимости и эффективности действия соответствующие механизмы *реабилитационных мер,* а также выполнить анализ осуществимости намечаемых проектов, - по сути означающий проведение модернизированного технико - экономического обоснования - МТЭО (принятая в США аббревиатура для обозначения этой процедуры - **RI/FSs**). Разработка **RAAS** методологии была начата в **PNNL, Battelle** 1990г. и завершилась в 1995г. Процесс разработки включал создание и испытание пяти последовательных прототипов. Каждый прототип был протестирован профессионалами из Минэнерго США и Агентства по охране окружающей среды, а также представителями частной промышленности. Основное внимание было обращено на то, чтобы методология оптимизировала труд *ЛПР* и специалистов по внедрению результатов *CO-AP* при выполнении *RI/FS процессов*, сводя к минимуму финансовые, материальные и временные затраты. Команда обученных работе с **HCRA** специалистов, выполняющая работы по **RI/FS**, может обоснованно рекомендовать оптимальные комбинации технологий для *реабилитационных мер*. При этом гарантируется, что такие рекомендации будут наиболее полно реализовывать нормы или требования, установленные федеральными законами, или отдельными ведомствами, или местными административными руководителями (все перечисленные относятся к *лицам*, уполномоченным *принимать* соответствующие решения, - *ЛПР*). Команда аналитиков должна будет оценить альтернативы *реабилитационных мер*, используя предоставляемые в их распоряжение системой **RAAS** данные по показателям эффективности, реальности внедрения, стоимости и приемлемости, а также методики работы с избранными мерами.

Remedial alternatives (варианты *реабилитационных мер* **[7,13]**) - анализируемые в *рамках RI/FS процесса* текущие варианты технологических процессов, которые являются основой и содержанием *реабилитационных мер*, и выбираются, оцениваются, сопоставляются, анализируются, корректируются по техническим, экологическим и экономическим показателям, после чего, наконец, рекомендуются методологией *CO-AP* к внедрению.

Remedial investigation, RI (исследования по мерам, снижающие *опасности и риск* [7,13]).

Remediation site (объект для потенциальных *реабилитационных мер в рамках RI/FS процесса*[7]) - включает идентифицированные опасные *источники* химических загрязнений и радиоактивности или облучений, а также их сочетания, которые размещены в границах промплощадки, промзоны, и т.д., в общем случае - в границах *ПНТ*, но иногда и вне этих границ, однако если они создают угрозу для *рецепторов*, расположенных внутри *ПНТ*, на ее границах и даже вне *ПНТ*, однако, в непосредственной близости от них.

Removal technologies (технологические процессы для *физического устранения.*/ *ликвидации* загрязненных сред, материалов или *источников* излучения [13]).

Residual human health effects (уменьшение *ущерба* для здоровья человека [7,13]) - понятие, производное от термина **human health effect.**

Residual source inventory (*инвентаризация* остаточных *источников* [7,13]) - составная часть 3-х процедур, которые выполняются при анализе параметров, определяющих пригодность загрязненных сред к обработке, при сопоставлении анализируемых вариантов *CO-AP*, и, конечно, при оценках значений *риска*, который достигается в результате проведения *реабилитационных мер*. К *дозообразующим средам* (**exposure media**) в **RI/FS process** относят загрязненные составляющие *Окр.Ср.* - воздух, воды, суша; опасные для здоровья человека вещества и материалы, в том числе *источники* излучений; загрязненную пищу, растительного и животного происхождения, и т.д. и т.п. - все, что в контактах с потенциальными реципиентами, - человеком, представителями флоры и фауны, становится причиной их *экспонирования*, а, следовательно, создает условия для получения доз и последующих ущербов, выражаемых *рисками* и *квотами опасности* облучения.

Rio 1992 Conference Recommendations (*Рекомендации Конференции РИО –1992)-*** Речь идет о рекомендациях Конференции ООН по Окружающей Среде и Развитию, состоявшейся в Рио-де-Жанейро в 1992г. Они сводятся, в главном, к системе постановлений и требований по так называемому "*Устойчивому развитию*" – *УР* **(Sustainable development)**.

Rio 1992 Conference Recommendations: Генезис. Оригинальным источником рекомендаций явилась "Концепция *УР,*" разработанная ранее Международной Комиссией ООН по окружающей среде и развитию, известной под именем "комиссии Брунтланд." Концепция в целом выражала оптимизм в принципиальной возможности согласования деятельности человека с законами природы. На этом базировался оптимизм в процветания человечества в будущем. Идея комиссии Брунтланд, по сути, основывается на натур философской схеме представления мира как некоего единого организма, который может развиваться в связи с растущими потребностями человечества и удовлетворять эти потребности. Целью такого организма или системы является "устойчивое развитие" с учетом существующих и нарастающих экологических, экономических, промышленных и ресурсных и прочих ограничений.

Rio 1992 Conference Recommendations: Математика. Идея *рекомендаций Конференции РИО –1992*, сформулированная в решениях "Повестка дня на XXI век," может быть формально сформулирована в терминах математической задачи нелинейного программирования, где оптимизируемыми параметрами целевой функции являются: качество жизни, уровень экономического развития и экологического благополучия, а в качестве ограничений используются показатели *состояния окружающей среды* **(state of environment)**, экосистем и охраняемых территорий. Однако такая задача столь фантастически сложна, что без значительных приближений и упрощений решить ее не представляется возможным. Кроме того, отсутствует положительный опыт решения подобных задач из прошлой истории науки, даже для гораздо более простых исходных условий, чем вышеупомянутые. До сих пор сохраняются не только различные взгляды, как на саму проблему, так и на способы ее решения. Принципиальным, на наш взгляд, является разная оценка состояния устойчивости биосферы. Можно встретить, например, *пессимистические оценки* [**Лосев, Россия**]. В соответствии с ними, биосфера Земли уже находится в условиях жесткого экологического кризиса, так как вступила в период катастрофы. Порог устойчивости человечество перешло в начале века, когда оно превысило величину потребления первичной биологической продукции, допустимую для крупных позвоночных, и составляющую около 1% от всей биопродукции. Кстати, этот порог можно выразить в единицах энергии, точнее мощности. Критическая мощность будет составлять, примерно, 1ТВт. Это та энергия (мощность), которую человек, не нарушая и не деформируя окружающую среду, может использовать в своих целях, это - критический показатель предельной мощности, которая отведена для развития земной цивилизации по - Лосеву. Однако, существуют и *оптимистические оценки* [Р. Макнамара. "Сохранение Земли. Стратегия поддерживаемого проживания," 1991]. Согласно им, критический порог несущей емкости Земли еще не нарушен. Следует лишь держаться в пределах несущей емкости Земли и все будет в порядке. Кстати, вышеупомянутая комиссия Брунтланд в целом также выразила оптимизм о будущем человечества, при условии согласования его деятельности с законами природы. Кроме того, понимание законов и показателей развития вообще, и *устойчивого* в частности, видимо не может быть единым для всех народов и стран [см. сборники "Пути Евразии, Русская интеллигенция и судьбы России," 1992 и "Русская философия собственности 18-20 вв.," 1993]. В соответствии этими и другими известными нам оценками этнологов, народы стран Запада имеют врожденное *экспансионистское миропонимание*, толкающее их к своеобразному осмыслению природы и нужд других народов. Как следствия этого миропонимания - индивидуалистический капитализм, идеи "мировой цивилизации" по западному образцу, стремление к неограниченному росту производства и потребления и т.п. Пока и поскольку Запад сохраняет такое миропонимание, он не способен поверить в реальность угрозы социально экологической катастрофы (отсюда и оптимизм Макнамары и Брунтланд). Однако такая катастрофа не неизбежна в странах Запада, а также, и может быть даже раньше, - в странах, подчиненных Западу экономически и идеологически. Вместе с тем известно, что другие этносы имеют иное врожденное миропонимание. Японцам, например, присуща нацеленность на совершенствование своего хозяйства и природного окружения в корпоративных, но не индивидуальных интересах. В этой связи Япония имеет более естественные этнокультурные предпосылки перехода к *устойчивому развитию*. Русская этническая идея также нацелена не на экспансию, а на *упорядочение хаоса общинными усилиями*. Поэтому идея *устойчивого, то есть более упорядоченного, чем до сих пор, развития* созвучна и русскому миропониманию. Важно иметь в виду, что этнокультурное миропонимание и отвечающее ему поведение народов и стран существует объективно, и не поддается произвольному "улучшению." С позиций объективного этнологического знания Запад таков, "каким он рожден," и его стремление всех и все "вестернизировать" также естественно, как естественны стремления и приоритеты других культур, и особенности других этносов. [Л. Шестов, и др. Подробнее см. сборники "Пути Евразии, Русская интеллигенция и судьбы России," 1992 и "Русская философия собственности 18-20 вв.," 1993.]

Rio 1992 Conference Recommendations:Комментарий. Понятно, что надеяться на реализацию *Рекомендаций Конференции РИО –1992* на практике на глобальном уровне наивно. Вместе с тем, ценность такой постановки столь же очевидна, в той мере,, в какой ее результаты могут быть использованы как задания, или "уставки" для ориентации и количественного планирования работ на нижележащих уровнях. Именно здесь они должны быть преобразованы в конкретные плановые показатели, и только на местных или локальных уровнях могут быть превращены в реальные результаты, соответствующие общим *Рекомендациям Конференции РИО –1992*. Значительную роль в этом может сыграть решение проблем ликвидации наследия прошлой холодной войны (**the past Cold War Legacy**), и применение для подготовки оптимальных решений *методологии КУР* (ООАР v COAP v OBP; *FCRA v HCRA v RPA*). Входы систем компьютерной поддержки (СКП), созданные на основе этой методологии, готовы воспринимать соответствующие *Рекомендациям Конференции РИО –1992* задания/уставки. На выходах СКП для лиц, принимающих ответственные решения по *устойчивому развитию* территорий и производств, будут формироваться соответствующие предложения. Использование этих СКП (**MEPAS, RAAS, ReOpt, etc.**) позволяет подготавливать и принимать обоснованные и качественные решения на местном и региональном уровнях, в рамках рекомендаций по *устойчивому развитию,* с учетом ограниченных ресурсов и требований общественности к повышению эффективности затрат на реализацию таких рекомендаций.

Risk (*Риск* [3,5,13,18]*)*, - понятие, идентифицированное ниже в рубриках **Risk Analysis** и **Risk estimation;** понятие *риск* в практике наиболее часто используется со следующими определениями: *оцененный (***assessed***); повседневный (***everyday***); повторяющийся (***recurrent***); повышенный, в сравнении с фоном (***elevated***); поддающийся количественной оценке (***quantifiable***); постоянно присутствующий (***ever-present***); потенциальный (***potential***); пренебрежимо малый (***negligible***); приемлемый (***acceptable***); неприемлемый (***unacceptable***); принимаемый (***accepted***); природный (***natural***); профессиональный (***occupational***); непрофессиональный (***non-occupational***); прямой или непосредственный (***direct***); расчетный (***calculated, estimated***); рутинный (***routine***); связанный с развитием техники (***technology-assodated risk***); совокупный (***integrated, overall***); суммарный (***total***);социальный (***social***); сравнительный (***comparative***); средний (***average***); существенный (***substantial***); технологический (***technological***); действительный, фактический или реальный (***actual***); статистический (***statistical***), прогнозный (***forecasting***), чрезвычайный (***extraordinary, emergency***); эксплуатационный (***operational***); якобы существующий или предполагаемый (***alleged***); для здоровья человека (***health risk***); для населения (***public risk***); для окружающей среды (***environmental risk***); смерти или смертельный (***mortality***); аварийный (***accidental***); воспринимаемый (***perceived***);генетический (***genetic***); гипотетический (***hypothetical***);добровольный (***voluntary***); недобровольный (***involuntary***); допустимый(***allowed***); излишний (***undue***); искусственный или антропогенный (***artificial***); контролируемый (***controlled***); краткосрочный (***short-term***); долгосрочный(***long-term***);накопленный (***cumulative***); отдаленных последствий (***delayed effect risk***); отложенный (***delayed***); нулевой (***zero***); относительный (***relative***).

Risk Analysis *(анализ риска*[3,7,13,18]*)* - базовое понятие специальной области знаний. Среди основных задач этой области, решаемых в рамках **RI/FS** процесса и методологии **HCRA:** 1) выявление, учет и *идентификация* радиационных и химических опасностей на данной территории (**inventory, identification**) до проведения восстановительных мероприятий; 2) расчет соответствующего им *риск*а (**baseline risk estimation**), и изучение его распределения в пространстве и во времени; 3) *ранжирование источников опасности* по исходным показателям *риск*а (**baseline risk analysis**); 4) повторный учет и *идентификация* радиационных и химических опасностей на данном предприятии (**facility, installation**) или данной загрязненной территории (**site, release site, waste site**) после проведения восстановительных мероприятий, наконец, *расчет риск*а от опасностей, оставшихся после проведения *реабилитационных мер* (**residual risk estimation**) и новых, которые могут появиться как побочный продукт внедрения некоторых восстановительных технологий (**secondary stream**); повторное изучение распределения *риск*а в пространстве и во времени, и ранжирование *остаточных источников* опасности по достигнутым показателям *риск*а (**residual risk analysis**).

Risk Analysis: Совет: Для пользователей методологией HCRA из стран Восточноевропейского региона, в том числе -из бывших республик СССР, могут быть полезны некоторые подробности из данной области знаний. Во-первых, под рубрикой **Risk** вы найдете перевод на русский язык большинства понятий, используемых в *ВАБ и ВАР*. Они, видимо, не требуют дальнейших комментариев. Во-вторых, ниже рассмотрены более содержательные категории риска, которые мы полагали целесообразным снабдить обширными комментариями.

Risk Analysis: Комментарий 1: Результаты анализа, выполняемые с применением показателей *риск*а, хорошо зарекомендовали себя в процедурах подготовки и оптимизации данных для принятия обоснованных решений. Они в особенности удобны при оценке и сопоставлении социально- экономических эффектов негативного воздействия техногенных опасностей на здоровье и жизнь, как отдельного человека, так и групп людей, в предположении позитивного влияния *реабилитационных мер*. Вместе с тем, использование методологии и показателей *риск*а целесообразно тогда, и, может быть, только там, где реализация *реабилитационных мер* априори предполагает затрат значительных финансовых, материальных и социальных ресурсов. Именно в таких случаях, *оценка и анализ риск*а позволяет сопоставить возможные социально- экономические последствия и принять единственно правильное решение о целесообразности действия, - то есть, применения *реабилитационных мер* или бездействия - отказа от них. Таким образом, *анализ риск*а можно рассматривать как основу максимизирующего поведения субъектов разных иерархических уровней в условиях рыночных отношений.

Risk Analysis: Математика 1 [11]. Недостаточно корректным с математической точки зрения, но хорошо отражающим логику большинства вычислительных процедур, может служить представление о расчете *индивидуальных рисков* через произведение трех компонент: $R = R1*R2*R3.$ В этом выражении из [8,9,10,11]: R - *уровень риск*а; R1 - вероятность (для уже свершившихся событий - частота) возникновения события или явления, которые обусловливают формирование и действие вредных факторов; **R2** - вероятность формирования определенных уровней физических полей, ударных нагрузок, полей концентрации вредных веществ в различных средах и их дозовых нагрузок, воздействующих на людей и другие объекты биосферы; **R3** - вероятность того, что указанные выше уровни полей и нагрузок приведут к определенному ущербу: ухудшению состояния здоровья и снижению жизнедеятельности людей, в том числе летальному поражению, поражению тех или иных популяций животных и растений, сдвигу равновесного состояния экосистем, экономическому ущербу и т. п.

Risk Analysis: Комментарий 2. В анализе оперируют тремя *видами рисков*: *индивидуальным, коллективным или суммарным* (**total**), *и социальным*. Различают также под - виды рисков, как *индивидуального, так и коллективного - это стохастический и детерминированный риски*. Все определения, приводимые ниже, будут относиться к первому подвиду, поскольку *в рамках **RI/FS** процесса*, а также в обеспечивающих этот процесс компьютерных программах (например, **MEPAS**), а также в методологии **HCRA**, оперируют только со *стохастическими рисками*. Однако формулы для расчета *детерминированных рисков* также будут приведены, в виде исключения.

Risk Analysis: Individual risk *(индивидуальный риск)* - вероятность, а во многих приложениях - частота возникновения, поражающих воздействий определенного вида для индивида: смерть, заболевание (**carcinogenic risk**), травма, потеря трудоспособности, возникающие при реализации определенных опасностей в определенной точке пространства - там, где находится индивид. При расчетах определяют значения *индивидуального риска* канцерогенного заболевания - радиационного, и - *индивидуального риска* канцерогенного заболевания - химического. Эти *виды риска* используются на терминалах системы **MEPAS** как базовые - стандартные выходы.

Risk Analysis: Individual risk & hazard quotients in MEPAS *(индивидуальный риск и квоты опасности в* **MEPAS [5,14]**) - в программе **MEPAS** *риск канцерогенный*, от радиационных воздействий, рассчитывается по формуле: **Rr,rp** = **H** * **Drp** * **2.555*10E+4,** (здесь и далее сохранены обозначения оригиналов), где: - знак произведения; **Rr,rp** - *риск в течение жизни* при мощности ежедневной *экспозиционной дозы* радиоактивного облучения - **Drp** рентг/день); а **H** - *конверсионный фактор риска*. В соответствии с рекомендаций Национальной Академией Наук США 1992г, **H** выражается цифрой заболеваемости раком в *10E-4 случаев /на 1 rem экспозиционной *дозы* радиоактивного облучения, получаемой в течение всей жизни (70 лет). Численный коэффициент 2.555*10E+4 - то всего на всего количество дней в 70 годах (дн) человеческой жизни. В той же программе **MEPAS** *канцерогенный риск,* но от химических воздействий, рассчитывается по иной формуле: **Rc,rp = 1 - exp(- D,rp * q,k),** где: * - знак произведения; **Rc,rp** - *верхняя граница риска* при ежедневной экспозиции человека *дозой* мощностью **D,rp** (безразмерная величина), в течение всей 70-летней жизни, а q,k - значение фактора **CPF** для данного **k**-того химического соединения (кг*день/мг). Далее, объединением *канцерогенного радиационного и канцерогенного токсического риска,* в вычислительной схеме **MEPAS,** формируется *критерий* (в оригинале - **parameter): Rmax,r = maximum (Rrp),** по величине которого судят о *канцерогенном суммарном эффекте* от воздействия на индивида исследуемой (-мых) опасности(-тей) и условий данного региона. В выражении *критерия:* **Rmax,r** - максимальный, приведенный к 70 - летней жизни, *риск для индивида,* принадлежащего к рассматриваемой группе населения, и оцененный по дозовым маршрутам "**r**" для всех *ТПЗ* (**usage locations**) данного района, и также по всем отрезкам времени. Как эквивалент *риска* в программе **MEPAS** используется понятие *квоты опасности* - меры опасности для здоровья человека от воздействия различных химических загрязнений и видов радиоактивности. Эта мера используется в **American RI/FS process** как эквивалент *риска,* но для не канцерогенных химических токсикантов. Оценка *квоты опасности* выполняется по формуле **In,rp = D,rp/RfD,k,** где: **In,rp** - количественный показатель индивидуального *ущерба* для здоровья от не канцерогенных химикатов (безразмерная величина); **D,rp** - мощность ежедневной экспозиционной *дозы;* **RfD,k** - минимально - опасная *доза* **k**-того токсического вещества.

Risk Analysis: Individual risk in Norms of Radiation Safety in Russia, 1996
(*индивидуальный риск в "Нормах радиационной безопасности РФ-1996г* [**Е, 15**])- в *НРБ-96 риск* предлагается рассчитывать по формуле*: r = p(E) * r(e) *E* (здесь и далее сохранены обозначения оригинала [**15**]). В этой формуле: **r** - *индивидуальный риск* возникновения стохастических эффектов; **Е** - индивидуальная эффективная *доза;* **p (E)** - вероятность событий, создающих *дозу* **E; r(e)** - коэффициенты *риска от смертельного рака,* а также от серьезных наследственных эффектов и *не смертельного рака* (приведенного по вреду к последствиям от смертельного рака). Коэффициент *риска* "**r**" предлагается полагать равным 5,6 *10*E-2 1/чел-Зв в случаях облучения профессионалов, и **r** = 7,3 *10 *E-2 1/чел -Зв, когда оценивается *риск для населения.*

Risk Analysis: Total risk in MEPAS (*коллективный или суммарный риск* в **MEPAS** [**5, 8, 18**]) - определяется для людей, находящихся в прилегающих к опасной промплощадке районах, и подвергающихся воздействию техногенных *источников опасности;* вычисляется в **MEPAS** как *сумма рисков* по всем дозовым маршрутам, с учетом числа людей, расположенных на этих маршрутах: **Rt= Sum[p=1,2...P]* Sum [r=1,2,.Rp]*Rrp*Ppr** (сохранены обозначения основного оригинала[**18**]). В этом выражении: **Rt** - *суммарный риск,* рассчитанный для ограниченного промежутка времени "**t**" для людей, подвергающихся данному опасному воздействию(безразмерная величина); **p** - индекс, или номер *ТПЗ* (**usage location**); **P** - общее количество *ТПЗ* в данном регионе (**site, installation**); **r** - индекс, или номер, дозового маршрута; **Rp** - количество *дозовых маршрутов* для "**р**"-той *ТПЗ*, **Ppr** - число людей в "**р**"-той *ТПЗ*, располагающихся на "**r**"-том *дозовом маршруте;* **Rrp** - *индивидуальный риск* для людей, расположенных в "**р**"-той *ТПЗ*, и находящихся на "**r**"-том дозовом маршруте. Вычисляемый показатель выбран для формирования критерия, характеризующего эффект воздействия исследуемой опасности и условий данного региона на население.

Risk Analysis: Total risk in Norms of Radiation Safety in Russia, 1996, (*коллективный или суммарный риск* в *НРБ-96* [15])- для оценок используется формула: **R= p (S(e)*r(e) *S(e)** (здесь и далее сохранены обозначения оригинала [15]). В этом выражении: **R** - *коллективный риск* возникновения стохастических эффектов; **S(e)** - коллективная эффективные *доза*; **p(S(e)** - вероятность событий, создающих дозу **S(e); r(e)** - коэффициент риска от смертельного рака, серьезных наследственных эффектов и не смертельного рака (приведенного по вреду к последствиям от смертельного рака). Так называемый коэффициент риска равен: г = 5,6 *10*Е-2 1/чел-3в для профессионального облучения, и **r** = 7,3 *10 *Е-2 1/чеп-Зв для населения. В *НРБ-96* для событий с тяжелыми последствиями от детерминированных эффектов консервативно принимается: **r = p (E), R = p(E) *N,** где: **N** - численность популяции, подвергающейся радиационному воздействию в дозе **E** > 0,5 Зв.; **r, R, p (E)** - *индивидуальный и коллективный риски* возникновения последствий от детерминированных эффектов, **E** - индивидуальная эффективная *доза*, а **p(E)** - вероятность событий, создающих дозу **E.**

Risk Analysis: Комментарий 3 [12]. Еще одна разновидность *риска* - *социальный* (**social risk**) определяется как зависимость вероятности нежелательных "событий," состоящих в поражении не менее определенного числа людей. При этом предполагается, что они подвергаются поражающим воздействиям определенного вида (смерти, заболеваниям, травмам), которые проявляются при реализации определенных опасностей. Величина *социального риск*а пропорциональна числу людей, находящихся под воздействием изучаемой опасности. Эта разновидность *риск*а применяется для характеристики масштабов опасности при авариях и катастрофах. В методологии **RI/FS** и **HCRA** не используется.

Risk Analysis: Комментарий 4 [7]. *Статистический (***statistical***), потенциальный (***potential***) и прогнозный* (**forecasting**), - следующие разновидности риска, которые также используются в анализе. Они различаются способами получения конкретных величин, имеют свои специфические области использования, и не могут заменить один другого в практических приложениях. Так, *статистический, или ретро* - *риск*, определяется статистическими методами на основе измеряемых и накапливаемых фактических данных (чаще всего так рассчитываются "*природные риски*"). При расчетах *риск статистический индивидуальный* интерпретируют как математическое ожидание *ущерба*, возникающего при авариях, катастрофах и опасных природных явлениях - **R(МО)**. Величина *статистического индивидуального риск*, то есть, - **R(МО)**, выражается произведением вероятности события **R1** на степень его тяжести, выраженную в иде *ущерба* **R3** того или иного рода. Однако при такой оценке условно полагают, то величина *ущерба* имеет *детерминированное* значение (**R3 = det = Y**), то-есть го вероятностная природа не учитывается. Обычно принимаются во внимание всевозможные виды опасных природных явлений, происшествий, аварий и катастроф, применительно к данному объекту, а оценку *индивидуального риска* производят как сумму произведений вероятностей указанных событий на соответствующие ущербы: **R(MO)= Sum (i=1,..n)[R1i*Yi].** Здесь: **R1i** - вероятности появления данного опасного i-го события, а **Yi** - величины ущербов при каждом i-ом событии. *Потенциальный риск* оценивается аналитическими или вычислительными методами, и, как правило, с помощью специально разработанных компьютерных программ, в том числе поддерживающих систем **RAAS** и **MEPAS**. *Прогнозный риск* рассчитывается посредством несложной алгебраической процедуры объединения частично измеренных и другой части вычисленных данных. В этом смысле, множество значений *прогнозного риск*а одновременно принадлежит двум не пересекающимся множествам значений *статистического и потенциального рисков*. Соответствующие системы иногда называют "гибридными": измерительно-компьютерными системами. В гибридных системах поддержки оценка последствий опасных воздействий базируется на алгоритмах определения *прогнозного риск*а. Примером такой категории и такой системы может быть *риск*, рассчитываемый на **MEPAS** в режиме подачи на его входы предварительно измеренных в *Окр.Ср.* концентраций загрязнителей.

Risk Analysis: Комментарий 4 [7,11]. Последние две из рассматриваемых здесь категорий риска - *рутинный* (**routine**) *и аварийный* (**accidental***)* - термины, используемые при соотнесении оценок *риска* с динамическими условиями задачи. *Рутинный риск* соответствует стационарным или квази- стационарным условиям, когда в математических описаниях изменениями во времени пренебрегают. Такие расчеты характерны для оценок *риска*, выполняемых в предположении нормального - повседневного использования загрязненных территорий или опасных техногенных *источников* по их прямому назначению. Таким образом, *рутинный риск* используется, когда надо оценить последствия *рутинной* (аналогии - нормальной, регламентной, штатной) эксплуатации или использования объектов радиационно- химической опасности. Ему наиболее близка категория *потенциального*, и, отчасти *прогнозного рисков*. Здесь событиями, которые могут обусловливать возникновение нежелательных последствий - *индивидуальным риск*ов, будут *выбросы и сбросы*, перевозка и *миграция* материалов или продуктов, содержащих вредные вещества. Периодичность и масштабы таких событий с хорошим приближением могут быть отнесены к детерминированным. Это же относится и к основным параметрам физических полей- радиационного, электромагнитного, теплового и др., которые негативно воздействуют на объекты живой природы. Как следствие принятых допущений (см. **Математика 1,** и **Комментарий 4**), первый множитель **R1** общей логической процедуры, записанной в виде произведения из трех сомножителей, может быть принят как некоторая постоянная величина, - коэффициент, а при выполнении несложного нормирования и равным единице. Тогда частная логика процедуры оценки будет упрощена и сведена к произведения двух сомножителей - случайных величин: **R = R2*R3.** В рамках такой частной логики определений построены алгоритмы вычислений *потенциального и прогнозного рисков* в рамках **RI/FS** *процесса* и в

НРБ-96, - те, которые приведены в начале данного пояснения (следует принять, конечно, во внимание разницу в обозначениях, то есть, что **R= Rr,rp** или **Rc,rp, R2 = Drp**, а **R3 = H** или **q**). Рассматривая формулы и тексты цитируемых определений легко заметить, что в практических приложениях, каковыми являются расчеты, производимые, например, в **MEPAS**, вводится еще одно упрощающее расчеты предположение. А именно - последний сомножитель общей логической процедуры **R3** выражается также детерминированными величинами **H** или **q**. *Аварийный риск* соответствует нестационарным условиям изучаемого объекта, когда в его математическом описании изменениями во времени пренебречь нельзя. Соответствующие оценки *риска* относят к динамическим. Последние чаще всего выполняют применительно к нештатным - аварийным состояниям загрязненных территорий или опасных техногенных *источников*. *Аварийный риск* используют для оценки последствий аварий на объектах радиационной и химической опасности. Ему также наиболее близка категория потенциального, и, отчасти прогнозного *рисков*, но в форме, существенно более сложной, чем в случае *рутинного риска*. Сложность отчасти связана с недопустимостью принятия допущения о детерминированности событий, - события здесь являются случайными величинами, отчасти - с многочисленными неопределенностями на всех этапах производимых оценок **[5,6,7,8,9,10,11]**.

Risk Analysis: Замечание 1 [7].. Не рассматривая более подробно последнюю из разновидностей оценки *риска*, поскольку к процессу **RI/FS** она не имеет отношения, сделаем только одно замечание. Первый множитель **R1** общей логической процедуры в этом случае не может быть принят равным единице. В соответствующих методиках его значения, а также сопутствующие расчетам величины неопределенностей, обосновываются и оцениваются достаточно сложными способами, а проведение расчетов обычно связано со значительными методическими трудностями. Кроме того, отметим, для полноты информации еще две категории *рисков*, которые не часто используются в процедурах **RI/FS**, однако могут встретиться в практической деятельности - *ранние* и *поздние риски*. Этими понятиями пользуются, если необходимо подчеркнуть время проявления последствий. Например, когда оценка относится ко времени после *экспонирования (облучения)*. Другой пример использования - когда надо подчеркнуть пороговый характер связи *доза* - эффект для ранних проявлений последствий. К *ранним рискам,* для радиационно обусловленных эффектов, относят смерти, а также пред - нормальные состояния (утомляемость, тошноты и рвоты после острого общего облучения поглощенными *дозами* более 1 Гр = 100 рад), фиброзы легких. К показателям *поздних рисков* - летальный рак, не летальные раки щитовидной железы, кожи и молочной железы, а также наследственные эффекты.

Risk Analysis Methodology for Environment and Health, RAMEH (общая методологии *анализа риска* для решения проблем *окружающей среды и здоровья*, МАРОСЗ **[1]**).

Risk assessment, RA (*оценка риска* **[7,13,18]**)

Secondary waste stream (потоки *вторичных отходов* [1,7,13]) - *источники потенциальных риск*а и *опасности* в форме загрязненных дозообразующих и геосред, загрязненных продуктов потребления и отходов, которые могут возникнуть после применения технологий **RI/FS.** Их отличие от *источников* первичных в том, что потоки вторичных отходов возникают как неизбежный побочный продукт применения некоторых восстановительных технологий. В результате продолжается распространение радиационно и токсически опасных газовых, жидких или твердых примесей, которые ранее содержались в исходных *выброс*ах или *сброс*ах. Следовательно, процесс оценок *риск*а, ранжирования вторичных *источников* по степени опасности, выбора и оптимизации новых технологий **American RI/FS process,** и т.д. может, а иногда и должен быть продолжен по второму кругу.

Site (*промышленная населенная территория - ПНТ* [7]) - чаще всего это - промплощадка, промзона, включающие "идентифицированные" *источники* опасности (**identification**)

Site: Генезис [7]: "**Site**" - пример одного из наиболее хорошо изученных в рамках **American RI/FS process** типовых индустриальных районов, однако включающих относительно хорошо развитый аграрный сектор. Обладает всеми признаками "*объекта для реабилитационных мер" как результат CO-AP* (**remediation object**). Может включать любую из шести групп предварительно выявленных (**inventory, identification**) *источников* опасных выбросов и сбросов, в общем случае – "отходов" (**waste**), или любые сочетания групп таких *источников* (**waste sites**). *ПНТ* может включать ряд *технопатогенных зон - ТПЗ* (**usage location**) и является наименьшей по площади территорией, для которой разрабатываются рекомендации **HCRA**. В границах *ПНТ* могут быть решены два вида задач по оптимизации *восстановительных мероприятий*. А именно, может быть найден глобальный оптимум для всей *ПНТ*, либо локальные оптимумы для каждого из предприятий, их комплексов, участков загрязненной территории, *ТПЗ*, и т.п., входящих в состав или представляющих данную *ПНТ*.

Site Conceptual Model, Typical (*концептуальная модель* промышленной населенной территории - *ПНТ*, участка территории, промзоны, промплощадки [13]) - расчетная модель идентифицированного единичного источника опасных *выбросов* или *сбросов*. Включает: опись запасов идентифицированного *загрязнителя* (**contaminant inventory**), перечень свойств источника опасности (**source and media properties**); характеристику потенциальных *дозовых путей* (**exposure pathways**).

Site-specific knowledge (сведения о загрязненной территории) - имеются в виду те, которые необходимо иметь и учитывать при моделировании.

Source (*источник* загрязнения *ОкрСр* и *дозообразующих сред* [7,18]).

Source properties (*свойства источника* [7,13]) - методологией **HCRA** рассматриваются два основных типа *источников* химических загрязнений и радиоактивности и радиоактивности и облучений, и, соответственно, два различных перечня их свойств. Первый тип - это *источники* первичных *выбросов* и *сбросов* и облучений. К ним относят *источники* разного уровня, и естественно отличающимися свойствами: 1.1. Объекты с установленными местоположениями *выбросов* или *сбросов*. К таковым принадлежат: загрязненные траншеи, пруды - отстойники, бассейны выдержки, и т.д., а также различного рода над- и подземные производственные здания, хранилища и коммуникации с возможными неплотностями в них, и, наконец, резервуары, могильники, вытяжные трубы и проч. Свойства *источников* определены пространственно - временными координатами позиций, из которых загрязнения могут поступать в *Окр.Ср.* 1.2. Объекты, эксплуатируемые по схожим регламентам (**operable unit).** Здесь *свойства источника* определены параметрами контроля этих объектов, или мониторинга состояния обслуживаемых хранилищ, либо параметрами контролируемых подземных вод, геологических пластов и проч.; 1.3. Разнообразные промышленные объекты, однако реализующие близкие по целям способам и средствам *реабилитационные меры*. Искомые свойства могут либо задаваться пользователем методологии **HCRA**, либо, напротив, определяться одной из составляющих этого комплекса, - например, системой **MEPAS**. К *источникам* также относят так называемые *"вторичные потоки отходов."* Итак, *источники* типа 2. возникают в результате естественного распространения в *Окр.Ср.* загрязнений, или искусственно - как следствие проведения *процессов переработки или обработки* газовых, жидких или твердых радиационно или токсически *загрязненных сред*, а также материалов и *источников* облучения. В соответствии с методологией **HCRA**, свойства *источников* второго типа выявляются при оценке эффективности *мероприятий по реабилитации*, и обязательно включаются в перечни выходных показателей *МТЭО*.

Source term (характеристика *источника выбросов /сбросов* [18])

Stages of "life cycle" of hazardous technologies: R&D, Project making, Manufacturing & Construction, Assembling, and also Decommissioning and Preserved (стадии *"жизненного цикла"* опасных технологий: "научно-исследовательская," "проектная," "производственная," "монтажная"; а также "снятия с эксплуатации" и "хранения демонтированных опасных объектов").

State of object/envoronment/territory (*состояние объекта, Окр.Ср., территории* 19])- выражается в теории идентификации, оптимального управления и специальном разделе высшей математики, называемом "алгеброй пространства состояний" через посредство так называемых *"переменных состояния"* (**variables of state**). Строгость и недвусмысленность определения понятия *"состояния"* прямо влияет в дальнейшем на обоснованность решения задачи выбора адекватных технологий для *реабилитационных мер в рамках RI/FS процесса*, поскольку такое решение базируется на математических процедурах оптимизации.

State of object: Математика [7,19]: Практически любые физические объекты могут быть описаны системами обыкновенных дифференциальных уравнений, или уравнений в частных производных или разностных уравнений на детерминистской или стохастической основе. Объекты, описываемые дифференциальными уравнениями в частных производных, в свою очередь, могут быть аппроксимированы обыкновенными дифференциальными уравнениями, которые содержат только производные по времени. И, наконец, любое из обыкновенных дифференциальных уравнений можно преобразовать в систему дифференциальных уравнений первого порядка. В свою очередь, система из "n" линейных дифференциальных уравнений первого порядка определяет полностью свое решение лишь в том случае, когда заданы все ее коэффициенты и известны "n" начальных условий. Начальные условия образуют "n"- мерный вектор, который полностью определяет *состояние* объекта в начальный момент времени t1 (предполагается, что все входные и возмущающие воздействия известны с момента t1 и далее). Указанный вектор называется *вектором состояния объекта* в момент времени t1, а его компоненты - *переменными состояния*.
Соответствующее векторное дифференциальное уравнение называется *уравнением состояния объекта*. Оно имеет вид: $X'(t) = AX(t) + ZF(t) + BU(t)$; где: $X(t) = \{xi(t)\}$ и $X'(t) = \{x'i(t)\}$ - транспонированные векторы состояния объекта размерности n и его первая производная, xi – "переменные состояния"; A, Z, B - матрицы коэффициенты размерностью $n*n$, $n*l$, и $n*m$ соответственно; $U = \{ui(t)\}$, $F = \{fh(t)\}$ - транспонированные векторы входных (управляющих и возмущающих) воздействий на объект размерностью $m =< n$, $w <> n$. Измеряемые выходы объекта в таком случае могут определяться в функции вектора состояния, как: $Y(t) = CX(t)$, где $Y(t) = \{yj(t)\}$ - транспонированный вектор выходных сигналов объекта размерностью $v =< n$; C - матрица коэффициентов размерностью $v*n$. Именно к такому виду математического представления могут быть сведены описания и математические основания последующих оценок и решений,, которые выполняются применительно к объектам, анализируемым *в рамках* **RI/FS** *процесса*.

Sustainable Development, SD (*Устойчивое развитие – УР*) – обеспечение сбалансированного решения социально-экономических задач и проблем сохранения благоприятной окружающей среды и природно-ресурсного потенциала в целях удовлетворения потребностей нынешнего и будущих поколений людей.

Sustainable Development: Комментарий. Необходимыми признаками перехода к *устойчивому развитию* является достижение сбалансированного функционирования триады: Природная среда – Население – Производство/Экономика. В этой фазе развития механизмы разработки и *принятия решений* должны быть ориентированы на соответствующие приоритеты, учитывать последствия реализации этих решений в экономической, социальной, экологической сферах, а также предусматривать наиболее полную *оценку затрат, выгод и рисков.* При этом рекомендуется руководствоваться следующими общими критериями: никакая хозяйственная деятельность не может быть оправдана, если выгода от нее не превышает вызываемого ущерба; ущерб окружающей среде должен быть на столь низком уровне, какой только может быть разумно достигнут с учетом экономических и социальных факторов.

Technology Analysis (*анализ технологий* - технологических процессов *в рамках RI/FS процесса* **[7,13]**) - производится по многим показателям, которые рассматриваются в других учебных блоках данного курса по методологии **HCRA**. Заключительной, подводящий итоги *анализа технологий*, является оценка эффекта их применения, который характеризуется по результатам обработки *первичных выбросов и сбросов*, а также по опасности *вторичных потоков загрязнителей и радиоактивности* - опосредованных *источников* канцерогенного *риска* и токсической опасности, в виде загрязненных геосред - атмосферного воздуха, вод, почв и "отходов." Критерий *анализа технологий* таков **[5,6,7]**: если показатели опасности *вторичных потоков загрязнителей и радиоактивности* удовлетворяют заданным целям очистки, то анализируемые технологии *реабилитационных мер* считаются приемлемыми. В случае же, когда заданных целей не удается достигнуть, то, в соответствии с методологией *RI/FS процесса* и **HCRA** нужно будет опробовать другие наборы технологий из имеющегося, например, в системе **ReOpt** перечня, и т.д. Так продолжают до тех пор, пока: либо будут удовлетворены поставленные требования, либо истощен перечень из 100 технологий *реабилитационных мер*, содержащихся в базе данных системы **RAAS.**

Technology Analysis: Комментарий [7]. К анализируемым в методологии **HCRA** относят более 100 технологий, которые классифицируют по ряду признаков: Первая группа признаков – "адреса применения," когда технологии различаются в зависимости от удаленности восстанавливаемых объектов от первичного источника опасности: на 1) *"местные"* (**in situ**) и 2) *"удаленные"* (**ex situ**). Вторая группа признаков - функция *"местных"* технологий: 3) *"удаления,"* *"перемещения,"* *"вывоза"* и т.д. загрязнителей или *загрязненных сред* (**removal**); 4) *"разрушения"* или *"уничтожения"* (**destruction**) загрязняющих веществ; 5) *"закрепления"* или *"связывания"* химических загрязнений и радиоактивности (**immobilization**); 6) создания *"защитных оболочек"* (**containment**); а также: 7) *"технологического контроля"* (**engineered control**); и 8) *"регламентного контроля и обслуживания"* (**institutional control**). Третья группа признаков - функции *"удаленных"* технологий: процессы 9) *"разделения"* загрязняющих веществ (**separation**); 10) *"разрушения или уничтожения"* загрязняющих веществ (**destruction**); 11) *"закрепления или связывания"* химических загрязнений и радиоактивности (**immobilization**); а также перевозка транспортом или иное подобное 12) *"перемещение"* загрязненных и радиоактивных материалов (**material handling**); 13) их *"хранение"* (**storage**); или 14) *"удаление материалов"* от мест плотного проживания людей, фауны и флоры, перемещение их в места безопасного *"размещения и хранения"* (**disposal**). Четвертая группа признаков -

избранные *местные* технологии для *реабилитационных мер*, применяемые в той или иной геологической формации: 15) в *"вадозных зонах"* (**vadose zone**);или 16) в *водоносных горизонтах (*grounwater aquifer); а также 17) в *"непроточных водоемах" (*surface water impoundment). Пятая группа признаков - классификация по возможностям и параметрам применения *"удаленны"* технологий к различным 18) агрегатным состояниям загрязняющих и радиоактивных веществ или материалов, находящихся в твердом, жидком или газообразном. Наконец, шестая группа признаков - по различным задачам *"очистки"* загрязненных территорий, например: 19) *"обработки"* или *"переработки"* загрязнителей в местах их сосредоточения; 20) *"перемещения,"* загрязнителей, *"заключенных в изолирующие оболочки,"* к пунктам обработки с последующим контролируемым размещением продуктов переработки, и т.д.

Technology feasibility analysis (*анализ осуществимости* проектов, выполняемый в рамках *RI/FS процесса* [5,8,13]) - одна из составляющих *МТЭО* - модернизированного технико-экономического обоснования (**Feasibility Studies - FS**), выполнение которого предписывается законодательством США (**CERCLA**). В соответствии с логикой *RI/FS процесса* и методологией **HCRA** используется при обосновании технологий *реабилитационных мер*.

Technology feasibility analysis: Комментарий [13]. Алгоритм данной составляющей *МТЭО* (**Feasibility Study**) включает количественную оценку возможных эффектов применения технологий *СО-АР*, выбранных для *обработки или переработки первичных выбросов* и *сбросов* радиоактивности или токсичности в *Окр.Ср.* (**primary releases**), а также оценку показателей опасности *вторичных потоков загрязнителей* (**secondary streams**). **Комментарий [E].** Если, например, степень *обработки или переработки* загрязнителей удовлетворяют *заданным целям очистки* территории (**cleanup objectives**), то выбранные для этих функций *обработки или переработки* технологии *реабилитационные меры* по правилам **RI/FS** считаются приемлемыми. Тот же подход используется и для других функций *реабилитационных мер*. В случаях, когда заданных результатов или характеристик таким образом не достигают, проверяются другие наборы технологий (из имеющегося, например, в БД системы **ReOpt** перечня), и т.д. Так продолжают, до тех пор, пока: либо будут удовлетворены поставленные требования, либо будет истощен перечень из более чем 100 технологий *реабилитационных мер*, используемых в **American RI/FS process**.

Technology selection and performance (выбор технологий *реабилитации* и оценка их эффективности [7,13]) - ключевая задача **American RI/FS process** и методологии **HCRA**. Процедура принятия решений включает 6 этапов: 1). Выбираются *стратегии реабилитационных мер* (**remediation strategy**); 2). Стратегии уточняются и утверждаются сопоставлением с нормативными требованиями (**allowed objectives**); 3). Оценивается потенциальное качество принятых в пп. 1). и 2). решений (**objective logic**); 4). Устанавливаются принципы выбора технологий (**technology screening**); 5). Производится *выбор первичных* *технологий* (**initial technology screening**); и, наконец, - 6). Принимается окончательное решение по выбору адекватных конкретным условиям данной территории технологий (**technology screening**).

Transport of contaminants (*распространение/миграция* химических загрязнений и радиоактивности и радиоактивности [5,8,18]) - удаление *сбросов/выбросов* радиоактивного или химического вещества от исходного местоположения источника. Рассматриваются процессы диффузии в *Окр.Ср.* растворенных или взвешенных в геосредах примесей, либо перенос примесей движущими геосредами, либо, наконец, перемещение опасных веществ транспортными средствами и людьми, работающими в контакте с ними. Параметры химических загрязнений и радиоактивности и радиоактивности в различных точках *Окр.Ср.* оцениваются с помощью моделей распространения (переноса, диффузии) химических загрязнений и радиоактивности и радиоактивности.

Transport of contaminants: Генезис [2]: Термин *"миграция"* введен стандартом 6107/6-86 Международной Организации по Стандартам (ИСО) как "самопроизвольное или принудительное перемещение растворенных или взвешенных веществ или организмов в водном объекте." **Transport pathway** (*пути переноса* химических или радиационных загрязнений)- к таковым относят компоненты окружающей среды: подземные и/или поверхностные воды, почву и воздух. По этим *путям* загрязнения могут перемешаться от *первичного источника выбросов/сбросов* к биоте и человеку.

Transport scenario (*сценарии переноса - С.П.* [7,18]) - включают разнообразные варианты перемещения загрязнения в природных средах.

Transport scenario: Комментарий [E].*С.П.* создаются при моделировании, объединением нескольких *путей переноса* в некоторую последовательность,, которая характеризует возможное перемещение загрязнения в окружающей среде. Пример *С.П.*: *Из почвы в ->подземные воды, из них в -> поверхностные воды.*

Typical Transport Analysis (типичный анализ распространения/*миграции* химических загрязнений и радиоактивности) - включает анализ [7]: 1). Распределения концентраций загрязняющих примесей у источника *выброс* а или *сброс*а, а также в местах расположения *рецепторов* (**contaminant concentrations at source and receptor locations**); 2). Распространения химических загрязнений и радиоактивности и радиоактивности, вызванное воздушным или водным переносом (**water-driven and wind-driven transport**). Мат. моделирование в **RI/FS process** проводится с учетом радиоактивного распада и естественного ослабления токсичности загрязнителей (**radioactive decay and natural attenuation**).

Umbrellas of safety (*"Зонтики" безопасности*)

Usage location (*технопатогенные зоны - ТПЗ* [5,8,19]) - участки территорий, на которых люди подвергаются опасности получения доз вследствие загрязненности составляющих окружающей среды: воды, почвы, воздуха, где пути переноса химических загрязнений и радиоактивности могут перекрещиваться или совпадать с *дозовыми маршрутами. ТПЗ* является наименьшей по площади территорией, на которой выполняются оценки **MEPAS**.

Usage location: Генезис [13,18,19]: Понятие технопатогеной зоны ассоциируется *в рамках RI/FS процесса* с последней средой из *сценария переноса.* В технопатогеной зоне должны быть учтены и рассмотрены, помимо обычно идентифицируемых *источников* опасности и *риск*а, например такие, как: 1) скважины для поднятия подземные вод, и др. водозаборные устройства, 2) рекреационные парки вдоль водоемов, 3) сельскохозяйственные угодья и т.п., 4) а также все населенные территории, расположенные в радиусе 80 км от каждого источника выбросов в атмосферу. Радиус зоны от источника сбросов, то есть граница опасности для людей от места поступления химических загрязнений и радиоактивности *в воду или почву,* и соответственно "дальнобойность" методик *RI/FS процесса* не ограничиваются.

Waste management (*управление отходами* [18]) - комплекс мер и процедур, направленных на: 1) снижение поступления в среду обитания человека отходов современной индустрии, - материалов и/или энергии, опасных для него и/ или способных нарушить функции окружающей среды (**source reduction**); 2) возвращение в индустриальный цикл для повторного использования переработки таких энергии и/или материалов (**recycling**); 3) уменьшение опасности поступающих в среду обитания человека материалов и/или энергии, в первую очередь,- понижением концентрации в среде обитания таких материалов и/или потенциалов энергетических потоков (**treatment**); 4) удаление опасных материалов и/или энергии от мест плотного проживания людей, фауны и флоры, перемещение их в места безопасного хранения и размещение в хранилищах (**disposal**).

References

1. Competitive Risk Management. Vitaly Eremenko. Training aids for the NATO ASI, Bourgas, Bulgaria, May 2-11, 2000.
2. ECOLOGY. Russian-English Dictionary.-1300 Terms. Moscow 1993. 120 pages.
3. English - Russian - English Terminological Dictionary of Nuclear Power.- 9000 Terms. Moscow, 1993, 196 p.
4. Eremenko V.A. Comparative risk assessment with due regard to human factor uncertainties,- IAEA-TECDOC-671, 1992.
5. Eremenko V.A.. Glossary of the MEPAS Terms. - ACTIS, 1995. - 7 pages - <Appendix to the Deliverable for the BOA-1994 with Battelle>.
6. Eremenko V.A., Andreev O.V., Sharov V.F. Generalized antropogeny methodology of risk assessment for developing countries. Proceeding of SRA - Europe Conference, Roma, 18 -20 October, 1993.
7. Eremenko V., Shilova O. English - Russian Terminological Glossaries of RAAS (The 1997 Deliverable for Task 1, BOA with Battelle), 1997.

8. Eremenko V.(RF), G.Whelan, J.Droppo (USA). Invitation to Introducing a Complex of Computer Systems for Data Preparation USA - MEPAS (СКОЗОС) for Taking the Optimum Deciding on Polluted Territories. - Moscow-Richland, 1995. - 32p.- <Digest for Seminar Material on MEPAS>

9. Guidelines for Organizational Self-Assessment of Safety Culture and for Reviews by the Assessment of Safety Culture in Organizations Team, IAEATECDOC-743, IAEA, Vienna, 1994

10. History Of Methods. Dennis Bley, Jim Droppo and Vitaly Eremenko. Training aids for the NATO ASI, Bourgas, Bulgaria, May 2-11, 2000.

11. Izmalkov V, and Izmalkov A. Safety and Risk under Technogene Influences, Parts 1 and 2. - Moscow - St-Petersburg, 1994. - 269p.

12. Kuzmin I, Mahutov N, etc.. Safety and Risk.- St-Petersburg, 1997. - 164p.

13. Michael K. White. Remedial Action Assessment System (RAAS): Decision support tools for performing streamlined feasibility studies.// Pacific Northwest Laboratory, 1995. - 12 ps.

14. Materials of reviewing a draft of risk assessment textbook for a NATO ASI held in Bulgaria in May 2000. Deliverable for Task 1 BOA -Contract 323725-A-R4, - IX.30, 2001, Battelle PNNL - ACTIS, March 31, 2001

15. Norms of Radiation Safety in Russia, 1996. (НРБ-96). Hygienic Standards GS 2.6.1.054-96. - Moscow,1996.- 127p.-< State Committee of Russia >.

16. Quantitative assessment problem of safety activities for organizations team in nuclear industry. V.A.Eremenko, O.V. Andreev. International Center of Educational Systems (ICES), Moscow. Final Program and Abstracts. Society for Risk Analysis, Annual Meeting. Baltimore, M, USA, Dec. 4-7, 1994, p.59

17. Risk Assessment Activities for the Cold War Facilities and Environmental Legacies. Vitaly Eremenko, Dennis Bley, James Droppo. Handout for the NATO ASI, Bourgas, Bulgaria, May 2-11, 2000.

18. Whelan G, Buck J., Droppo J, etc. Overview of the Multimedia Environmental Pollutant Assessment System (MEPAS) // Hazardous waste & hazardous materials, Vol.9, No.2, 1992 Mary Ann Liebert, Inc.,Publishers, pp. 191-208.

19. В.А.Еременко "Повышение радиационной и химической безопасности населения и среды обитания ограниченной административной территории с использованием компьютерных наблюдателей и модального управления ее состоянием." Комплексная радиационная и химическая безопасность населения и Среды обитания Московской области: Материалы МО НПК. - М., 1996.-с. 94- 99

Index